Recent Progress in Plant Taxonomy and Floristic Studies

Recent Progress in Plant Taxonomy and Floristic Studies

Editors

Alessio Papini
Mushtaq Ahmad
Fazal Ullah
Wajid Zaman

MDPI • Basel • Beijing • Wuhan • Barcelona • Belgrade • Manchester • Tokyo • Cluj • Tianjin

Editors

Alessio Papini
Department of Biology
University of Florence
Florence
Italy

Mushtaq Ahmad
Department of Plant Sciences
Quaid-i- Azam University
Islamabad
Pakistan

Fazal Ullah
School of Life Sciences
Lanzhou University
Lanzhou
China

Wajid Zaman
Department of Life Sciences
Yeungnam University
Gyeongsan
Korea, South

Editorial Office
MDPI
St. Alban-Anlage 66
4052 Basel, Switzerland

This is a reprint of articles from the Special Issue published online in the open access journal *Agronomy* (ISSN 2073-4395) (available at: www.mdpi.com/journal/agronomy/special_issues/plantaxonomy_floristic).

For citation purposes, cite each article independently as indicated on the article page online and as indicated below:

LastName, A.A.; LastName, B.B.; LastName, C.C. Article Title. *Journal Name* **Year,** *Volume Number*, Page Range.

ISBN 978-3-0365-7677-0 (Hbk)
ISBN 978-3-0365-7676-3 (PDF)

Contents

Preface to "Recent Progress in Plant Taxonomy and Floristic Studies"

There is a relentless human need to understand life by tracing the historical development of the discipline of plant sciences, the part of natural science that covers traditionally treated plants.

The history of plant taxonomy and floristic studies—the botanical classification of plants into different groups and their distribution in different natural habitats and evolution—stretches from the work of ancient Greeks to modern evolutionary botanists and plant scientists. As an area of science, plant systematics came into being slowly, with early plans usually being considered as part of the research on medicine or drugs. Later on, classification, description, and evolution were driven through natural biology and natural history. Until the discovery of the theory of evolution, almost all classifications and descriptions were based on natural history and natural biology. Botany in the 18th and 19th centuries made significant advancements toward a more holistic classification methodology, eventually based on evolutionary relationships.

Alessio Papini, Mushtaq Ahmad, Fazal Ullah, and Wajid Zaman
Editors

agronomy

MDPI

Article

Morphological, Anatomical and Chemical Characterization of *Ricinus communis* L. (Euphorbiaceae)

Iman H. Nour [1,*], Khadiga Alhadead [2], Faten Y. Ellmouni [3], Reem Badr [1], Tamannouha I. Saad [4], Ahmed EL-Banhawy [5] and Salwa M. Abdel Rahman [1]

1 Botany and Microbiology Department, Faculty of Science, Alexandria University, Alexandria 21511, Egypt; reem.badr@alexu.edu.eg (R.B.); salwam218@gmail.com (S.M.A.R.)
2 Environmental Sciences Department, Faculty of Natural Resources and Environmental Sciences, Omar Al-Mukhtar University, El-Beida P.O. Box 919, Libya; khadiga.alhadead@omu.edu.ly
3 Botany Department, Faculty of Science, Fayoum University, Fayoum 63514, Egypt; fyl00@fayoum.edu.eg
4 Botany Department, Faculty of Science, University of Derna, Qubbah Branch, El-Qubbah 11595, Libya; t.saad@uod.edu.ly
5 Botany and Microbiology Department, Faculty of Science, Suez Canal University, Ismailia 41522, Egypt; ahmedbanhawy@science.suez.edu.eg
* Correspondence: iman.nour@alexu.edu.eg

Abstract: *Ricinus communis* L. (Euphorbiaceae, Acalyphoideae) is a highly variable species known as the castor oil plant. This study aimed to describe *R. communis* using several methodologies, such as vegetative morphometry, leaf surface ultrastructure, soil analysis, and gas chromatography-mass spectrometry (GC-MS) analysis, to understand the diversity of this species. The morphological analysis revealed that some samples had purple stems while others were grayish-green. The purple-stemmed *R. communis* phenotype reflects the intra-specific diversity of the species. The multivariate analysis of 25 *R. communis* samples based on 34 vegetative morphometric characteristics revealed that they belonged to three main groups (morphotypes). Each group attained some specific characteristics discriminating it from the other groups. Selected samples from each group were investigated using SEM, soil analysis, and GC-MS. The performed GC-MS technique revealed that six major compounds were detected in the chromatograms of the studied samples. The highest percentages of n-Hexadecanoic acid and 9,12,15-Octadecatrienoic acid were recorded. *Ricinus communis* demonstrated adaptive growth capability, where plants inhabiting coastal sites are salt-sensitive, while inland plants are relatively drought-tolerant species. The intra-specific variation between *R. communis* morphotypes indicated the possibility of the direct and indirect use of these varieties in genetic improvement programs of the species.

Keywords: Gas chromatography-mass spectrometry (GC-MS); leaf; morphometry; morphotype; *Ricinus communis*; SEM; stomata

Citation: Nour, I.H.; Alhadead, K.; Ellmouni, F.Y.; Badr, R.; Saad, T.I.; EL-Banhawy, A.; Abdel Rahman, S.M. Morphological, Anatomical and Chemical Characterization of *Ricinus communis* L. (Euphorbiaceae). *Agronomy* **2023**, *13*, 985. https://doi.org/10.3390/agronomy13040985

Academic Editors: Alessio Papini, Mushtaq Ahmad, Fazal Ullah, Wajid Zaman and Teresa Navarro

Received: 9 February 2023
Revised: 8 March 2023
Accepted: 17 March 2023
Published: 27 March 2023

1. Introduction

Ricinus communis L. is known as the castor oil plant [1]. It is a perennial oilseed shrub that belongs to the Euphorbiaceae family and subfamily Acalyphoideae [2–4]. The species originated in Africa and is currently cultivated in many tropical and subtropical regions around the world [5]. The plant is a high protein source for animal feedstock and can be cultivated as a garden ornament [6]. The castor bean plant has earned increasing attention due to its commercial castor oil production, pharmacological activities, and agricultural applications [7]. The species was reported to possess therapeutic properties, such as antiasthmatic, antidiabetic, anticancer, antioxidant, antimicrobial, anti-inflammatory, antiulcer, wound healing, and laxative effects [8–14]. According to Abbes et al. [15], castor seeds contain a toxic glycoprotein called ricin, which is regarded as a potent poison. The leaf and seed extracts of *R. communis* have larvicidal activity against mosquitoes [16]. It contains ricinine in all parts of the plant which is considered a strong insecticide [17]. Leaf extracts of *R. communis* were effective against the replication of the hepatitis A virus [18]. Mboyazi et al. [19] reported that extracts of *R. communis* leaf have proved to be a safe source of therapeutic agents. Since *R. communis* oil has a high ricinoleic fatty acid content, it is now used in biofuel production [20].

This species can be self- and cross-pollinated, containing a large amount of pollen biomass per flower, which contributes generously to the plant's distribution [21]. However, Allan et al. [22] reported that castor bean germplasm had limited genetic diversity. In contrast, Anjani [23] mentioned that many wild and semi-wild types with wide genotypic and phenotypic diversity resulted from natural selection in various agro-climatic regions. Nevertheless, the plant's morphological characteristics vary greatly depending on the environment. This highly variable species can be classified into wide varieties and forms based on different vegetative and floral morphological characteristics [24–30]. Despite that, limited leaf characteristics of *R. communis* were reported in those previous works. Although there are several forms with distinctive characteristics, they are closely related through intermediate forms and hybridization [29,31]. Investigating the leaf's micromorphological characteristics—epidermal cells and stomata—explains the plant responses to abiotic and biotic stresses [32]. Stomata mediate interactions and act as channels between plants and the environment [33].

Analyzing soil in different sites identifies the soil's chemical, physical, and biological characteristics and determines the concentration of soil nutrients [34]. Soil pollutants released by anthropogenic activities endanger the stability of biological systems [35]. Nazzal et al. [36] considered copper, iron, manganese, and zinc as major soil contaminants. Plants growing in such contaminated soils have developed defense mechanisms for toxic metal ions [37]. Heavy metal precipitation and mobilization in soil may be due to the uncontrolled discharge of wastes during industrialization or agricultural activities that affects the metabolism of plants [38–41]. Tyagi et al. [42] studied the impact of industrial pollution on *R. communis* compared to a plant grown in unpolluted natural areas. *Ricinus communis* can thrive in heavy-metal-contaminated soils [43] and function as a biosensor of environmental quality because of its massive growth and large leaf area, which aids in pollution detection [44]. Thus, it is a possible candidate for environmental restoration and bioremediation [45].

The essential oil isolated from *R. communis* leaves could be used in the formulation of natural remedies [46]. Mboyazi et al. [19] mentioned that harvesting *R. communis* at different geographic locations could yield various phytochemicals with potential therapeutic significance. Major compounds have been detected in castor leaf extract with the help of GC-MS analysis, such as octadecanoic acid, n-hexadecanoic acid, 1-hexadecanol, triethyl citrate, 2-Methyl, diethyl phthalate, 3-octadecene, α-thujone, and 1,8-cineole [17,46]. Meanwhile, long-chain fatty acids and their derivatives were detected in the castor seed oil of some cultivars, such as oleic acid, palmitic acid, and linoleic acids [17,47].

Leaf traits are frequently measured to predict how anthropogenic pressures will affect ecosystems [48]. The leaf oil composition of *R. communis* has been insufficiently studied compared to its seeds, which was extensively studied [6,49–53]. Accordingly, this study aimed to describe its leaves using different approaches to understand the diversity of this species and how environmental conditions may affect its morphological traits and chemical composition. Thus, several methodologies were used to characterize this plant in different locations in Egypt, such as vegetative morphometry, leaf surface ultrastructure, soil analysis, and gas chromatography-mass spectrometry (GC-MS) analysis.

2. Materials and Methods

2.1. Plant Materials and Vegetative Morphometry

Twenty-five *R. communis* samples were gathered from eleven sites in the western Mediterranean coastal desert of Egypt (Table 1, Supplementary Figure S1). From each shrub, ten leaves were collected from different branches. Measurements of the third and fourth healthy undamaged leaves from the top of the stem were carried out to minimize variation due to different stages of leaf growth. Voucher specimens were deposited in ALEX University Herbarium, Alexandria, Egypt. ImageJ program (version 1.51j8) was used to measure the quantitative characters [54]. Leaves with a number of lobes above ten were disregarded in this study. The central lobe (first lobe) was measured, while the lateral lobes were represented as an average of the two similar lobes, as well as the lobe sinus (depth between every two lobes) (Figure 1). The terminology of Singh [55] and Beth et al. [56] was used, the categories of leaf area were described according to Ash [57], and the plant stature was ranked based on the study of Silva et al. [27].

Table 1. Localities and coordinates of *Ricinus communis* samples from Egypt.

Site	Sample No	Coordination	Location
1	1–4	31°02′40.3″ N 29°42′13.9″ E	Mehwar Al-Ta'ameir, about 700 m from kilo 26 wastewater treatment plant, and 26 km from West Alexandria.
2	5–7	31°00′00.0″ N 29°36′35.2″ E	36 km from Alexandria–Matrouh International Coastal Road, and 200 m from Amoun Resort, Sidi Kirayr.
3	8–9	30°59′19.3″ N 29°35′27.2″ E	39 km from Alexandria–Matrouh International Coastal Road, Sidi Kirayr.
4	10–11	30°49′19.5″ N 29°11′47.8″ E	Omayed Biosphere Reserve, Alexandria Desert Road.
5	12–13	30°49′16.4″ N 29°11′49.9″ E	1 km from East Omayed biosphere reserve, Alexandria Desert Road.
6	14–15	30°52′07.4″ N 29°20′20.3″ E	International Coastal Road, 5 km west of El Hammam Central Hospital, El Hammam.
7	16	30°52′50.7″ N 29°22′07.9″ E	International Coastal Road, 2 km west of El Marwa Resort, El Hammam, Egypt.
8	17–19	31°00′43.6″ N 29°38′04.7″ E	175 m from South Sidi Kirayr Bridge.
9	20–21	30°56′57.6″ N 29°41′27.5″ E	3 km from Borg-El Arab International Airport, Borg El-Arab.
10	22–23	30°56′50.7″ N 29°50′31.9″ E	Cairo–Alexandria Desert Road, 7 km south of Amaria General Hospital, second Al Amaria.
11	24–25	30°56′14.3″ N 29°50′57.9″ E	Cairo–Alexandria Desert Road, 11 km south of Amaria General Hospital, second Al Amaria.

Figure 1. Description of leaf morphological traits in *Ricinus communis*. The yellow arrow indicates the lobe sinus (depth between two lobes).

2.2. Leaf Surface Ultrastructure (SEM)

The leaf surface micromorphology was investigated for six *R. communis* samples (4, 8, 11, 12, 17, and 22). These samples were selected from the three major subclusters that resulted from the multivariate analysis of the macromorphological data. Both the abaxial (AB) and adaxial (AD) leaf surfaces were mounted on the copper sample holder with double-sided adhesive tape. The samples' coating was completed with gold in a Polaron JFC-1100E coating unit for 5 min. Firstly, the leaves were examined using a stereomicroscope and directly observed under a JEOL JSM-IT200 Scanning Electron Microscope (Tokyo, Japan) at the Electron Microscopy Unit of the Faculty of Science, Alexandria University, Alexandria, Egypt. Twenty-seven quantitative and qualitative characteristics were studied for both leaf surfaces. The quantitative traits included stomatal count at 2560 × 1920 μm^2 as well as stomatal and epidermal size parameters. Measurements were made by using ImageJ (1.51j8) [54]. The terminology of Barthlott et al. [58] was used.

2.3. Soil Sampling and Elemental Analysis

A total of thirty-three soil samples were collected for the eleven studied sites (three subsamples per site). According to Piper's method [59], every three subsamples were mixed as a composite sample. Six samples corresponding to six sites (1, 3–5, 8, and 10) were selected based on the multivariate analysis of the vegetative traits. Then, the samples were crushed into small pieces using a mechanical crusher and dried overnight at 35 °C to a constant weight. The dried samples were ground to fine powder, and then the powdered samples were sieved using a standard set of sieves to diameters <125 and >63 μm. Every powdered sample was shaken using an electric shaker to be sure that the sample was homogenized. The saturation percentage (SP) was identified as the amount of water in milliliters needed to saturate 100 g of soil. Soil pH was measured in a soil-water ratio of 1:2.5 by using a pH meter (PB-21, Sartorius, Göttingen, Germany). Soil organic matter (SOM) was examined using the potassium dichromatic oxidation titration technique, according to Walkely and Black [60].

The available nitrogen (N) was determined using the Kjeldahl method with a Kjeldahl analyzer (Kjeltec TM8200, FOSS, Shanghai, China) [61]. The available phosphorus (P) was digested using perchloric and sulfuric acids, then was analyzed using the molybdenum antimony blue colorimetric method [62]. According to Hanway and Heidel [63], the available potassium (K) was digested using ammonium acetate and analyzed via atomic adsorption spectrometry. In order to determine the metal soil contents, the solutions were subjected to inductively coupled plasma-optical emission spectroscopy (ICP-OES; Agilent 5100 VDV, Santa Clara, CA, USA). The contents of Zn^{++}, Ca^{++}, Cu^{++}, K^{+}, Mg^{++} and Na^{+}, were computed as parts per million (ppm). Flow rates of plasma, auxiliary, and nebulizer of ICP-OES were kept at 12, 1, and 0.7 mL min^{-1}, respectively. The sample uptake and stabilization time was 10 s for each sample.

2.4. Gas Chromatography-Mass Spectrometry (GC-MS) Analysis

2.4.1. Preparation of Plant Extracts

The fresh *R. communis* leaves were shade-dried until the samples became well-dried for grinding. After drying, the plant materials were ground well into a fine powder using a mechanical blender, and then were dried under shade and ground again. The dried plant material was soaked in hexane and then concentrated to dryness using a rotary evaporator at 40 °C [64].

2.4.2. GC-MS Analysis Conditions

The GC-MS analysis was carried out for the six selected samples (1, 3–5, 8, and 10) by using gas chromatography-mass spectrometry with the following specifications: a TRACE GC Ultra Gas Chromatographs (THERMO Scientific Corp., Waltham, MA, USA) coupled with a Thermo mass spectrometer detector (ISQ Single Quadrupole Mass Spectrometer). The GC-MS system was equipped with a TR-5 MS column (30 m × 0.32 mm i.d., 0.25 μm film thickness). The column oven temperature was initially maintained at 150 °C for 2 min, and then increased at 1 °C/min to 210 °C, and next was sustained at 220 °C for 40 min. The injector and detector temperatures were 275 °C and 220 °C, respectively. The carrier gas was helium, with a flow rate of 1.5 mL/min [65]. The quantities of the phytochemicals detected in the extracts were expressed as percentages from the total extract (area %). The identification of the compounds was de-convoluted using AMDIS software and identified by its retention indices (relative to n-alkanes C8–C22), and mass spectrum matching to the Wiley spectral library collection and NSIT mainlib and replib databases.

2.5. Data Analysis

Different software was used to analyze the data, including Excel 365, Minitab version 20, and R software (Vienna, Austria); the latter had the necessary packages loaded [66]. The mean and standard deviation were computed as descriptive statistics for each component. Based on several measurable qualities, interpretive multivariate statistics were employed to divide the 25 samples into many homogenous groups.

The similarity and dissimilarity between and among samples were visualized using the "pheatmap" and "ggplot2" packages [67,68]. The color scale varied depending on how much the studied readings differed. According to Viscosi and Cardini [69], a red color implies a high degree of similarity between accessions, whereas a blue color denotes a low degree of similarity. Following the normalization and scaling of various variables using R-software, agglomerative cluster analysis was performed using Ward's linkage approach and Euclidean distance measurement. Installing the "factoextra" and "ggplot2" packages in R allowed for the creation of a scatter diagram using principal component analysis (PCA) to display the distance matrices [70].

The correlation coefficients for the association between the two variables were obtained and displayed using the "Corrplot" package [71]. White with a 0 shows no link between the two variables, whereas blue with a 1 suggests a high positive correlation. Red with a −1 indicates a significant negative correlation.

Utilizing Minitab 20, the optimal approach was used to apply the Box–Cox transformation for dependent variables that are not normal. Under the general linear model, several comparisons were made using one-way and two-way analysis of variance (ANOVA). Results demonstrated a satisfactory fit for several models, and after data transformation, normal residual probability plots displayed a linear attitude for all studies. *p*-values were measured as significant at ($p \leq 0.05$). Tukey's test for pairwise comparisons was used to conduct a post hoc analysis of all group interactions. The post hoc analysis results are shown as letters, with the same letter indicating no significant difference between groups and different letters indicating significant differences between groups.

3. Results

3.1. Vegetative Morphometry

The descriptive and qualitative data of the vegetative morphometry are illustrated in Table 2 and Supplementary Table S1. *Ricinus communis* is a shrubby plant, and its height varied from 1.02 m in sample 5 to 3.72 m in sample 24. The plant's stature was categorized as very tall, ">2.50 m" (Figure 2a); tall, "2.0–2.5 m"; medium, "1.51–2.0 m"; and short, "1.0–1.5 m" (Figure 2b). The stems were smooth, glabrous, and with divergent branches sometimes covered by wax (Figure 2a,b). Stems were mostly greyish-green (Figure 2a), except for samples 1–4, 12–13, and 20–21, which had purple ones (Figure 2b). Each shrub had both condensed and elongated internodes. The petiole was green or purple and inserted sub-basally to the leaf at 9.85% to 26.14%. The leaf petiole length ranged from 5.58 cm to 39.8 cm (in samples 10 and 19, respectively), and the petiole length ratio varied approximately from half to about one and a half times the leaf length.

Table 2. Descriptive quantitative data of *Ricinus communis* vegetative morphometry.

Variable	Mean ± StDev	SE Mean	Min.	Q1	Median	Q3	Max.	IQR
Plant length (m)	2.144 ± 0.725	0.145	1.02	1.57	2.03	2.7	3.72	1.13
Petiole length/leaf length ratio	0.9756 ± 0.2785	0.0557	0.49	0.755	1.01	1.155	1.54	0.4
Petiole attachment (%)	18.302 ± 2.838	0.568	11.76	16.31	17.65	20.805	23	4.495
Leaf area (cm^2)	178.7 ± 90.9	18.2	52.9	112.1	151.8	251	352.9	138.9
Leaf incision (%)	63.796 ± 4.454	0.891	54.82	61.525	63.77	65.755	74.4	4.23
Leaf length (cm)	19.351 ± 4.722	0.944	11.6	15.805	18.42	23.38	27.49	7.575
Leaf width (cm)	20.57 ± 5.33	1.07	12.3	16.82	19.62	25.38	30.52	8.55
Leaf length/width ratio	0.9444 ± 0.03318	0.00664	0.88	0.92	0.94	0.97	1	0.05
First lobe length (cm)	13.51 ± 3.366	0.673	8.22	10.775	13.63	16.495	20.06	5.72
First lobe width (cm)	4.579 ± 1.217	0.243	2.17	3.615	4.53	5.515	7	1.9
First lobe length/width ratio	3.0224 ± 0.3283	0.0657	2.43	2.785	3	3.205	3.97	0.42
Second lobe length (cm)	12.505 ± 3.161	0.632	7.55	10.09	11.83	15.04	18.84	4.95
Second lobe width (cm)	3.864 ± 1.019	0.204	1.97	3.26	3.69	4.62	5.51	1.36
Second lobe length/width ratio	3.3036 ± 0.2736	0.0547	2.75	3.15	3.3	3.46	4.15	0.31
Third lobe length (cm)	10.855 ± 2.924	0.585	6.64	8.765	10.03	13.445	16.62	4.68
Third lobe width (cm)	3.075 ± 0.893	0.179	1.72	2.52	2.92	3.865	4.72	1.345
Third lobe length/width ratio	3.6092 ± 0.29	0.058	2.94	3.39	3.66	3.825	4.17	0.435
Fourth lobe length (cm)	8.916 ± 2.588	0.518	4.93	7.14	8.34	11.095	13.69	3.955
Fourth lobe width (cm)	2.39 ± 0.757	0.151	1.25	1.93	2.26	2.77	3.78	0.84
Fourth lobe length/width ratio	3.8452 ± 0.378	0.0756	3.08	3.54	3.79	4.16	4.59	0.62
Fifth lobe length (cm)	6.933 ± 2.187	0.437	3.5	5.045	6.46	9.135	10.21	4.09
Fifth lobe width (cm)	1.739 ± 0.583	0.117	0.88	1.225	1.72	2.195	2.8	0.97
Fifth lobe length/width ratio	4.103 ± 0.508	0.102	3.29	3.725	4.06	4.44	5.33	0.715
Depth between first and second lobe (cm)	4.948 ± 1.807	0.207	1.5	3.682	4.617	6.029	9.19	2.348
Depth between second and third lobe (cm)	4.68 ± 1.69	0.194	1.559	3.524	4.435	5.472	8.902	1.948
Depth between third and fourth lobe (cm)	3.922 ± 1.489	0.171	1.313	2.697	3.643	4.797	7.609	2.099
Depth between fourth and fifth lobe (cm)	3.028 ± 1.182	0.136	0.98	2.043	2.817	3.848	6.204	1.805

Figure 2. (**a**,**b**) Photographs of *Ricinus communis*: (**a**) very tall grayish-green stem showing both condensed and elongated internodes (sample no 7), (**b**) short purple-waxy stem (sample no 4). (**c–e**) Dried herbarium leaves of the plant: (**c**) mesophyll blade with lanceolate lobes (sample no 11) and (**d**,**e**) macrophyll blade. (**d**) First, second and third lobes are oblong and the fifth is lanceolate (sample no 24). (**e**) First, second and third lobes are lanceolate and the fifth is linear-lanceolate (sample no 17). Scale bar = 30 cm (**a**,**b**); 4 cm (**c–e**).

Leaves were alternate, green, glabrous, more or less flat, and elliptic, and their size was 7.10–32.57 × 7.89–34.39 cm (Table 2). The leaf area ranged from 225.14 to 5000.19 mm^2, where the leaf blade classes fell into the mesophyll (225–1820 mm^2) and macrophyll (<1820 mm^2) categories (Table 2). *Ricinus communis* leaves are symmetric or asymmetric, with a variable number of lobes in the same shrub; 7 to 13 lobes. Leaves are simple and lobed with a palmatipartite incision. The incision percentage of the leaf represents 52.28 to 80.64% (in samples 1 and 21, respectively), with a curved sinus.

The first lobe (central lobe) size was 4.86–24.41 × 1.49–7.83 cm, the second pair size was 4.60–21.49 × 1.26–7.29 cm, and the third pair size was 3.86–17.84 × 1.01–6.15 cm. The latter lobes had oblong or lanceolate shapes. Generally, the fourth and fifth pairs attained lanceolate lobes; their sizes were 2.97–14.61 × 0.58–4.62 cm and 2.35–12.61 × 0.51–3.47 cm, respectively (except sample 17 which had linear-lanceolate fifth lobe). Each lobe had a prominent vein, acuminate apex, and serrate margin. The leaf teeth were straight or curved with 4 to 7 orders.

3.2. Multivariate Analysis

Twenty-five samples of *R. communis* were subjected to cluster analysis based on 34 quantitative and qualitative vegetative morphological characteristics. The agglomerative cluster analysis revealed two major clusters (Figure 3a). The first cluster represented group 1, which contained nine samples collected from sites 1, 3, 6, 7, 8 and 11. The second major cluster was divided into two subclusters; groups 2 and 3. Group 2 also comprised nine samples from sites 1, 2, 5, 8, 9 and 10, while group 3 contained the remaining samples from sites 2, 4, 6, 9, 10 and 11. The heatmap (Figure 3b) exemplified the overall variations between the investigated samples.

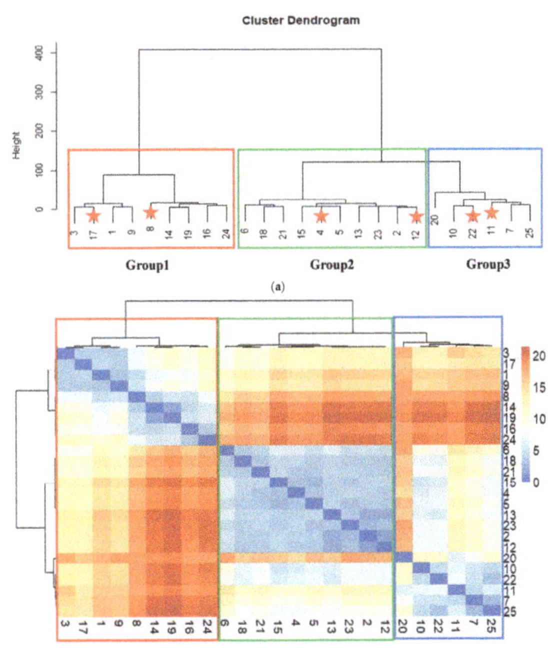

Figure 3. (**a**) Agglomerative cluster analysis of *Ricinus communis* based on 34 vegetative morphometric characters. Red asterisks show the six investigated samples using SEM. (**b**) Heatmap of *Ricinus communis* based on 34 vegetative morphometric characters.

The principal component analysis (PCA) performed for the twenty-five samples (Figure 4), based on 27 quantitative characters, revealed that the first axis (Dimension 1) scored 56.5%, followed by the second axis (Dimension 2) that accounted for 12.8% of the total variation. The first axis separated group 1 and sample 6 (belonging to group 2) from groups 2 and 3. The second axis split samples of group 1 (except 16 and 24), samples 4, 5, and 18 of group 2, and samples 10, 11, and 20 from group 3. The significant characteristics attributed to the ordination of dimensions 1 and 2 and their correlation values are summarized in Table 3. For dimension 1, the highest contribution was for the leaf area (0.987), and the lowest was for the leaf length/width ratio (0.425). In contrast, dimension 2 recorded a higher correlation for the third lobe length/width ratio (0.818), and the lowest was the fifth lobe length/width ratio (0.463).

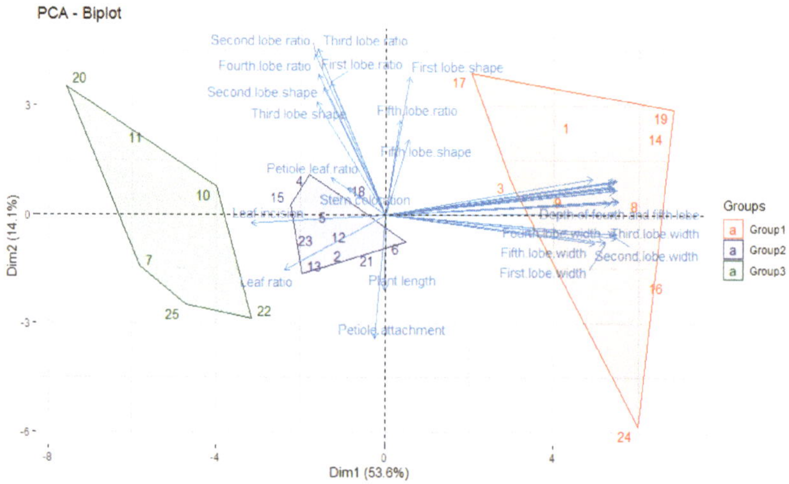

Figure 4. Principal Component Analysis (PCA) of 25 *Ricinus communis* samples based on 27 quantitative morphometric characters with the correlation between different variables and the first two components. Length of the variable arrow representing the importance of different variables where longer arrows contribute the most to the discrimination of the three groups.

Table 3. Principal component analysis (PCA) with their Eigenvalues and variance for *Ricinus communis* based on 27 quantitative morphometric traits.

Dimension 1	r	Dimension 2	r
Leaf area (cm^2)	0.987	Third lobe length/width ratio	0.818
Depth of third and fourth lobe (cm)	0.984	Second lobe length/width ratio	0.793
Second lobe width (cm)	0.980	Fourth lobe length/width ratio	0.691
Third lobe width (cm)	0.980	First lobe shape	0.679
Fourth lobe length (cm)	0.980	First lobe length/width ratio	0.651
Third lobe length (cm)	0.978	Second lobe shape	0.626
Depth of fourth and fifth lobe (cm)	0.9745	Third lobe shape	0.555
Leaf width (cm)	0.973	Fifth lobe length/width ratio	0.463
Leaf length (cm)	0.970	Petiole attachment (%)	−0.613
First lobe length (cm)	0.970		
Second lobe length (cm)	0.970		
Fourth lobe width (cm)	0.967		
Depth of second and third lobe (cm)	0.966		
Depth of first and second lobe (cm)	0.953		
Fifth lobe length (cm)	0.948		
First lobe width (cm)	0.932		

Table 3. *Cont.*

Dimension 1	r	Dimension 2	r
Fifth lobe width (cm)	0.887		
Blade class	0.878		
Leaf incision (%)	−0.568		
Leaf length/width ratio	−0.425		
Eigenvalue	17.68	Eigenvalue	4.640
Variance %	53.59	Variance %	14.06
Cumulative variance %	53.59	Cumulative variance %	67.66

According to the cluster analysis, a one-way ANOVA was performed for the three resulting groups to indicate the statistically significant morphological characters (Table 4). There was a significant difference between the three studied groups for the following traits: leaf length, width and area, first lobe width, second lobe length and width, third lobe length and width, fourth lobe length and width, fifth lobe length and width, depth between first and second lobe, depth between second and third lobe, depth between the third and fourth lobe and depth between fourth and fifth.

Table 4. One-way ANOVA results and pairwise comparisons between the three groups of *Ricinus communis* showing the significant morphological traits. Groups that share same letters are non-significant, while different letters represent significantly different groups.

Variable	Group 1	Group 2	Group 3
Plant length (m)	2.207 ± 0.721 ^{a}	1.931 ± 0.72 ^{a}	2.403 ± 0.764 ^{a}
Petiole length/leaf length ratio	0.91 ± 0.2776 ^{a}	0.962 ± 0.262 ^{a}	1.097 ± 0.316 ^{a}
Petiole attachment (%)	17.824 ± 2.417 ^{a}	19.165 ± 2.671 ^{a}	17.58 ± 3.74 ^{a}
Leaf area (cm^2)	283.7 ± 50.2 ^{a}	145.57 ± 17.9 ^{b}	76.61 ± 22.31 ^{c}
Leaf incision (%)	60.75 ± 3.64 ^{b}	64.74 ± 4.19 ^{ab}	66.8 ± 3.62 ^{a}
Leaf length (cm)	24.523 ± 1.79 ^{a}	18.393 ± 1.409 ^{b}	13.19 ± 1.676 ^{c}
Leaf width (cm)	26.493 ± 2.318 ^{a}	19.281 ± 1.459 ^{b}	13.813 ± 1.622 ^{c}
Leaf length/width ratio	0.92333 ± 0.01936 ^{a}	0.958 ± 0.02974 ^{a}	0.9533 ± 0.0427 ^{a}
First lobe length (cm)	17.168 ± 1.489 ^{a}	12.832 ± 1.016 ^{b}	9.152 ± 1.081 ^{b}
First lobe width (cm)	5.801 ± 0.575 ^{a}	4.412 ± 0.549 ^{b}	3.023 ± 0.624 ^{c}
First lobe length/width ratio	2.991 ± 0.303 ^{a}	2.968 ± 0.2523 ^{a}	3.16 ± 0.475 ^{a}
Second lobe length (cm)	15.961 ± 1.587 ^{a}	11.839 ± 0.699 ^{b}	8.432 ± 0.861 ^{c}
Second lobe width (cm)	4.937 ± 0.496 ^{a}	3.672 ± 0.319 ^{b}	2.577 ± 0.526 ^{c}
Second lobe length/width ratio	3.2778 ± 0.2954 ^{a}	3.275 ± 0.1203 ^{a}	3.39 ± 0.426 ^{a}
Third lobe length (cm)	14.174 ± 1.402 ^{a}	10.009 ± 0.645 ^{b}	7.285 ± 0.769 ^{c}
Third lobe width (cm)	4.047 ± 0.541 ^{a}	2.832 ± 0.243 ^{b}	2.023 ± 0.334 ^{c}
Third lobe length/width ratio	3.57 ± 0.358 ^{a}	3.607 ± 0.2132 ^{a}	3.672 ± 0.33 ^{a}
Fourth lobe length (cm)	11.871 ± 1.122 ^{a}	8.121 ± 0.689 ^{b}	5.81 ± 0.845 ^{c}
Fourth lobe width (cm)	3.198 ± 0.517 ^{a}	2.183 ± 0.1863 ^{b}	1.523 ± 0.315 ^{c}
Fourth lobe length/width ratio	3.801 ± 0.403 ^{a}	3.812 ± 0.3083 ^{a}	3.967 ± 0.48 ^{a}
Fifth lobe length (cm)	9.416 ± 0.586 ^{a}	6.249 ± 1.029 ^{b}	4.35 ± 0.681 ^{c}
Fifth lobe width (cm)	2.303 ± 0.395 ^{a}	1.641 ± 0.338 ^{b}	1.055 ± 0.142 ^{c}
Fifth lobe length/width ratio	4.252 ± 0.63 ^{a}	3.923 ± 0.415 ^{a}	4.178 ± 0.426 ^{a}
Depth between first and second lobe (cm)	6.744 ± 0.866 ^{a}	4.494 ± 0.488 ^{b}	3.052 ± 0.494 ^{c}
Depth between second and third lobe (cm)	6.322 ± 0.772 ^{a}	4.3113 ± 0.3129 ^{b}	2.88 ± 0.459 ^{c}
Depth between third and fourth lobe (cm)	5.403 ± 0.567 ^{a}	3.5332 ± 0.2429 ^{b}	2.373 ± 0.394 ^{c}
Depth between fourth and fifth lobe (cm)	4.2 ± 0.512 ^{a}	2.7431 ± 0.1295 ^{b}	1.766 ± 0.303 ^{c}

3.3. Leaf Surface Ultrastructure (SEM)

Six *R. communis* samples were selected from the three subclusters (groups) that resulted from the cluster analysis based on vegetative morphology, illustrated with red asterisks in Figure 3a. The quantitative leaf ultrastructure characters of the six selected

R. communis samples are represented in Supplementary Table S2. *Ricinus communis* has an amphistomatous leaf; however, the stomatal count on the abaxial surface is higher than that of the adaxial one at a unit area (2560 × 1920 μm^2). The one-way ANOVA demonstrated that the stomatal length, width, and area at the closed state and the epidermal cell length/width ratio significantly differed between the abaxial and the adaxial surfaces (Supplementary Table S3).

The epidermal cell outline was undistinguishable (Figure 5a), isodiametric, pentagonal, hexagonal (Figure 5b), or oblong, and more or less arranged (Figure 5c). The anticlinal wall was straight, and the relief of the cell boundary was channeled. A convex periclinal wall was observed in the examined samples (Figure 5b,c). Generally, the fine relief of the periclinal wall was striate with different densities. A densely striate periclinal wall was detected in sample 22 on both surfaces and sample 12 on the AD surface (Figure 5d). Moderate striations were found on the AB surface of sample 11 (Figure 5e). Conversely, the remaining samples exhibited sparsely striated elements (Figure 5f). Furthermore, the AB surface of sample 4 showed rodlets of epicuticular crystalloid wax (Figure 5g), where the average length of the rodlets was 2.18 ± 1.5 μm, and their average width was 0.42 ± 0.2 μm (Figure 5h). Rarely, a sessile gland was found on the AB surface near the tips of the margin's teeth but not associated with every tooth (Figure 5i). The average gland size was 311.038 ± 9.69 × 354.8495 ± 111.99 μm. The paracytic stomata had smooth guard cell surfaces and elliptic pore shapes (Figure 5a–g).

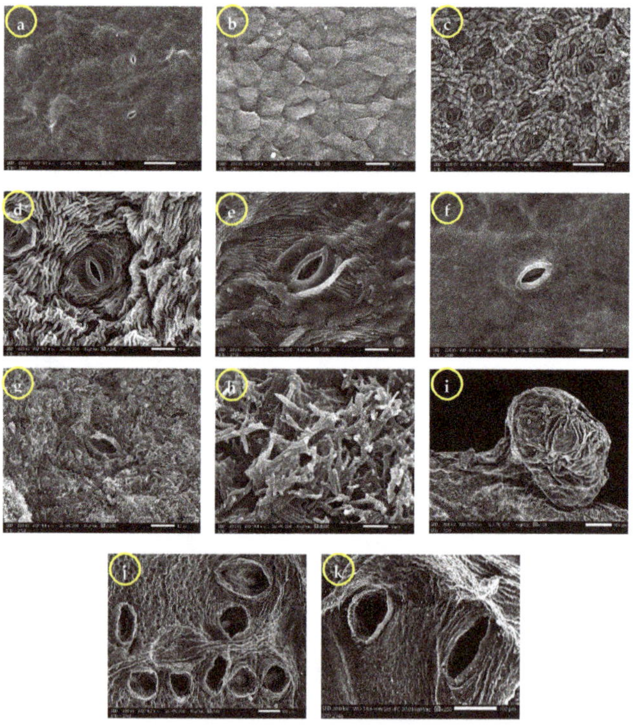

Figure 5. Scanning electron microscope (SEM) photomicrographs of *Ricinus communis* leaf surfaces. (**a**,**c**) General view showing the epidermal cell outline, anticlinal wall, and curvature of the outer periclinal wall. (**d**–**f**) Showing the striation density of the fine relief of the periclinal wall: (**d**) dense, (**e**) moderate, (**f**) sparse. (**g**,**h**) Rodlets of epicuticular crystalloid wax. (**i**–**k**) Anomalous stomata. Abaxial leaf surface (**a**,**f**–**i**). Adaxial leaf surface (**b**–**e**,**j**,**k**). Scale bar = 100 μm (i–k); 50 μm (**a**,**c**); 10 μm (**b**,**d**–**g**); 2 μm (**h**).

Anomalous stomata were noticed in their opened state on both leaf surfaces of samples 8 and 12 (Figure 5j,k, respectively). At the AB surface, the stomatal size was 166.47–286.33 × 94.33–175.33 µm, and their area was 15,004.28–39,840.30 µm^2. The stomatal pore size was 143.68–250.68 × 70.55–133.05 µm, and its area ranged from 8122.01 to 21,468.05 µm^2, while at the AD surface, the measured stomatal size was 139.60–333.94 × 88.83–150.85 µm and its area was 10,707.04–35,927.39 µm^2. Its pore size was 120.99–235.04 × 54.20–122.28 µm, and the pore area was 6063.53 to 20,388.87 µm^2.

For the AB surface, the stomatal size with the closed pore was 7.66–27.74 × 2.94–12.05 µm, and its area varied from 27.01 to 246.55 µm^2 (Figure 6a), showing the most variation between all samples, while the stomatal size with an opened pore was 11.24–24.54 × 4.81–13.20 µm with a pore size is 6.99–19.98 × 2.23–8.55 µm. The stomatal and pore areas were in the range of 42.04–207.88 µm^2 and 12.58–90.10 µm^2, respectively (Figure 6c), showing the low variation between all samples. The epidermal cell parameters (in terms of mean ± SD) were as follows: the length was 28.85 ± 7.04 µm, the width was 14.12 ± 4.75 µm, and the area was 427.52 ± 246.17 µm^2 (Figure 6e), showing the significant variation between all samples.

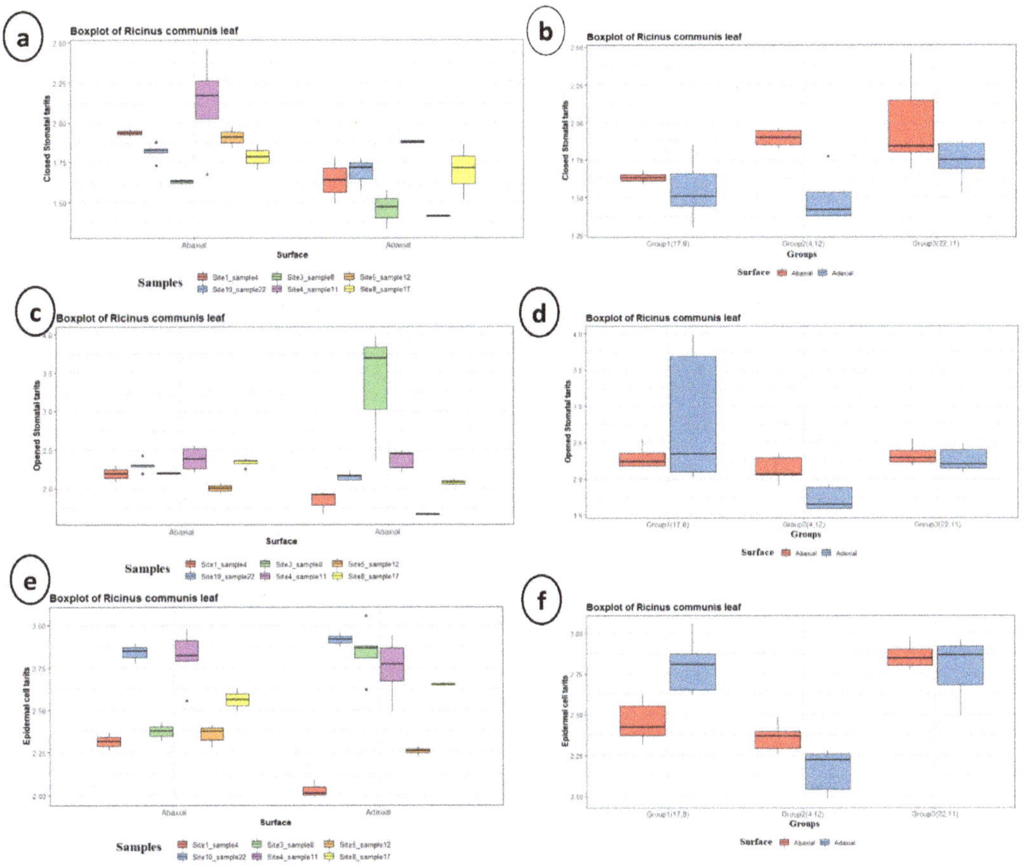

Figure 6. Boxplots of the leaf ultrastructure traits of *Ricinus communis* for all samples and for groups, respectively. (**a**,**b**) closed stomatal traits, (**c**,**d**) opened stomata, (**e**,**f**) epidermal cell size traits.

For the AD surface, showing a significant difference between all samples, the stomatal size with the closed pore was 5.82–13.17 × 2.66–6.30 μm, and its area varied from 10.24 to 56.13 μm^2 (Figure 6a), while the stomatal size with an opened pore was 6.76–132.49 × 3.38–65.73 μm with a pore size is 4.13–116.03 × 1.13–33.07 μm. The stomatal and pore areas were in the range of 16.95–6780.36 μm^2 and 3.30–2403.10 μm^2, respectively (Figure 6c). The epidermal cell length was 26.36 ± 9.19 μm, the width was 16.04 ± 6.00 μm, and the area was 451.23 ± 287.98 μm^2 (Figure 6e).

The two-way ANOVA (Table 5) showed a significant difference in the epidermal cell width and area of group 1 (samples 8 and 17), between their abaxial and adaxial surfaces (Figure 6f). Group 2 (samples 4 and 12) exhibited significant variation for the following characteristics: closed stomata length and area; opened stomata length, width and area; stomatal pore length and area; epidermal cell length and length/width ratio (Figure 6b,d). In contrast, group 3 (samples 11 and 22) showed no significant difference between the two surfaces. The number of stomata, closed stomata width, and the opened stomata length/width ratio represented non-significant variation for each group (Figure 6b,d,f).

Table 5. Two-way ANOVA analysis of *Ricinus communis* abaxial and adaxial leaf ultrastructure traits. Different letters represent significant differences.

Characters	Group 1 (Samples 8 and 17)		Group 2 (Samples 4 and 12)		Group 3 (Samples 11 and 22)	
	Abaxial	Adaxial	Abaxial	Adaxial	Abaxial	Adaxial
Number of stomata	14.67 ± 5.77 ^{a}	13.75 ± 1.258 ^{a}	16.5 ± 0.707 ^{a}	15 ± 1.41 ^{a}	25 ± 13 ^{a}	17.67 ± 7.23 ^{a}
Closed stomata length (μm)	9.095 ± 1.337 ^{bc}	8.567 ± 1.934 ^{c}	13.5 ± 1.238 ^{ab}	8.28 ± 3.21 ^{c}	16.68 ± 4.9 ^{a}	12.22 ± 1.315 ^{abc}
Closed stomata width (μm)	4.719 ± 0.395 ^{a}	3.889 ± 1.345 ^{a}	5.525 ± 0.602 ^{a}	3.737 ± 1.469 ^{a}	6.37 ± 3.1 ^{a}	4.478 ± 1.261 ^{a}
Closed stomata length/width ratio	1.948 ± 0.406 ^{b}	2.288 ± 0.44 ^{ab}	2.451 ± 0.1606 ^{ab}	2.2152 ± 0.0771 ^{ab}	2.957 ± 1.096 ^{a}	2.87 ± 0.635 a
Closed stomata area (μm^2)	29.4 ± 2.26 a ^{b}	26.18 ± 15.42 ^{b}	60.51 ± 9.43 ^{a}	21.81 ± 12.36 ^{b}	88.2 ± 72.9 ^{a}	40.45 ± 14.81 ^{ab}
Opened stomata length (μm)	16.78 ± 1.516 ^{a}	48.9 ± 53.7 ^{a}	14.26 ± 3.61 ^{a}	8.652 ± 2.046 ^{b}	19.263 ± 3.033 ^{a}	16.922 ± 3.069 ^{a}
Opened stomata width (μm)	9.323 ± 2.19 ^{ab}	26 ± 25.6 ^{a}	6.919 ± 1.473 ^{b}	4.567 ± 1.076 ^{c}	9.177 ± 1.37 ^{ab}	8.577 ± 1.765 ^{ab}
Opened stomata length/width ratio	1.859 ± 0.337 ^{a}	1.666 ± 0.294 ^{a}	2.076 ± 0.358 ^{a}	1.921 ± 0.321 ^{a}	2.114 ± 0.276 ^{a}	2.006 ± 0.322 ^{a}
Opened stomata area (μm^2)	122.3 ± 37.4 ^{a}	2099 ± 2996 ^{a}	80.6 ± 36.9 ^{a}	27.31 ± 11.71 ^{b}	132.6 ± 38 ^{a}	110 ± 43.2 ^{a}
Stomatal pore length (μm)	11.07 ± 2.99 ^{a}	41.2 ± 48.1 ^{a}	9.91 ± 2.86 ^{a}	5.212 ± 1.202 ^{b}	13.27 ± 2.975 ^{a}	11.766 ± 2.614 ^{a}
Stomatal pore width (μm)	5.004 ± 2.176 ^{a}	12.97 ± 12.99 ^{a}	4.021 ± 1.12 ^{ab}	2.319 ± 1.255 ^{b}	3.611 ± 0.937 ^{ab}	4.127 ± 1.374 ^{a}
Stomatal pore length/width ratio	2.411 ± 0.663 ^{b}	2.621 ± 0.735 ^{ab}	2.491 ± 0.336 ^{b}	2.555 ± 0.816 ^{b}	3.808 ± 0.894 ^{a}	3 ± 0.631 ^{ab}
Stomatal pore area (μm^2)	38.8 ± 26.9 ^{a}	743 ± 1063 ^{a}	29.06 ± 14.69 ^{a}	8.29 ± 5.11 ^{b}	33.36 ± 18.13 ^{a}	35.25 ± 18.73 ^{a}
Epidermal cell length (μm)	25.4 ± 3.07 ^{bc}	30.57 ± 5.12 ^{ab}	22.42 ± 2.7 ^{c}	13.909 ± 1.818 ^{d}	35.32 ± 4.51 ^{a}	31.94 ± 5.65 ^{ab}
Epidermal cell width (μm)	11.47 ± 3.68 ^{bc}	21.26 ± 5 ^{a}	10.491 ± 2.224 ^{c}	10.75 ± 2.97 ^{c}	17.83 ± 3.96 ^{a}	17.06 ± 5.43 ^{ab}
Epidermal cell length/width ratio	2.369 ± 0.663 ^{a}	1.494 ± 0.393 ^{ab}	2.224 ± 0.54 ^{a}	1.389 ± 0.453 ^{b}	2.054 ± 0.428 ^{ab}	1.977 ± 0.52 ^{ab}
Epidermal cell area (μm^2)	260.4 ± 85.1 ^{b}	600.9 ± 246.8 ^{a}	199.8 ± 44.1 ^{bc}	123.7 ± 38.5 ^{c}	674.6 ± 121.5 ^{a}	623.7 ± 200.5 ^{a}

3.4. Soil Analysis

The soil physicochemical properties of the studied sites (1, 3–5, 8, and 10) are illustrated in Table 6. The mechanical analysis revealed that the studied sites have calcareous loamy sand with a saturation percentage between 39 (site 10) and 59 (site 5). The highest pH was recorded for site 3 (7.91), and the lowest was 7.3 for site 8. The electrical conductivity ranged from 4.05 dS/m (site 8) to 65 dS/m (site 1). Site 1 showed the highest EC (65.55 dS/m) and the highest concentration of Ca^{++}, Mg^{++}, Na^{+}, K^{+}, HCO_3^{-}, Cl^{-}, SO_4^{--} (464, 246, 315, 4.7, 16, 278, and 735.7 meq/L, respectively), SAR (16.7), available K (348 ppm), and $CaCO_3$ (25%), while site 8 showed the lowest EC (4.05 dS/m) and the lowest concentration of Ca^{++}, Mg^{++}, Na^{+}, K^{+}, HCO_3^{-}, Cl^{-}, SO_4^{--} (28.0, 14.6, 19.6, 2.1, 5.2, 20.2, and 38.9 meq/L, respectively), SAR (4.2), and $CaCO_3$ (15.7%). The lowest values of organic matter (0.06%), available nitrogen (0.56 ppm), and available phosphorus (6 ppm) were recorded at site 1. The highest concentration of available phosphorus (16 ppm), and the lowest available potassium (137 ppm) was detected at site 10. Minor concentrations of iron, zinc, manganese, and organic matter were recognized in the studied samples. Bar plots and boxplots of soil traits per site and group are illustrated in Figure 7. The one-way ANOVA of the soil traits showed non-significant variation for the three studied groups (Supplementary Table S4).

Table 6. Physicochemical soil properties of *Ricinus communis*. Where SP refers to Saturation Percentage, and SAR refers to Sodium Adsorption Ration. Different letters represent significant differences.

Variable	Mean ± StDev	SE Mean	Min.	Q1	Median	Q3	Max.	IQR
Gravel (%)	3.533 ± 1.317 [c]	0.538	1.7	2.375	3.75	4.25	5.6	1.875
Sand (%)	73.5 ± 13.05 [abc]	5.33	60	61.5	72	86.25	90	24.75
Silt (%)	19 ± 9.25 [c]	3.78	8	9.5	19.5	27.75	30	18.25
Clay (%)	7.5 ± 4.04 [c]	1.65	2	4.25	7.5	11.25	12	7
SP (%)	50.67 ± 6.98 [abc]	2.85	39	45.75	51	56.75	59	11
pH	7.5483 ± 0.2046 [c]	0.0835	7.3	7.39	7.55	7.6475	7.91	0.2575
EC (dS/m)	17.3 ± 23.75 [c]	9.7	4.05	5.96	8.25	24.71	65.55	18.75
Ca^{++} (meq/L)	118.2 ± 169.8 [abc]	69.3	28	43.8	55	159.5	464	115.8
Mg^{++} (meq/L)	63.1 ± 89.8 [abc]	36.7	14.6	23.1	29.5	86.3	246	63.1
Na^{+} (meq/L)	79.1 ± 115.8 [abc]	47.3	19.6	25.9	36.5	108	315	82.1
K^{+} (meq/L)	3.217 ± 0.999 [c]	0.408	2.1	2.475	2.9	4.25	4.7	1.775
HCO_3^{-} (meq/L)	9.18 ± 3.8 [c]	1.55	5.2	6.17	8.45	11.88	16	5.7
Cl^{-} (meq/L)	73 ± 100.6 [abc]	41.1	20.2	28.3	35.3	98.4	278	70.1
SO_4^{--} (meq/L)	181 ± 272 [ab]	111	39	61	80	248	736	187
SAR	7.1 ± 4.75 [c]	1.94	4.2	4.5	5.65	8.52	16.7	4.02
N (ppm)	0.93 ± 0.275 [c]	0.112	0.56	0.643	0.975	1.15	1.3	0.508
P (ppm)	9.65 ± 3.7 [c]	1.51	6	6.97	8.3	13	16	6.03
K (ppm)	192.2 ± 78.5 [a]	32.1	137	146	167	225.8	348	79.8
Fe (ppm)	0.643 ± 0.279 [c]	0.114	0.28	0.408	0.62	0.907	1.02	0.5
Zn^{++} (ppm)	0.3217 ± 0.0679 [c]	0.0277	0.27	0.2775	0.295	0.3675	0.45	0.09
Mn^{++} (ppm)	0.2883 ± 0.1158 [c]	0.0473	0.18	0.2175	0.23	0.405	0.48	0.1875
Cu^{++} (ppm)	0.2133 ± 0.1724 [c]	0.0704	0.03	0.06	0.175	0.3925	0.46	0.3325
O.M. (%)	0.1267 ± 0.0758 [c]	0.0309	0.06	0.06	0.1	0.215	0.23	0.155
$CaCO_3$ (%)	19.67 ± 3.41 [c]	1.39	15.7	16.9	19	22.75	25	5.85

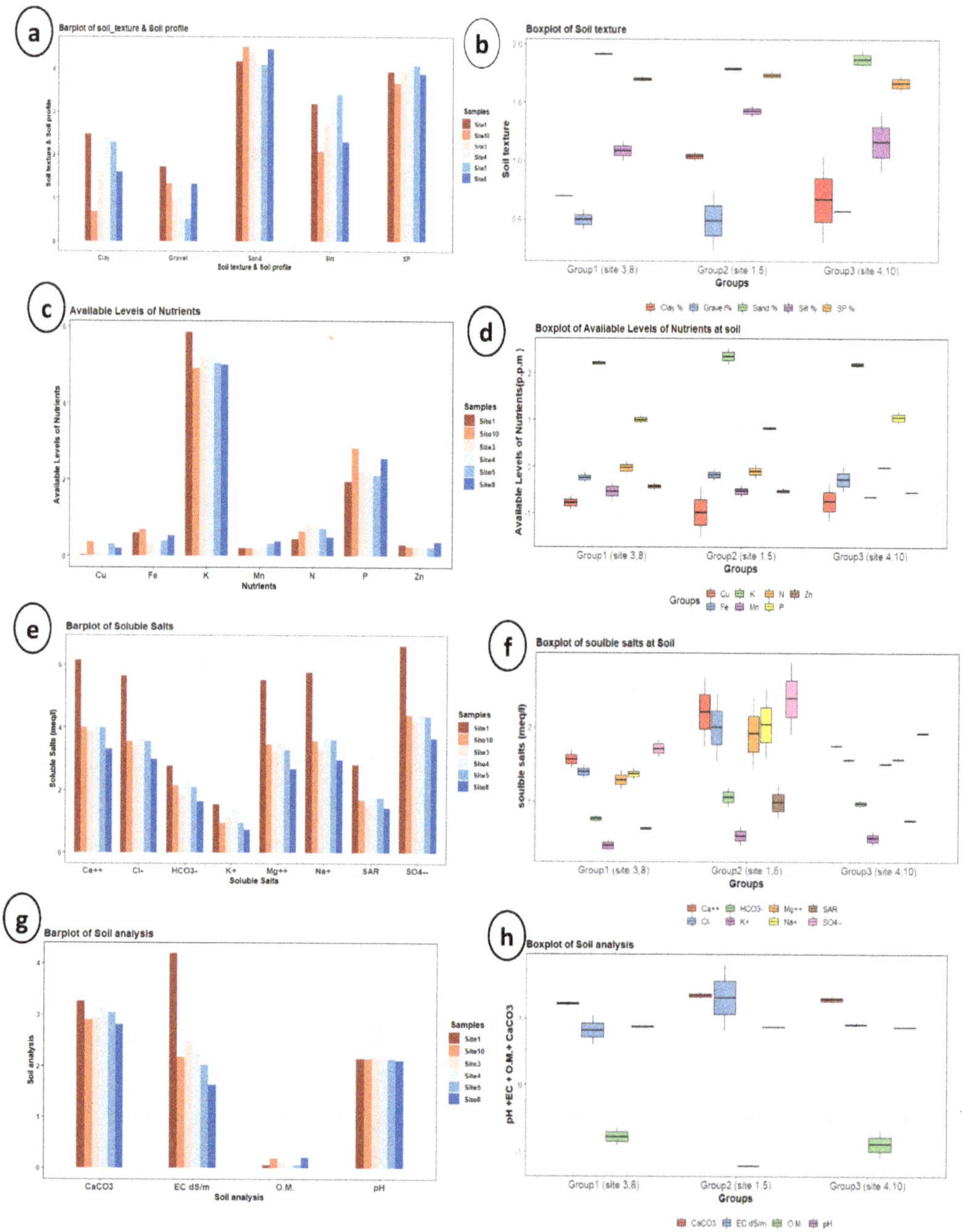

Figure 7. Bar plots and boxplots for soil analysis per site and for groups, respectively: (**a**,**b**) soil texture and profile, (**c**,**d**) nutrient concentrations, (**e**,**f**) soluble salts concentration, and (**g**,**h**) $CaCO_3$ (%), EC (ds/m), organic matter (O.M %), and pH.

3.5. Gas Chromatography-Mass Spectrometry (GC-MS)

The hexane extracts of *R. communis* samples using GC-MS are presented in Table 7. The total ion chromatograms are shown in Figure 8. The peaks were distributed at a retention time of 6.36–34.31. Comparing the analytical data revealed differences in the quantitative composition of phytochemicals between different samples.

Table 7. Major phytochemical compounds identified in hexane extract of *Ricinus communis*. RT: Retention time (in minutes); MF: molecular formula; MW: molecular weight. Different letters represent significant differences.

Rt	Phytochemical Compounds	M.F	M.W	Mean ± StDev	SE Mean	Min.	Q1	Median	Q3	Max.	IQR
10.97	Isophytol	$C_{20}H_{40}O$	296	1.383 ± 0.965c	0.394	0.42	0.495	1.3	2.03	3.08	1.535
13.53	n-Hexadecanoic acid	$C_{16}H_{32}O_2$	256	29.78 ± 6.15a	2.51	22.04	23.66	30.67	35.21	36.25	11.55
16.88	9,12,15-Octadecatrienoic acid	$C_{18}H_{30}O_2$	278	22.65 ± 3.92b	1.6	17.31	18.98	23.26	25.97	26.86	6.99
15.27	Oleic acid	$C_{18}H_{34}O_2$	282	0.22 ± 0.1361c	0.0556	0.13	0.1375	0.17	0.2875	0.49	0.15
17.10	Octadecanoic acid	$C_{18}H_{36}O_2$	284	3.63 ± 0.676c	0.276	3	3.203	3.5	3.912	4.94	0.71
19.24	Tributyl acetylcitrate	$C_{20}H_{34}O_8$	402	1.552 ± 1.54c	0.629	0.15	0.165	1.145	3.25	3.55	3.085

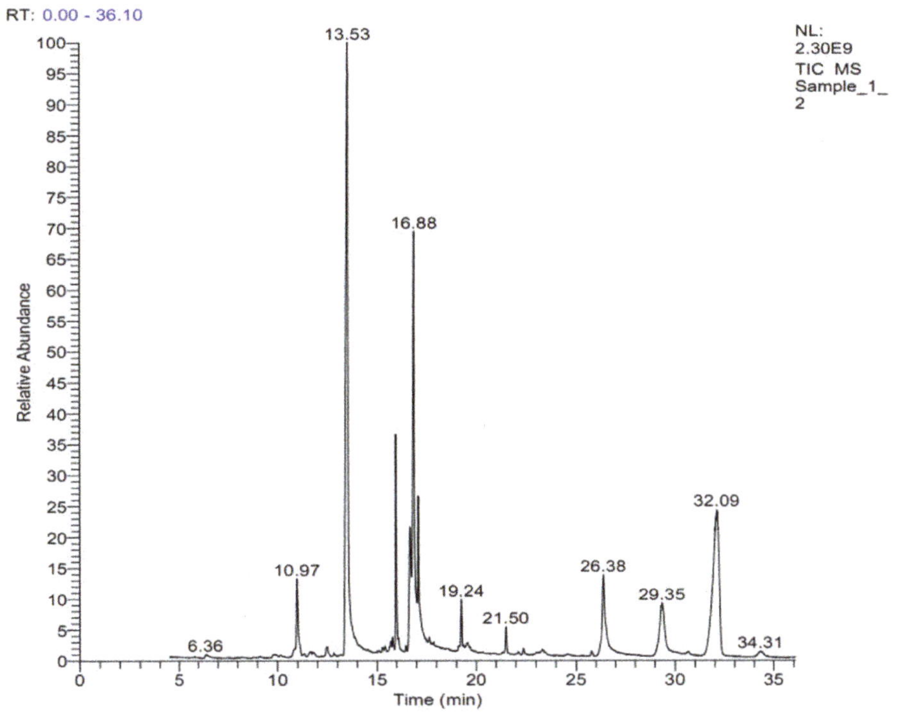

Figure 8. GC-MS chromatogram of *Ricinus communis* hexane extract.

Six major common compounds were detected in the chromatograms of the studied samples. Oleic acid represented the lowest values for the studied samples. The highest percentages for n-Hexadecanoic acid (36.25%) and 9,12,15-Octadecatrienoic acid (26.86%) were recorded in the hexane extract of site 5, whereas their lowest values were recorded for sites 10 and 1, respectively. The highest concentration of Isophytol was detected for site 1 (3.08%), and the lowest was site 10 (0.42%). Octadecanoic acid had its highest value (4.94%) for site 8 and the lowest (3.0%) for site 4. The highest concentration of Tributyl Acetylcitrate was found in site 10 (3.55%), and the lowest was 0.15 in site 8. Bar plots and boxplots of the identified phytochemical compounds concentrations per site and group are illustrated in Figure 9. The one-way ANOVA of the phytochemical concentrations exhibited non-significant variation for the three studied groups (Supplementary Table S5).

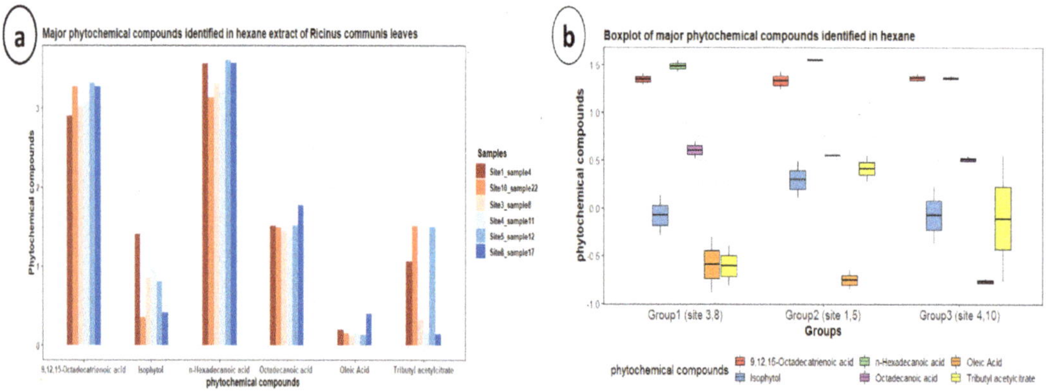

Figure 9. (**a**,**b**) Bar plots and boxplots for major phytochemical compounds identified in hexane extract of *Ricinus communis* per site and for groups.

3.6. Correlation Analysis

Pearson's correlation analysis was performed to explore the relationship and correlation between examined parameters (the macro- and micromorphological characters, soil analysis, and chemical composition) (Figure 10a,b). The results showed a significant positive correlation ($p \leq 0.05$) between vegetative morphological traits, which confirms the discrimination of the resulting groups (morphotypes).

Plant length showed a positive correlation with leaf size parameters (length, width, and area), the length and width of each lobe, depth between first and second lobe, depth between second and third lobe, depth between third and fourth lobe, depth between fourth and fifth lobe, pH, and N. The leaf size parameters and lobes dimensions (first and second lobe dimensions, and the third, fourth and fifth lobe lengths) positively correlate with the soil pH and nitrogen content. A positive correlation exists between petiole length/leaf length ratio and petiole attachment against EC, soluble salts (Ca^{++}, Mg^{++}, Na^{+}, K^{+}, HCO_3^{-}, Cl^{-}, and SO_4^{--}), $CaCO_3$, SAR, and isophytol. Phosphorus negative correlated with EC, soluble salts (Ca^{++}, Mg^{++}, Na^{+}, K^{+}, HCO_3^{-}, Cl^{-}, and SO_4^{--}), $CaCO_3$, SAR, N, Fe, Zn^{++}, and Mn^{++}. A negative correlation was found between EC, soluble salts (Ca^{++}, Mg^{++}, Na^{+}, K^{+}, HCO_3^{-}, Cl^{-}, and SO_4^{--}) and SAR against Cu^{++}, Mn^{++}, N, and P (Figure 10a).

The analysis revealed that the leaf length, width, and area positively correlated with the opened stomata and the pore lengths, widths, and areas at the AD surface. pH and nitrogen are positively correlated with the opened stomata and the pore length, width, and area in the AD leaf surface, while leaf size parameters negatively correlate with the closed stomata and the epidermal cell lengths, widths, and areas at both leaf surfaces. The soil-soluble HCO_3^{-} positively correlated with length, width, and area of the closed stomata at both leaf surfaces. Oleic and octadecanoic acids are positively correlated with the stomatal pore size parameters at the AB surface of the leaf (Figure 10b).

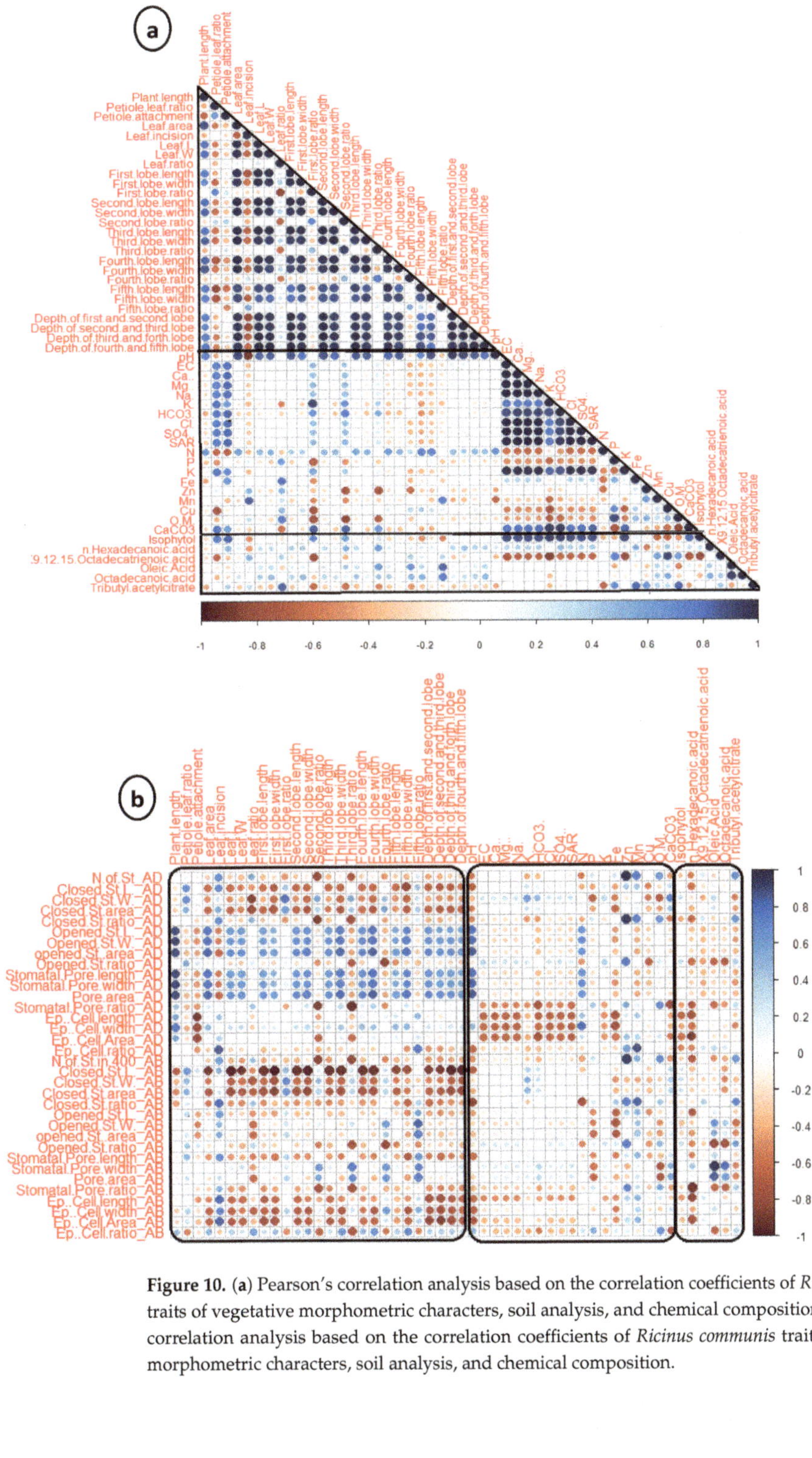

Figure 10. (**a**) Pearson's correlation analysis based on the correlation coefficients of *Ricinus communis* traits of vegetative morphometric characters, soil analysis, and chemical composition. (**b**) Pearson's correlation analysis based on the correlation coefficients of *Ricinus communis* traits of vegetative morphometric characters, soil analysis, and chemical composition.

4. Discussion

Plant growth is affected by stressful environmental conditions such as soil salinity and drought [72]. One of the highly adaptable plant traits to environmental change is leaf size [32,73]. Soil nitrogen is essential for plant development and growth [74]. Some researchers have reported that plant growth parameters, such as leaf size and area, are affected by many abiotic factors, including soil pH and nitrogen supply [75,76]. The latter outcomes are congruent with the present study, where the leaf and lobe size parameters positively correlate with the soil nitrogen concentration (Figure 10a). The plant length demonstrated a positive correlation with leaf parameters, and a similar finding was reported by Liu et al. [77]. Phosphorus negatively correlated with EC, soluble salts (Ca^{++}, Mg^{++}, Na^+, K^+, HCO_3^-, Cl^-, and SO_4^{--}), $CaCO_3$, SAR, N, Fe, Zn^{++}, and Mn^{++}. The phosphorus uptake decreased due to the precipitation of phosphate ions with Ca^{++} in salt-stressed soil [78]. A positive correlation exists between petiole length/leaf length ratio and petiole attachment against EC, soluble salts (Ca^{++}, Mg^{++}, Na^+, HCO_3^-, Cl^-, SO_4^{--}), $CaCO_3$, SAR, and isophytol. Hammad et al. [72] stated that environmental factors influenced castor oil composition.

The physiology and anatomical properties of a plant's stomata are highly plastic and are affected by various environmental factors, such as the availability of nutrients [79]. The correlation analysis examined whether the macromorphological characteristics were associated with leaf ultrastructure. The analysis revealed that the leaf length, width, and area at the adaxial surface positively correlated with the opened stomata and the pore lengths, widths, and areas. In turn, the latter-mentioned traits and the epidermal cell length and area positively correlate with soil pH and nitrogen concentration (Figure 10b). Generally, the ionic form in which the element is located in the soil is affected by pH [80]. However, these leaf size parameters negatively correlate with the closed stomata and the epidermal cell lengths, widths, and areas at both leaf surfaces (Figure 10b). This is incompatible with Murphy et al. [73], who tested a field-grown woody species and proposed that differences in leaf stomatal traits are induced by sun and shade but not regulated by leaf size. Oleic and Octadecanoic acids correlated positively with the stomatal pore. Mohamed et al. [81] reported that the stomatal aperture and fatty acid composition are highly correlated.

Ricinus communis growing along roadsides and wastelands of Egypt is affected by several abiotic stresses [82]. Anomalous large stomata were observed on the leaf surfaces of samples 8 (collected from site 3, 39 km from Alexandria–Matrouh International Coastal Road, Sidi Kirayr) and 12 (collected from site 5, 1 km from East Omayed Biosphere Reserve, Alexandria Desert Road). Plants absorb minerals such as magnesium, calcium, and copper from the soil. Such anomalies in the stomata may be ascribed to the accumulation and transportation of Mg^{++} ions, which may interfere with the functions and mobilization of other mineral ions and cause leaf distortion [83]. Otherwise, copper deficiency in the studied sites may lead to morphological alterations in the leaf architecture, where it is a vital element with different functions in plant development and metabolism [84,85]. Moreover, the distortion of stomata may be attributed to the existence of some environmental pollutants or phytotoxic gases at sites 3 and 5. Angeles et al. [86] reported that the diffusion of these pollutants through the stomatal opening according to different concentration gradients caused lesions and a distorted stomatal complex on leaves of *Mangifera indica* L. Plants located in the study area, the northwestern coastal desert of Egypt, are under pressure resulting from anthropogenic disturbance, agricultural practices, pollution, and urbanization [87]. Even the Omayed Biosphere Reserve has recently encountered new human-induced disturbances [88,89].

Important morphological markers of *R. communis* were previously used in the identification, and to verify the genuineness, of different castor varieties. These markers included plant stature, stem color, the presence/absence of anthocyanin pigmentation, stem wax, type of internode, bloom, branching habits, leaf shape, and adaxial and abaxial leaf surface traits [26–29]. The multivariate (clustering and PCA) analysis of 25 *R. communis* samples based on 34 vegetative morphometric characters revealed three subclusters (groups). A heatmap of all the combined traits, vegetative morphometry, leaf micromorphology, and soil and GC-MS analysis (Figure 11) confirmed the discrimination of the three groups or morphotypes resulting from multivariate vegetative morphological characteristics analysis. Each group attained some specific characteristics discriminating it from the other groups (Tables 4 and 5). Samples 1–4, 12–13, and 20–21 collected from sites 1, 5, and 9, respectively, had purple stems. These samples had no more specific characteristics; thus, they were scattered across the three groups. Wahibah et al. [28] reported that the purple-stemmed *R. communis* variety with an anthocyanin coloration reflected the intra-specific diversity of the species. According to Shankar et al. [90] and Santha et al. [91], this phenotype is resistant to the severe wilt syndrome caused by *Fusarium oxysporum* f. sp. *ricini*, which leads to up to 77% yield loss based on the degree of infection. Its ability to resist fungal growth indicates the existence of a specific resistance mechanism in this genotype. This variety provides valuable genetic resources to produce wilt-resistant genotypes.

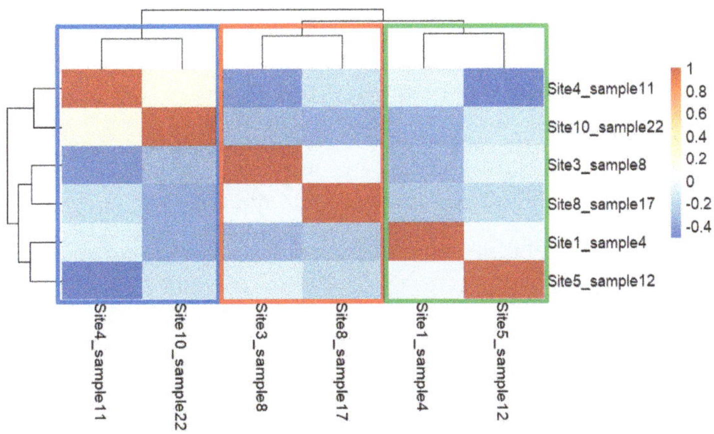

Figure 11. Heatmap analysis of *Ricinus communis* combined traits (vegetative morphometry, leaf micromorphology, soil analysis, and GC-MS analysis).

Morphological traits are useful indicators of environmental stress [92]. For the vegetative morphometry, group one represented the highest range of leaf length (22.73–27.49 cm) and width (23.88–30.52 cm), first lobe length and width (15.13–20.06 cm and 5.05–7.00 cm, respectively), second lobe length (14.10–18.84 cm) and width (4.13–5.51 cm), third lobe length (12.50–16.62 cm), fourth lobe length (10.76–13.69 cm), fifth lobe length (8.62–10.21 cm), depth between first and second lobe (5.43–8.04 cm), depth between second and third lobe (5.13–7.39 cm), depth between third and fourth lobe (4.53–6.14 cm), depth between fourth and fifth lobe (3.71–5.10 cm), and macrophyll leaf blade with leaf area (207.04–352.89 cm^2). For SEM leaf characteristics, this group was characterized by the lowest number of closed stomata at the AB surface. For the soil chemical characteristics, this group (sites 3 and 8) exhibited the lowest values for soluble salts (Ca^{++}, Mg^{++}, Na^{+}, HCO_3^{-}, Cl^{-}, SO_4^{--}) and SAR. Members of this group are located at coastal sites, where plants grown along coastal areas are exposed to salt spray [92]. The macrophyll leaf blade characterizing this group reveals the ability of its individuals to retain a sizeable transpiring surface area, which is considered a sort of acclimation to salt stress [93,94].

Group two exemplified the medium range of leaf length and width (16.19–20.35 cm and 17.58–21.63 cm, respectively), first lobe length and width (10.79–14.08 cm and 3.53–5.44 cm, respectively), second lobe length (10.53–12.70 cm) and width (3.17–4.14 cm), third lobe length (9.19–11.26 cm), fourth lobe length (7.13–9.28 cm), depth between first and second lobe (3.60–5.24 cm), depth between second and third lobe (3.83–4.85 cm), depth between third and fourth lobe (3.10–3.83 cm), depth between fourth and fifth lobe (2.54–2.91 cm), and mesophyll leaf blade with leaf area (117.31–181.24 cm^2). For SEM leaf characteristics, group 2 (samples 4 and 12) showed smaller values than the other two groups. For the adaxial leaf surface, the opened stomata width and area was 3.38–5.83 µm and 16.95–40.28 μm^2, respectively, pore area was 3.30–14.98 μm^2, and epidermal cell length and area was 11.77–16.44 µm and 74.51–163.22 μm^2, respectively. Sample 4 specifically (collected from site 1 at Mehwar Al-Ta'ameir, about 700 m from kilo 26 wastewater treatment plant, and 26 km from West Alexandria) had a small epidermal cell size at both leaf surfaces. This is in accordance with Cookson et al. [95], who mentioned that area of the leaf is affected by environmental factors because of the cell number and/or cell size differences. Additionally, rodlets of epicuticular crystalloid wax were observed on its AB surface. The presence of epicuticular waxes reveals how plants interact with their environment [96]. According to the soil chemical characters, this group (sites 1 and 5) displayed an example of wasteland areas. It recorded the lowest values for the available potassium, copper, and organic matter, whereas the highest was for $CaCO_3$. Furthermore, this group demonstrated a minor nutrient concentration of nitrogen and phosphorus. The depletion of nutrients, especially nitrogen, and phosphorus, characterizes wasteland areas [97]. Plants growing in such areas demonstrated a reduction in their growth and productivity due to some physiological and biochemical changes [98]. Nevertheless, site 1 represented the highest EC and concentrations of Ca^{++}, Mg^{++}, Na^+, K^+, HCO_3^-, Cl^-, SO_4^{--}, and SAR, copper, and organic matter. Jeschke and Wolf [99] mentioned that higher concentrations of NaCl enhanced the reduction in *R. communis* growth and leaf size, and suppressed shrub branching.

The third group had the lowest range of leaf length (11.60–15.42 cm) and width (12.30–16.06 cm), first lobe length and width (8.22–10.76 cm and 2.17–3.63 cm, respectively), second lobe length (7.55–9.65 cm) and width (1.97–3.35 cm), third lobe length (6.64–8.34 cm), fourth lobe length (4.93–7.15 cm), depth between first and second lobe (2.25–3.67 cm), depth between second and third lobe (2.11–3.38 cm), depth between third and fourth lobe (1.85–2.89 cm), depth between fourth and fifth lobe (1.44–2.22 cm), and mesophyll leaf blade with leaf area (52.94–106.81 cm^2). For SEM leaf characteristics, the highest number of closed stomata, and the highest scores for the epidermal cell area were noticed at the abaxial leaf surface (547.82–889.71 μm^2) of group 3 (samples 11 and 22). This group includes individuals growing in inland sites. Our findings are in accordance with Papazoglou [100], who reported that *R. communis* showed a tolerance capability under water stress conditions, although the leaf area decreased due to water scarcity. As an adaptive response to drought stress, plants reduce their leaf area and keep their stomata closed to minimize water losses through transpiration [101,102]. The lowest percent of n-Hexadecanoic acid is noticed for group 3 (22.04–24.2%), in comparison with the other two groups (26.86–36.25%). The Palmitic acid concentration is affected by drought [103,104]. n-Hexadecanoic acid is also called Palmitic acid and functions as an anti-inflammatory agent to treat rheumatic symptoms [105].

The results confirmed that *R. communis* could adapt to stresses in coastal lands or inland habitats, causing phenotypic and chemical variations to cope with the different environmental stresses in the study area. However, the soil heterogeneity may be ascribed to the soil's different nature and geomorphologic characteristics. A similar adaptive response was detected for some xerophytic taxa on the Al-Alamein–Alexandria International Desert Road of Egypt [87].

5. Conclusions

Ricinus communis samples were subjected to different analyses, such as analyses of vegetative morphometry, leaf surface ultrastructure, soil analysis, and GC-MS analysis, to assess the varietal diversity of the species. Multivariate analysis based on 34 quantitative and qualitative morphological traits distinguished 25 castor samples into three main groups (morphotypes). The results of UPGMA cluster analysis and PCA were consistent with each other. There was a discrimination of the three groups: the first one enclosed individuals inhabiting coastal habitats and was characterized by macrophyll leaf blades with larger leaf length, width, and area. The third group's individuals that occupied inland sites had the smallest leaf size parameters. In contrast, the second group was in the middle of the other two groups. The present study revealed that the following traits are important for studying the varietal identification of *R. communis*; leaf size parameters, leaf area, blade class, lobes size and shapes, stomatal and pore size, and epidermal cell size parameters. Some *R. communis* samples had purple stems, while the others were grayish-green. The purple-stemmed phenotype has no more specific characteristics; however, it may reflect the intra-specific diversity of the species. *Ricinus communis* revealed an adaptive growth capability, where plants inhabiting coastal sites are salt-sensitive, while inland plants are a relatively drought-tolerant species. The intra-specific variation between *R. communis* morphotypes indicated the possibility of the direct and indirect use of these varieties in genetic improvement programs of the species. Further molecular studies should be implemented to evaluate the genetic diversity among these morphotypes.

Supplementary Materials: The following are available online at https://www.mdpi.com/article/10.3390/agronomy13040985/s1, Figure S1: Map of Egypt indicating the sampling sites of *Ricinus communis*; Table S1: Qualitative characters of *Ricinus communis* vegetative morphometry. The plant's stature ranks: very tall ">2.50 m", tall "2.0–2.5 m", medium "1.51–2.0 m", and short "1.0–1.5 m"; Table S2: Quantitative characteristics of abaxial leaf (AB) and adaxial leaf (AD) surface ultrastructure of *Ricinus communis*; Table S3: One-way ANOVA of abaxial leaf and adaxial leaf surface ultrastructure characters of *Ricinus communis*; Table S4: The one-way ANOVA of the soil traits showed non-significant variation for the three studied groups of *Ricinus communis*; Table S5: The one-way ANOVA of the phytochemical concentration exhibited non-significant variation for the three studied groups of *Ricinus communis*.

Author Contributions: Conceptualization, I.H.N.; methodology, I.H.N., K.A., F.Y.E., R.B., T.I.S., A.E.-B. and S.M.A.R.; software, I.H.N. and F.Y.E.; validation, I.H.N., K.A., F.Y.E., R.B., T.I.S., A.E.-B. and S.M.A.R.; formal analysis, I.H.N., F.Y.E. and S.M.A.R.; investigation, I.H.N., R.B. and S.M.A.R.; resources, I.H.N., R.B. and S.M.A.R.; data curation, I.H.N., F.Y.E. and S.M.A.R.; writing—original draft preparation, I.H.N., K.A., F.Y.E., R.B., T.I.S., A.E.-B. and S.M.A.R.; writing—review and editing, I.H.N., F.Y.E. and A.E.-B.; visualization, I.H.N., K.A., F.Y.E., R.B., T.I.S., A.E.-B. and S.M.A.R.; Funding acquisition: I.H.N., K.A., F.Y.E., R.B., T.I.S., A.E.-B. and S.M.A.R. All authors have read and agreed to the published version of the manuscript.

Funding: This research received no external funding.

Data Availability Statement: The data presented in this study are available in this article.

Acknowledgments: Not applicable.

Conflicts of Interest: The authors declare no conflict of interest.

References

1. Kim, H.; Lei, P.; Wang, A.; Liu, S.; Zhao, Y.; Huang, F.; Yu, Z.; Zhu, G.; He, Z.; Tan, D.; et al. Genetic Diversity of Castor Bean (*Ricinus communis* L.) Revealed by ISSR and RAPD Markers. *Agronomy* **2021**, *11*, 457. [CrossRef]
2. Sahney, M.; Rahi, S.; Kumar, A.; Jaiswal, R. Melissopalynological Studies on Winter Honeys from Allahabad, Uttar Pradesh, India. *Palynology* **2018**, *42*, 540–552. [CrossRef]
3. Yang, S.; Mao, L.; Zheng, Z.; Chen, B.; Li, J. Pollen Atlas for Selected Subfamilies of Euphorbiaceae from Southern China: A Complementary Contribution to Quaternary Pollen Analysis. *Palynology* **2020**, *44*, 659–673. [CrossRef]
4. Kiran, B.R.; Prasad, M.N.V. *Ricinus communis* L. (Castor Bean), a Potential Multi-Purpose Environmental Crop for Improved and Integrated Phytoremediation. *EuroBiotech J.* **2017**, *1*, 101–116. [CrossRef]
5. Chan, A.P.; Crabtree, J.; Zhao, Q.; Lorenzi, H.; Orvis, J.; Puiu, D.; Melake-Berhan, A.; Jones, K.M.; Redman, J.; Chen, G.; et al. Draft Genome Sequence of the Oilseed Species *Ricinus communis*. *Nat. Biotechnol.* **2010**, *28*, 951–956. [CrossRef]
6. Perea-Flores, M.J.; Chanona-Pérez, J.J.; Garibay-Febles, V.; Calderón-Dominguez, G.; Terrés-Rojas, E.; Mendoza-Pérez, J.A.; Herrera-Bucio, R. Microscopy Techniques and Image Analysis for Evaluation of Some Chemical and Physical Properties and Morphological Features for Seeds of the Castor Oil Plant (*Ricinus communis*). *Ind. Crops Prod.* **2011**, *34*, 1057–1065. [CrossRef]

7. Enan, M.; Al-Deeb, M.; Fawzy, N.; Amiri, K. DNA Barcoding of *Ricinus communis* from Different Geographical Origin by Using Chloroplast Matk and Internal Transcribed Spacers. *Am. J. Plant Sci.* **2012**, *3*, 1304–1310. [CrossRef]
8. Taur, D.J.; Waghmare, M.G.; Bandal, R.S.; Patil, R.Y. Antinociceptive Activity of *Ricinus communis* L. Leaves. *Asian Pac. J. Trop. Biomed.* **2011**, *1*, 139–141. [CrossRef]
9. Ohishi, K.; Toume, K.; Arai, M.A.; Sadhu, S.K.; Ahmed, F.; Mizoguchi, T.; Itoh, M.; Ishibashi, M. Ricinine: A Pyridone Alkaloid from *Ricinus communis* That Activates the Wnt Signaling Pathway through Casein Kinase 1α. *Bioorg. Med. Chem.* **2014**, *22*, 4597–4601. [CrossRef]
10. Rakesh, M.R.; Kabra, M.P.; Rajkumar, V.S. Evaluation of Antiulcer Activity of Castor Oil in Rats. *Int. J. Res. Ayurveda Pharm.* **2011**, *2*, 1349–1353.
11. Tunaru, S.; Althoff, T.F.; Nüsing, R.M.; Diener, M.; Offermanns, S. Castor Oil Induces Laxation and Uterus Contraction via Ricinoleic Acid Activating Prostaglandin EP3 Receptors. *Proc. Natl. Acad. Sci. USA* **2012**, *109*, 9179–9184. [CrossRef] [PubMed]
12. Iqbal, J.; Zaib, S.; Farooq, U.; Khan, A.; Bibi, I.; Suleman, S. Antioxidant, Antimicrobial, and Free Radical Scavenging Potential of Aerial Parts of *Periploca aphylla* and *Ricinus communis*. *Int. Sch. Res. Not.* **2012**, *2012*, 563267. [CrossRef]
13. Shokeen, P.; Anand, P.; Murali, Y.K.; Tandon, V. Antidiabetic Activity of 50% Ethanolic Extract of *Ricinus communis* and Its Purified Fractions. *Food Chem. Toxicol.* **2008**, *46*, 3458–3466. [CrossRef] [PubMed]
14. Worbs, S.; Köhler, K.; Pauly, D.; Avondet, M.A.; Schaer, M.; Dorner, M.B.; Dorner, B.G. *Ricinus communis* Intoxications in Human and Veterinary Medicine-a Summary of Real Cases. *Toxins* **2011**, *3*, 1332–1372. [CrossRef] [PubMed]
15. Abbes, M.; Montana, M.; Curti, C.; Vanelle, P. Ricin Poisoning: A Review on Contamination Source, Diagnosis, Treatment, Prevention and Reporting of Ricin Poisoning. *Toxicon* **2021**, *195*, 86–92. [CrossRef]
16. Sogan, N.; Kapoor, N.; Singh, H.; Kala, S.; Nayak, A.; Nagpal, B. Larvicidal Activity of *Ricinus communis* Extract against Mosquitoes. *J. Vector Borne Dis.* **2018**, *55*, 282–290. [CrossRef] [PubMed]
17. Hussein, A.O.; Hameed, I.H.; Jasim, H.; Kareem, M.A. Determination of Alkaloid Compounds of *Ricinus communis* by Using Gas Chromatography-Mass Spectroscopy (GC-MS). *J. Med. Plants Res.* **2015**, *9*, 349–359. [CrossRef]
18. Elkousy, R.H.; Said, Z.N.A.; Abd El-Baseer, M.A.; Abu El wafa, S.A. Antiviral Activity of Castor Oil Plant (*Ricinus communis*) Leaf Extracts. *J. Ethnopharmacol.* **2021**, *271*, 113878. [CrossRef]
19. Mboyazi, S.N.; Nqotheni, M.I.; Maliehe, T.S.; Shandu, J.S. In Vitro Antibacterial and in Silico Toxicity Properties of Phytocompounds from *Ricinus communis* Leaf Extract. *Pharmacogn. J.* **2020**, *12*, 977–983. [CrossRef]
20. Osorio-González, C.S.; Gómez-Falcon, N.; Sandoval-Salas, F.; Saini, R.; Brar, S.K.; Ramírez, A.A. Production of Biodiesel from Castor Oil: A Review. *Energies* **2020**, *13*, 2467. [CrossRef]
21. Layek, U.; Manna, S.S.; Karmakar, P. Pollen Foraging Behaviour of Honey Bee (*Apis mellifera* L.) in Southern West Bengal, India. *Palynology* **2020**, *44*, 114–126. [CrossRef]
22. Allan, G.; Williams, A.; Rabinowicz, P.D.; Chan, A.P.; Ravel, J.; Keim, P. Worldwide Genotyping of Castor Bean Germplasm (*Ricinus communis* L.) Using AFLPs and SSRs. *Genet. Resour. Crop Evol.* **2008**, *55*, 365–378. [CrossRef]
23. Anjani, K. Purple-Coloured Castor (*Ricinus communis* L.)-A Rare Multiple Resistant Morphotype. *Curr. Sci.* **2005**, *88*, 215–216.
24. Kallamadi, P.R.; Nadigatla, V.P.R.G.R.; Mulpuri, S. Molecular Diversity in Castor (*Ricinus communis* L.). *Ind. Crops Prod.* **2015**, p *66*, 271–281. [CrossRef]
25. Purseglove, J.W. Castor, Sesame & Safflower By E. A. Weiss London: Leonard Hill Books (1971), pp. 901. *Exp. Agric.* **1972**, *8*, 282. [CrossRef]
26. Chaudhari, B.A.; Patel, M.P.; Patel, J.A.; Makwana, R.R.; Patel, A.M.; Patel, M.K. The Characterization of Castor (*Ricinus communis* L.) Genotypes for Morphological Traits. *Int. J. Curr. Microbiol. Appl. Sci.* **2019**, *8*, 2482–2492. [CrossRef]
27. Da Silva, A.R.; Silva, S.A.; Dos Santos, L.A.; De Souza, D.R.; De Melo Araujo, G.; Dantas, J.L.L.; Da Silva Leite, E.; Dantas, A.C.V.L. Characterization and Performance of Castor Bean Lineages and Parents at the UFRB Germplasm Bank. *PLoS ONE* **2019**, *14*, e0209335. [CrossRef]
28. Wahibah, N.N.; Fitmawati; Yahya, V.J.; Perdana, M.A.; Budiono, R. Morphological Variation of Castor Bean (*Ricinus communis* L.) on Peatland Area in Kepulauan Meranti Riau Indonesia. *J. Phys. Conf. Ser.* **2020**, *1655*, 012028. [CrossRef]
29. Shaheen, A.M. Morphological Variation within *Ricinus communis* L. Egypt: Fruit, Leaf, Seed and Pollen. *Pak. J. Biol. Sci.* **2002**, *5*, 1202–1206. [CrossRef]
30. Pandey, A.; Agrawal, S.; Patel, A.D.; Pendarkar, D.; Bambhaniya, J.D. Evergreen Conventional and Modern Microscopic Methods for Solving Various Cases of Forensic Botany by Morphological and Histological Study. *Pharmacogn. J.* **2019**, *11*, 171–176. [CrossRef]
31. CSIR. *The Wealth of India: A Dictionary of Indian Raw Materials and Industrial Products (Industrial Products—Part I)*; Council of Scientific & Industrial Research: New Delhi, India, 1972.
32. Fakhr, M.A.; Mazrou, Y.S.A.; Ellmouni, F.Y.; Elsaied, A.; Elhady, M.; Elkelish, A.; Nour, I.H. Investigating the Phenotypic Plasticity of the *Invasive Weed Trianthema portulacastrum* L. *Plants* **2022**, *11*, 77. [CrossRef] [PubMed]
33. Lin, P.; Chen, Y.; Ponce, G.; Acevedo, F.E.; Anderson, C.T.; Ali, J.G.; Felton, G.W.; Lynch, J.P. Stomata-Mediated Interactions between Plants, Herbivores, and the Environment. *Trends Plant Sci.* **2022**, *27*, 287–300. [CrossRef] [PubMed]
34. Fernandes, M.F.; Cardoso, D.; de Queiroz, L.P. An Updated Plant Checklist of the Brazilian Caatinga Seasonally Dry Forests and Woodlands Reveals High Species Richness and Endemism. *J. Arid Environ.* **2020**, *174*, 104079. [CrossRef]

35. Duruibe, J.O.; Ogwuegbu, M.O.C.; Egwurugwu, J.N. Heavy Metal Pollution and Human Biotoxic Effects. *Int. J. Phys. Sci.* **2007**, *2*, 112–118. [CrossRef]
36. Nazzal, Y.; Rosen, M.A.; Al-Rawabdeh, A.M. Assessment of Metal Pollution in Urban Road Dusts from Selected Highways of the Greater Toronto Area in Canada. *Environ. Monit. Assess.* **2013**, *185*, 1847–1858. [CrossRef] [PubMed]
37. Sameena, P.; Puthur, J.T. Heavy Metal Phytoremediation by Bioenergy Plants and Associated Tolerance Mechanisms. *Soil Sediment Contam.* **2021**, *30*, 253–274. [CrossRef]
38. Huang, J.; Guo, S.; Zeng, G.; Li, F.; Gu, Y.; Shi, Y.; Shi, L.; Liu, W.; Peng, S. A New Exploration of Health Risk Assessment Quantification from Sources of Soil Heavy Metals under Different Land Use. *Environ. Pollut.* **2018**, *243*, 49–58. [CrossRef]
39. Li, C.; Zhou, K.; Qin, W.; Tian, C.; Qi, M.; Yan, X.; Han, W. A Review on Heavy Metals Contamination in Soil: Effects, Sources, and Remediation Techniques. *Soil Sediment Contam.* **2019**, *28*, 380–394. [CrossRef]
40. Farooq, M.A.; Ali, S.; Hameed, A.; Bharwana, S.A.; Rizwan, M.; Ishaque, W.; Farid, M.; Mahmood, K.; Iqbal, Z. Cadmium Stress in Cotton Seedlings: Physiological, Photosynthesis and Oxidative Damages Alleviated by Glycinebetaine. *South Afr. J. Bot.* **2016**, *104*, 61–68. [CrossRef]
41. Siregar, A.S.; Sulistyo, I.; Prayogo, N.A. Heavy Metal Contamination in Water, Sediments and Planiliza Subviridis Tissue in the Donan River, Indonesia. *J. Water L. Dev.* **2020**, *45*, 157–164. [CrossRef]
42. Tyagi, K.; Sharma, S.; Kumar, S.; Ayub, S. Cytological, Morphological and Anatomical Studies of *Ricinus communis* Linn. Grown under the Influence of Industrial Effluent—A Comparative Study. *J. Pharm. Res.* **2013**, *7*, 454–458. [CrossRef]
43. Bauddh, K.; Singh, K.; Singh, B.; Singh, R.P. *Ricinus communis*: A Robust Plant for Bio-Energy and Phytoremediation of Toxic Metals from Contaminated Soil. *Ecol. Eng.* **2015**, *84*, 640–652. [CrossRef]
44. Kammerbauer, J.; Dick, T. Monitoring of Urban Traffic Emissions Using Some Physiological Indicators in *Ricinus communis* L. Plants. *Arch. Environ. Contam. Toxicol.* **2000**, *39*, 161–166. [CrossRef]
45. Boda, R.K.; Majeti, N.V.P.; Suthari, S. *Ricinus communis* L. (Castor Bean) as a Potential Candidate for Revegetating Industrial Waste Contaminated Sites in Peri-Urban Greater Hyderabad: Remarks on Seed Oil. *Environ. Sci. Pollut. Res.* **2017**, *24*, 19955–19964. [CrossRef] [PubMed]
46. Zarai, Z.; Chobba, I.B.; Mansour, R.B.; Békir, A.; Gharsallah, N.; Kadri, A. Essential Oil of the Leaves of *Ricinus communis* L.: In Vitro Cytotoxicity and Antimicrobial Properties. *Lipids Health Dis.* **2012**, *11*, 102. [CrossRef] [PubMed]
47. Warra, A.A. Physico-Chemical and GC/MS Analysis of Castor Bean (*Ricinus communis* L.) Seed Oil. *Chem. Mater. Res.* **2015**, *7*, 56–60.
48. Hammad, H.H.; El-Kateb, H.; Bahnasy, M.I. Growth, Yield and Fatty Acids Composition of Castor Bean (*Ricinus communis* L.) Cultivars. *Middle East J. Agric. Res.* **2019**, *8*, 638–653.
49. Firn, J.; McGree, J.M.; Harvey, E.; Flores-Moreno, H.; Schütz, M.; Buckley, Y.M.; Borer, E.T.; Seabloom, E.W.; La Pierre, K.J.; MacDougall, A.M.; et al. Leaf Nutrients, Not Specific Leaf Area, Are Consistent Indicators of Elevated Nutrient Inputs. *Nat. Ecol. Evol.* **2019**, *3*, 400–406. [CrossRef] [PubMed]
50. Perdomo, F.A.; Acosta-Osorio, A.A.; Herrera, G.; Vasco-Leal, J.F.; Mosquera-Artamonov, J.D.; Millan-Malo, B.; Rodriguez-Garcia, M.E. Physicochemical Characterization of Seven Mexican *Ricinus communis* L. Seeds & Oil Contents. *Biomass Bioenergy* **2013**, *48*, 17–24.
51. Hussein, H.M.; Ubaid, J.M.; Hameed, I.H. Inscticidal Activity of Methanolic Seeds Extract of *Ricinus communis* on Adult of *Callosobruchus maculatus* (Coleopteran: Brauchidae) and Analysis of Its Phytochemical Composition. *Int. J. Pharmacogn. Phytochem. Res.* **2016**, *8*, 1385–1397.
52. Salimon, J.; Noor, D.A.M.; Nazrizawati, A.; Noraishah, A. Fatty Acid Composition and Physicochemical Properties of Malaysian Castor Bean *Ricinus communis* L. Seed Oil. *Sains Malays.* **2010**, *39*, 761–764.
53. Nangbes, J.G.; Nvau, J.B.; Buba, W.M.; Zukdimma, A.N. Extraction and Characterization of Castor (*Ricinus communis*) Seed Oil. *Int. J. Eng. Sci.* **2013**, *2*, 105–109.
54. Yusuf, A.K.; Mamza, P.A.P.; Ahmed, A.S.; Agunwa, U. Extraction and Characterization of Castor Seed Oil from Wild *Ricinus communis* Linn. *Int. J. Sci. Environ. Technol.* **2015**, *4*, 1392–1404.
55. Schneider, C.A.; Rasband, W.S.; Eliceiri, K.W. NIH Image to ImageJ: 25 Years of Image Analysis. *Nat. Methods* **2012**, *9*, 671–675. [CrossRef] [PubMed]
56. Singh, G. *Plant Systematics: An Integrated Approach*; CRC Press: Boca Raton, FL, USA, 2019; ISBN 1000576779.
57. Ellis, B.; Daly, D.C.; Hickey, L.J.; Johnson, K.R.; Mitchell, J.D.; Wilf, P.; Wing, S.L. *Manual of Leaf Architecture*; Ithaca: New York, NY, USA, 2009; ISBN 9780801475184.
58. Ash, A. *Manual of Leaf Architecture: Morphological Description and Categorization of Dicotyledonous and Net-Veined Monocotyledonous Angiosperms*; Smithsonian Institution: Washington, DC, USA, 1999.
59. Barthlott, W. Epidermal and Seed Surface Characters of Plants: Systematic Applicability and Some Evolutionary Aspects. *Nord. J. Bot.* **1981**, *1*, 345–355. [CrossRef]
60. Piper, C.S. *Soil and Plant Analysis*; Scientific Publishers: Jodhpur, India, 2019; ISBN 9386237539.
61. Walkley, A.J.; Black, I.A. Estimation of Soil Organic Carbon by the Chromic Acid Titration Method. *Soil Sci.* **1934**, *37*, 29–38. [CrossRef]
62. Pansu, M.; Gautheyrou, J. *Phosphorus*; Springer: Berlin/Heidelberg, Germany, 2006; ISBN 978-3-540-31211-6.

63. Olsen, S.R.; Sommers, L.E. Phosphorus. Agronomy Monographs. In *Methods of Soil Analysis*; Academic Press: Madison, WI, USA, 1982; pp. 403–430. ISBN 9780891189770.
64. Hanway, J.J.; Heidal, H. Soil Analysis Methods as Used in the Iowa State College Soil Testing Laboratory. *Iowa State Coll. Agric. Bull.* **1952**, *57*, 1–31.
65. Tabassum, S.; Amin, F.; Erum, S.; Javed, H.; Kazmi, F.; Nisar, M.F.; Ullah, I.; Murtaza, I.; Ashraf, M. Effect of Hexane and Ethanol Extracts of Ten Basil Genotypes on the Growth of Selected Bacterial Strains. *Int. J. Agric. Biol.* **2016**, *18*, 735–740. [CrossRef]
66. Sbihi, H.M.; Nehdi, I.A.; Mokbli, S.; Romdhani-Younes, M.; Al-Resayes, S.I. Hexane and Ethanol Extracted Seed Oils and Leaf Essential Compositions from Two Castor Plant (*Ricinus communis* L.) Varieties. *Ind. Crops Prod.* **2018**, *122*, 174–181. [CrossRef]
67. Team, R.C. R: A Language and Environment for Statistical Computing. R Foundation for Statistical Computing, Vienna, Austria. 2013. Available online: https://www.gbif.org/tool/81287/r-a-language-and-environment-for-statistical-computing (accessed on 30 January 2023).
68. Wickham, H. *Data Analysis*; Springer: Berlin/Heidelberg, Germany, 2016; ISBN 331924275X.
69. Kolde, R. *Pheatmap: Pretty Heatmaps*, Version 1.0.8; R Package; R Foundation: Vienna, Austria, 2012.
70. Viscosi, V.; Cardini, A. Leaf Morphology, Taxonomy and Geometric Morphometrics: A Simplified Protocol for Beginners. *PLoS ONE* **2011**, *6*, e25630. [CrossRef] [PubMed]
71. Kassambara, A.; Mundt, F. Factoextr: Extract and Visualize the Results of Multivariate Data Analyses. 2020. Available online: https://cran.r-project.org/web/packages/factoextra/index.html (accessed on 30 January 2023).
72. Soetewey, A. Correlation Coefficient and Correlation Test in R. Available online: https://www.statsandr.com/blog/correlationcoefficient-and-correlation-test-in-r/ (accessed on 20 February 2022).
73. Murphy, M.R.C.; Jordan, G.J.; Brodribb, T.J. Differential Leaf Expansion Can Enable Hydraulic Acclimation to Sun and Shade. *Plant Cell Environ.* **2012**, *35*, 1407–1418. [CrossRef] [PubMed]
74. Xiong, D.; Wang, D.; Liu, X.; Peng, S.; Huang, J.; Li, Y. Leaf Density Explains Variation in Leaf Mass per Area in Rice between Cultivars and Nitrogen Treatments. *Ann. Bot.* **2016**, *117*, 963–971. [CrossRef]
75. James, J.J.; Tiller, R.L.; Richards, J.H. Multiple Resources Limit Plant Growth and Function in a Saline-Alkaline Desert Community. *J. Ecol.* **2005**, *93*, 113–126. [CrossRef]
76. Meier, I.C.; Leuschner, C. Leaf Size and Leaf Area Index in *Fagus sylvatica* Forests: Competing Effects of Precipitation, Temperature, and Nitrogen Availability. *Ecosystems* **2008**, *11*, 655–669. [CrossRef]
77. Liu, F.; Yang, W.; Wang, Z.; Xu, Z.; Liu, H.; Zhang, M.; Liu, Y.; An, S.; Sun, S. Plant Size Effects on the Relationships among Specific Leaf Area, Leaf Nutrient Content, and Photosynthetic Capacity in Tropical woody Species. *Acta Oecologica* **2010**, *36*, 149–159. [CrossRef]
78. Bano, A.; Fatima, M. Salt Tolerance in *Zea mays* (L). Following Inoculation with *Rhizobium* and *Pseudomonas*. *Biol. Fertil. Soils* **2009**, *45*, 405–413. [CrossRef]
79. Sathee, L.; Jain, V. Interaction of Elevated CO2 and Form of Nitrogen Nutrition Alters Leaf Abaxial and Adaxial Epidermal and Stomatal Anatomy of Wheat Seedlings. *Protoplasma* **2022**, *3*, 703–716. [CrossRef]
80. Bravo, S.; Amorós, J.A.; Pérez-De-Los-Reyes, C.; García, F.J.; Moreno, M.M.; Sánchez-Ormeño, M.; Higueras, P. Influence of the Soil PH in the Uptake and Bioaccumulation of Heavy Metals (Fe, Zn, Cu, Pb and Mn) and Other Elements (Ca, K, Al, Sr and Ba) in Vine Leaves, Castilla-La Mancha (Spain). *J. Geochem. Explor.* **2017**, *174*, 79–83. [CrossRef]
81. Mohamed, I.A.A.; Shalby, N.; MA El-Badri, A.; Saleem, M.H.; Khan, M.N.; Nawaz, M.A.; Qin, M.; Agami, R.A.; Kuai, J.; Wang, B. Stomata and Xylem Vessels Traits Improved by Melatonin Application Contribute to Enhancing Salt Tolerance and Fatty Acid Composition of *Brassica napus* L. Plants. *Agronomy* **2020**, *10*, 1186. [CrossRef]
82. Galal, T.M.; Essa, B.; Al-Yasi, H. Heavy Metals Uptake and Its Impact on the Growth Dynamics of the Riparian Shrub *Ricinus communis* L. along Egyptian Heterogenic Habitats. *Environ. Sci. Pollut. Res.* **2021**, *28*, 37158–37171. [CrossRef]
83. Azmat, R.; Saleem, A.; Ahmed, W.; Qayyum, A.; El-Serehy, H.A.; Ali, S. The Investigation of the Impact of Toxicity of Metals on Oxygen-Evolving Complex in *Spinacia oleracea*. *Antioxidants* **2022**, *11*, 1802. [CrossRef] [PubMed]
84. Abdel Latef, A.A.H.; Zaid, A.; Abo-Baker, A.B.A.E.; Salem, W.; Abu Alhmad, M.F. Mitigation of Copper Stress in Maize by Inoculation with *Paenibacillus polymyxa* and *Bacillus circulans*. *Plants* **2020**, *9*, 1513. [CrossRef] [PubMed]
85. Yruela, I. Copper in Plants. *Braz. J. Plant Physiol.* **2005**, *17*, 145–156. [CrossRef]
86. delos Angeles, M.D.; Serino, C.B.M.; Granada, K.B. Leaf Morpho-Anatomical Responses of *Mangifera indica* L. and *Ficus benjamina* L. to Air Pollution in Selected Areas of Cebu City, Philippines. *J. Nat. Stud.* **2018**, *17*, 44–53.
87. Morsy, A.A.; Youssef, A.M.; Mosallam, H.A.M.; Hashem, A.M. Assessment of Selected Species along Al-Alamein-Alexandria International Desert Road, Egypt. *J. Appl. Sci. Res.* **2008**, *4*, 1276–1284.
88. Marzouk, R.I.; El-darier, S.M.; Nour, I.H.; Kamal, S.A. Numerical Taxonomic Study of *Marrubium* L. (Lamiaceae) in Egypt. *Catrina* **2015**, *13*, 25–35.
89. Halmy, M.W.A.; Fawzy, M.; Ahmed, D.A.; Saeed, N.M.; Awad, M.A. Monitoring and Predicting the Potential Distribution of Alien Plant Species in Arid Ecosystem Using Remotely-Sensed Data. *Remote Sens. Appl. Soc. Environ.* **2019**, *13*, 69–84. [CrossRef]
90. Shankar, V.G.; Venkata Ramana Rao, P.; Reddy, A.V. Inheritance of Certain Morphological Characters and Fusarium Wilt Resistance in Castor, *Ricinus communis* L. *Sabrao J. Breed. Genet.* **2010**, *42*, 57–64.
91. Santha, E.B.M.; Prasad, L.; Yadav, P.; Bee, H. Defense Responses to *Fusarium oxysporum* f. sp. *ricini* Infection in Castor (*Ricinus communis* L.) Cultivars. *Indian Phytopathol.* **2019**, *72*, 647–656. [CrossRef]

92. Toscano, S.; Branca, F.; Romano, D.; Ferrante, A. An Evaluation of Different Parameters to Screen Ornamental Shrubs for Salt Spray Tolerance. *Biology* **2020**, *9*, 250. [CrossRef]
93. Lacerda, C.F.; Assis Júnior, J.O.; Lemos Filho, L.C.A.; de Oliveira, T.S.; Guimarães, F.V.A.; Gomes-Filho, E.; Prisco, J.T.; Bezerra, M.A. Morpho-Physiological Responses of Cowpea Leaves to Salt Stress. *Braz. J. Plant Physiol.* **2006**, *18*, 455–465. [CrossRef]
94. Matesanz, S.; Horgan-Kobelski, T.; Sultan, S.E. Phenotypic Plasticity and Population Differentiation in an Ongoing Species Invasion. *PLoS ONE* **2012**, *7*, e44955. [CrossRef]
95. Cookson, S.J.; Radziejwoski, A.; Granier, C. Cell and Leaf Size Plasticity in Arabidopsis: What Is the Role of Endoreduplication? *Plant Cell Environ.* **2006**, *29*, 1273–1283. [CrossRef] [PubMed]
96. Barthlott, W.; Neinhuis, C.; Cutler, D.; Ditsch, F.; Meusel, I.; Theisen, I.; Wilhelmi, H. Classification and Terminology of Plant Epicuticular Waxes. *Bot. J. Linn. Soc.* **1998**, *126*, 237–260. [CrossRef]
97. Reddy, P. *Influence of Nutrient Rich Organic Wastes in Wastelands Reclamation*; GRIN Verlag: München, Germany, 2014.
98. Öncel, I.; Keleş, Y.; Üstün, A.S. Interactive Effects of Temperature and Heavy Metal Stress on the Growth and Some Biochemical Compounds in Wheat Seedlings. *Environ. Pollut.* **2000**, *107*, 315–320. [CrossRef]
99. Jeschke, W.; Wolf, O. Effect of NaCI Salinity on Growth, Development, Ion Distribution, and Ion Translocation in Castor Bean (*Ricinus communis* L.). *J. Plant Physiol.* **1988**, *132*, 45–53. [CrossRef]
100. Papazoglou, E.G.; Alexopoulou, E.; Papadopoulos, G.K.; Economou-Antonaka, G. Tolerance to Drought and Water Stress Resistance Mechanism of Castor Bean. *Agronomy* **2020**, *10*, 1580. [CrossRef]
101. Zhou, H.; Zhou, G.; He, Q.; Zhou, L.; Ji, Y.; Zhou, M. Environmental Explanation of Maize Specific Leaf Area under Varying Water Stress Regimes. *Environ. Exp. Bot.* **2020**, *171*, 103932. [CrossRef]
102. Elliott-kingston, C.; Haworth, M.; Yearsley, J.M.; Batke, S.P.; Lawson, T.; Mcelwain, J.C.; Elliott-kingston, C. Does Size Matter? Atmospheric CO_2 May Be a Stronger Driver of Stomatal Closing Rate than Stomatal Size in Taxa That Diversified under Low CO_2. *Front. Plant Sci.* **2016**, *7*, 1253. [CrossRef] [PubMed]
103. Ullah, S.; Khan, M.N.; Lodhi, S.S.; Ahmed, I.; Tayyab, M.; Mehmood, T.; Din, I.U.; Khan, M.; Sohail, Q.; Akram, M. Targeted Metabolomics Reveals Fatty Acid Abundance Adjustments as Playing a Crucial Role in Drought-Stress Response and Post-Drought Recovery in Wheat. *Front. Genet.* **2022**, *13*, 972696. [CrossRef] [PubMed]
104. Yaghoubian, I.; Antar, M.; Ghassemi, S.; Modarres-Sanavy, S.A.M.; Smith, D.L. The Effects of Hydro-Priming and Colonization with *Piriformospora indica* and *Azotobacter chroococcum* on Physio-Biochemical Traits, Flavonolignans and Fatty Acids Composition of Milk Thistle (*Silybum marianum*) under Saline Conditions. *Plants* **2022**, *11*, 1281. [CrossRef] [PubMed]
105. Aparna, V.; Dileep, K.V.; Mandal, P.K.; Karthe, P.; Sadasivan, C.; Haridas, M. Anti-Inflammatory Property of n-Hexadecanoic Acid: Structural Evidence and Kinetic Assessment. *Chem. Biol. Drug Des.* **2012**, *80*, 434–439. [CrossRef]

agronomy

MDPI

Article

Pollen Visualization of Turkish Flora of Selected Plant Species under Light, Scanning, and Transmission Microscopy

Azize Demirpolat [1,*] and Pelin Yılmaz Sancar [2]

[1] Vocational School of Food, Agriculture and Livestock, Bingol University, Bingöl 12000, Turkey
[2] Department of Biology, Faculty of Science, Fırat University, Elazığ 23119, Turkey; peyilmaz@firat.edu.tr
* Correspondence: ademirpolat@bingol.edu.tr

Abstract: This study aims to examine pollen morphologically via studies of 16 plant taxa from Turkish flora. The pollen structures of the taxa examined, and their microscopic relevance, was discussed and studied with the help of microscopic visualization using scanning and transmission microscopies. Pollen grains were first acetolyzed, and then quantitative and qualitative pollen features were used to evaluate the species. The pollen grains were prolate, spherical, elliptic, tricolporate, 3-colpate, and hexacolpate. Sculpturing patterns of surfaces vary from reticulate, micro-reticulate, and striate regulate. As the findings reveal, palynological data can aid in the taxonomic classification of Turkish floral species. Microscopic implications can be made via micromorphological examination to correctly identify the species. While the pollen morphology of 16 taxa collected from the study area was studied for the first time from this region, the palynological research of some taxa was introduced to the literature for the first time with this study. Pollen morphology and photographic and statistical data of the taxa in our study were determined. This study contributed to bee plant research, melisapalinological studies, and systematic botanical flora studies.

Keywords: Turkish flora; pollen morphology; surface; clustering analysis (CA); principle component analysis (PCA)

Citation: Demirpolat, A.; Sancar, P.Y. Pollen Visualization of Turkish Flora of Selected Plant Species under Light, Scanning, and Transmission Microscopy. *Agronomy* **2023**, *13*, 962. https://doi.org/10.3390/agronomy13040962

Academic Editors: Alessio Papini, Mushtaq Ahmad, Fazal Ullah, Wajid Zaman and Rodolfo Gentili

Received: 1 February 2023
Revised: 20 March 2023
Accepted: 22 March 2023
Published: 24 March 2023

1. Introduction

Palynological features provide ancillary knowledge and indulge the systematic stand of species with its individual family. Pollen morphology can help support the taxonomic approach [1]. The word palynology was first confined by Hyde and William [2]. Since then, the botanical sciences have become an advanced sub-division with distinct applications. The microscopic investigation of spores, pollen, and other imperceptible structures is known as palynology [3]. Previous studies revealed that pollen morphology shows advancement in taxonomic characteristics like pollen type, exine and sexine appearance, position, measurement, number of apertures, type, and ornamentation [4]. Pollen grains are considered advantageous in morphological and phylogenetic studies [5–10]. Angiosperm might possess pollen apertures that are categorized based on their type; position and number vary from simple to composite [11].

Among angiosperm, pollen morphology plays an important role in advanced classification, which is carried out by using the maximum number of possible characteristics [12]. The division of families into subfamily tribes and subtribes is facilitated by palynology [13]. Different families have been divided into subfamilies and tribes on the bases of pollen morphology [14,15]. Some characteristics are specific to the family, which could include the size of pollen, type of sculpturing, aperture type, etc., which become a key to identifying that family. Asteraceae angiosperms show the most diverse types of pollen, showing the maximum number of pollen characteristics compared to any other taxa [7,16–18].

Palynology is also helpful for various uses, such as atmospheric pollen inspection, spore production, and the archaeological excavation of shipwrecks. Primarily, the investigation of morphological diversity in pollen is directly concerned with the classification of plants [18,19]. Palyno-morphological characteristics are fruitful for solving various taxonomic problems at species, genus, and family levels [7,20–22].

The primary objective of this research is to offer detailed insight into micromorphological features via SEM and TEM among selected plants from Turkey. In our study, taxa belonging to Lamiaceae, Boraginaceae, Fabaceae, Amaryllidaceae, Scrophulariaceae, Apiaceae, and Convolvulaceae families were selected and palynologically analyzed. The pollen morphology of 16 taxa collected from the study area was studied for the first time from this region, and palynological research of some taxa was introduced to the literature for the first time with this study. The current study provides a comprehensive analysis of the palyno-morphology for determining the extent to which these microscopic techniques can be utilized to discriminate among the taxa analyzed.

2. Materials and Methods

2.1. Specimen Collection and Plant İdentification

The 16 taxa belonging to different families were collected from Turkey (Figure 1, Table 1). The collection period runs from April 2020 to November 2021, including two seasons. Localities, times, and collector numbers are given in Table 1. After collecting the plants, they were then identified with the help of flora from Turkey. Plant material was identified, and voucher specimens were deposited in a Turkish herbarium (name it and insert its acronym). Flowering specimens were collected. Anthers were removed to create the specimens. Pollen was extracted directly and investigated with microscopic tools.

Figure 1. Map of the collection sites of the study area (Source: Google). The red zone represents the working area.

2.2. Light Microscopy (LM)

Pollen obtained from the samples was prepared according to the methods of Ertman and Wodehouse [23,24]. Some changes were made to the acetolysis procedure. Plant matter was marked by applying a drop of glycerin jelly. Each specimen received 4–5 prepared slides in total. Pollen samples were measured from at least 20 completely evolved grains according to the pattern beneath a Euromex CMEX- 10 PRO light microscope (Papenkamp 20, Holland) (100×). These measurements are reported in Tables 2 and 3 and in Figures 2–7. The terminology used is mainly from Faegri and Iversen, Ertman, and Kılıç et al. [25–27].

Table 1. Plant sampling, vouchering, and endemism.

No	Plant Name	Locality	Endemic Situation	Voucher Specimen
1.	*Ajuga reptans* L.	Bingöl-Tunceli	-	AD-257
2.	*Micromeria cremnophila* Boiss. Et Heldr. subsp. *anatolica* P. H. Davis	Elazig-Harput castle	Endemic	AD-1011
3.	*Onosma bornmuelleri Hausskn.*	Bingol	Endemic	AD-200
4.	*Onosma sericeum* Wild.	Bingol	-	AD-176
5.	*Rindera caespitosa* (A. DC.) Bunge	Bingol	Endemic	AD-78
6.	*Paracaryum* Schreber Boiss. subsp. *cristatum* Schreber Boiss.	Elazig	Endemic	AD-1012
7.	*Paracaryum strictum* (Koch) Boiss.	Elazig	Endemic	AD-1051
8.	*Astragalus aucheri* Boiss.	Elazig	Endemic	AD-1034
9.	*Astragalus aureus* Wild.	Bingol	-	AD-155
10.	*Astragalus dactylocarpus* Boiss.	Bingol	-	AD-230
11.	*Ixiolirion tataricum* Pallas Herbert subsp. *montanum* Labill. Takht.	Bingöl	-	AD-158
12.	*Linaria kurdica* Boiss. Et HOHEN. subsp. *kurdica* Boiss. Et Huet	Elazig	-	AD-1098
13.	*Linaria grandiflora* Desf.	Bingol	-	AD-253
14.	*Foeniculum vulgare* Miller	Bingol	-	AD-543
15.	*Convolvulus carduchorum* Davis	Elazig	Endemic	AD-123
16.	*Convolvulus reticulatus* Choisy subsp. *reticulatus* Choisy	Bingol	-	AD-112

Table 2. Qualitative micromorphological pollen characteristics of selected plants.

		Plant Name	Shape	Type and Number of Aperture	Exine Ornamentation	Colpi Orientation
1.	Lamiaceae	*A. reptans*	Oblate-spheroidal	3-colporate	Tectatae-granulatae	Sunken, angular
2.		*M. cremnophila* ssp. *anatolica*	Sub-oblate	6-colpate	Reticulate	Sunken
3.	Boraginaceae	*O. bornmuelleri*	Sub-prolate	3-colporate	Scabrate-reticulate	Sunken
4.		*O. sericea*	Sub-prolate	3-colporate	Scabrate-reticulate	Sunken
5.		*R. caespitosa*	Sub-prolate	3-colporate	Granulate	Pseudocolpus, margo
6.		*P. cristatum* ssp. *cristatum*	Prolate	6-colporate	Rugulate	Pseudocolpus, margo
7.		*P. strictum*	Prolate	6-colporate	Rugulate	Pseudocolpus, margo

Table 2. *Cont.*

		Plant Name	Shape	Type and Number of Aperture	Exine Ornamentation	Colpi Orientation
8.	Fabaceae	*A. aucheri*	Prolate	3-colporate	Reticulate	Sunken, angular, margins distinct
9.		*A. aureus*	Prolate	3-colporate	Reticulate	Sunken, angular, margins distinct
10.		*A. dactylocarpus*	Prolate	3-colporate	Reticulate	Sunken, angular, margins distinct
11.	Amaryllidaceae	*I. tataricum* subsp. *montanum*	Sub-prolate	3-colporate	Perforate	Sunken, angular, margins distinct
12.	Scrophulariaceae	*L. kurdica* subsp. *kurdica*	Oblate-spheroidal	3-colporate	Reticulate	Sunken
13.		*L. grandiflora*	Oblate-spheroidal	3-colporate	Reticulate	Sunken
14.	Apiaceae	*F. vulgare*	Per-prolate	3-colporate	Scabrate	Sunken
15.	Convolvulaceae	*C. carduchorum*	Oblate-spheroidal	3-colporate	Perforate	Sunken
16.		*C. reticulatus* subsp. *reticulatus*	Oblate-spheroidal	3-colporate	Perforate	Sunken

Table 3. Quantitative pollen features via microscopic measurements.

	Plant Name	Exine Thickness	Polar Diameter	Equatorial Diameter	P/E Ratio	Colpi Length	Colpi Width	Pores Length	Pores Width
Lamiaceae	*A. reptans*	1.1 ± (0.1)	22.4 ± (2.0)	22.7 ± (2.1)	0.98	16.7 ± (1.2)	7.6 ± (1)	-	-
	M. cremnophila ssp. *anatolica*	1.5 ± (1.2)	40 ± (2.3)	46.6 ± (1.3)	0.85	23.5 ± (1.3)	8.3 ± (1.6)	-	-
Boraginaceae	*O. bornmuelleri*	1.0 ± (0.5)	17.3 ± (1.0)	14.7 ± (1)	1.17	13.9 ± (0.3)	-	3 ± (0.5)	3.29 ± (0.5)
	O. sericea	1.0 ± (0.5)	17.8 ± (1.0)	14.9 ± (0.1)	1.19	13.8 ± (0.1)	-	2.83 ± (0.3)	2.29 ± (0.5)
	R. caespitosa	0.5 ± (0.3)	13 ± (1.0)	10 ± (0.1)	1.30	7.1 ± (1.0)	2.5 ± (1.2)	3.8 ± (0.5)	5.3 ± (0.3)
	P. cristatum ssp. *cristatum*	0.24 ±(0.2)	9.1 ± (0.3)	5.6 ± (0.1)	1.62	5.3 ± (1.3)	1.5 ± (1.5)	2.9 ± (0.3)	4.3 ± (0.7)
	P. strictum	0.23 ± (0.1)	10.1 ± (0.5)	5.2 ± (0.2)	1.94	6.5 ± (1.3)	1.9 ± (1.1)	2.8 ± (0.5)	4.7 ± (0.3)
Fabaceae	*A. aucheri*	0.45 ± (0.1)	35.5 ± (1.9)	23.6 ± (1.28)	1.50	29.2 ± (1.3)	4.2 ± (0.3)	6.3 ± (0.2)	4.2 ± (0.5)
	A.aureus	0.5 ± (0.1)	35.8 ± (1.2)	23.5 ± (1.0)	1.52	27.5 ± (0.3)	3.9 ± (0.9)	5.9 ± (0.6)	4.2 ± (0.7)
	A. dactylocarpus	0.5 ± (0.1)	38.7 ± (2.2)	25.3 ± (1.5)	1.52	29.8 ± (1.5)	5.5 ± (1.2)	7.4 ± (1.2)	5.7 ± (1.1)
Amaryllidaceae	*I. tataricum* subsp. *montanum*	1.8 ± (0.7)	51.2 ± (4.3)	41.6 ± (3.3)	1.23	38.9 ± (1.9)	15.0 ± (1.8)	-	-
Scrophulariaceae	*L. kurdica* subsp. *kurdica*	0.9 ± (0.2)	10.9 ± (1.2)	11.2 ± (1.1)	0.97	8.01 ± (2.3)	2.4 ± (2.3)	4.4 ± (1.4)	4.5 ± (1.8)
	L. grandiflora	0.82 ± (0.2)	10.0 ± (1.2)	10.3 ± (1.3)	0.97	7.85 ± (1.2)	3.8 ± (0.9)	4.23 ± (1.2)	3.98 ± (1.5)
Apiaceae	*F. vulgare*	1.8 ± (0.5)	26.8 ± (1.9)	10.0 ± (1.8)	2.68	14.9 ± (2.9)	4.0 ± (2.3)	3.8 ± (0.9)	4.1 ± (0.3)
Convolvulaceae	*C.carduchorum*	2.1 ± (0.4)	42.3 ± (1.3)	43 ± (1)	0.97	18.2 ± (1.9)	13.7 ± (1.1)	-	-
	C. reticulatus subsp. *reticulatus*	1.9 ± (0.3)	41.9 ± (0.9)	43.5 ± (1.2)	0.97	17.3 ± (1.2)	12.4 ± (1.3)	-	-

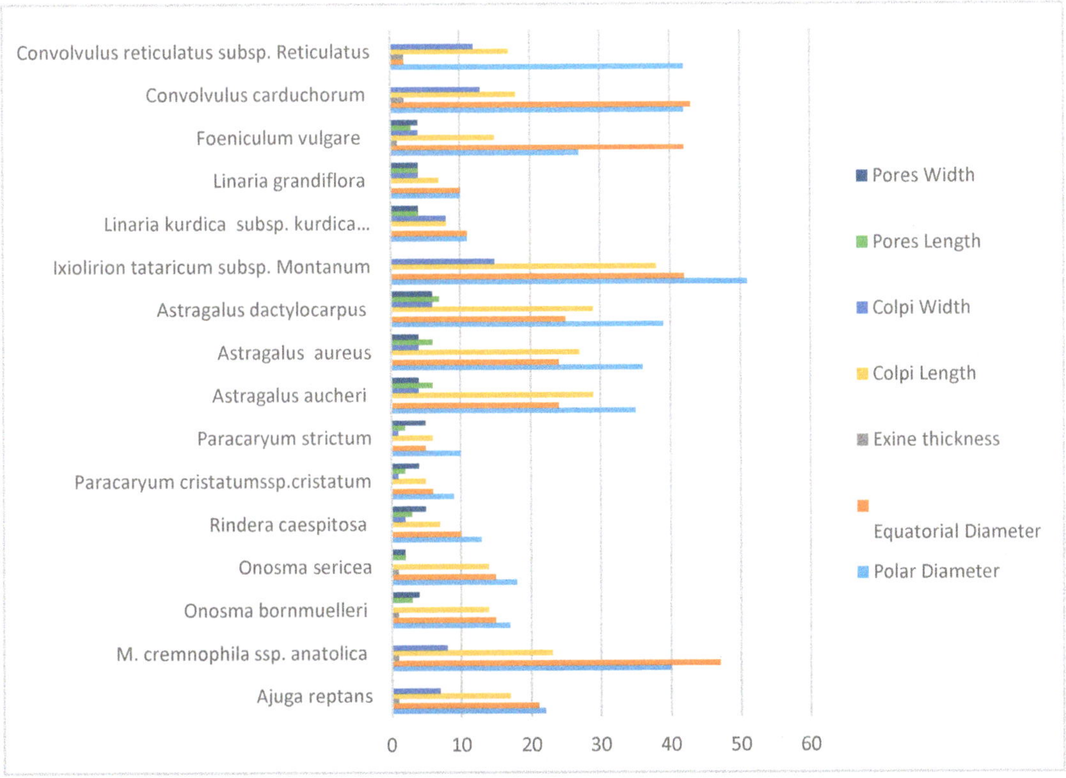

Figure 2. Graph of pollen's morphological measurements of the studied taxa.

2.3. *Scanning Electron Microscopy (SEM)*

Pollens belonging to each taxon were placed on metal carriers covered with double-sided adhesive tape under a binocular microscope and later coated with gold on a JEOL JSM 6510 (Akishima, Tokyo) (operating voltage range: 500 V–30 kV). The pollen appearances and detailed surface ornamentations of the studied pollen were obtained from Bingöl University Central Laboratory. Microphotographs at 1500×–10,000× magnification were taken for each taxon.

2.4. *Transmission Electron Microscopy (TEM)*

A transmission electron microscope (Beijing, P.R. China) (TEM, Tecnai G2 F20 S-TWIN, 200 kV) was used to analyze the sculpturing of pollen exine. The pollen sampled for TEM analysis was prepared by placing a drop of pollen on a carbon-coated copper TEM grid and allowed to evaporate (150 × 30 Mx).

2.5. *Cluster Analysis and PCA (Principal Component Analysis)*

The pollen data were subjected to cluster analysis from numerical taxonomic methods for the 16 taxa. For this analysis, XLSTAT 2022 (Denver, CO, USA) software and the UPGMA statistical method were used. The results of these analyses were illustrated using dendrograms and evaluated in terms of numerical chemotaxonomic relationships. In the resulting cluster analysis tree, the relationships of the species are indicated (Figure 8).

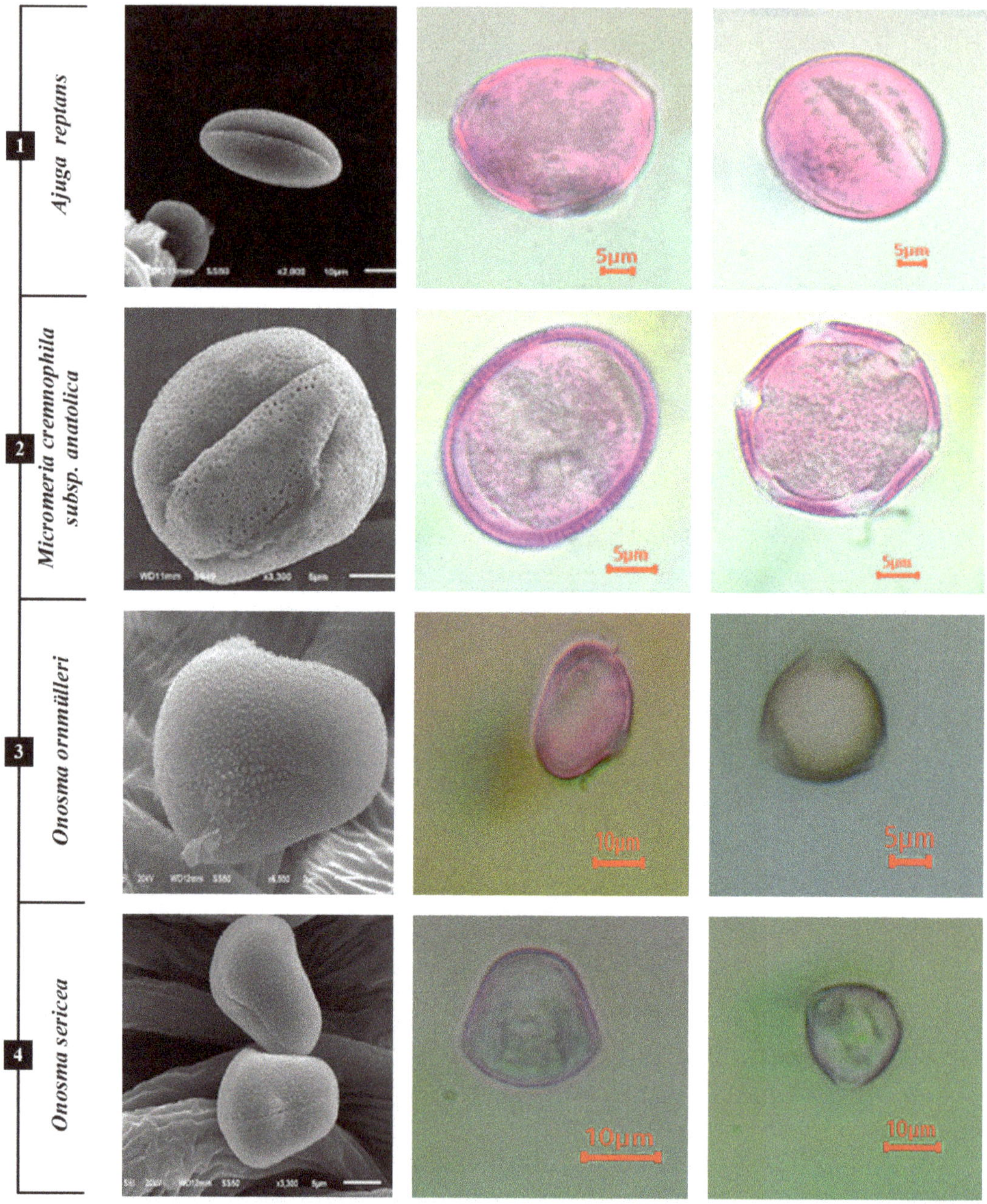

Figure 3. Scanning and light microscopic images of *A. reptans*, *M. cremnophila* subsp. *anatolica*, *O. bornmülleri*, and *O. sericea*.

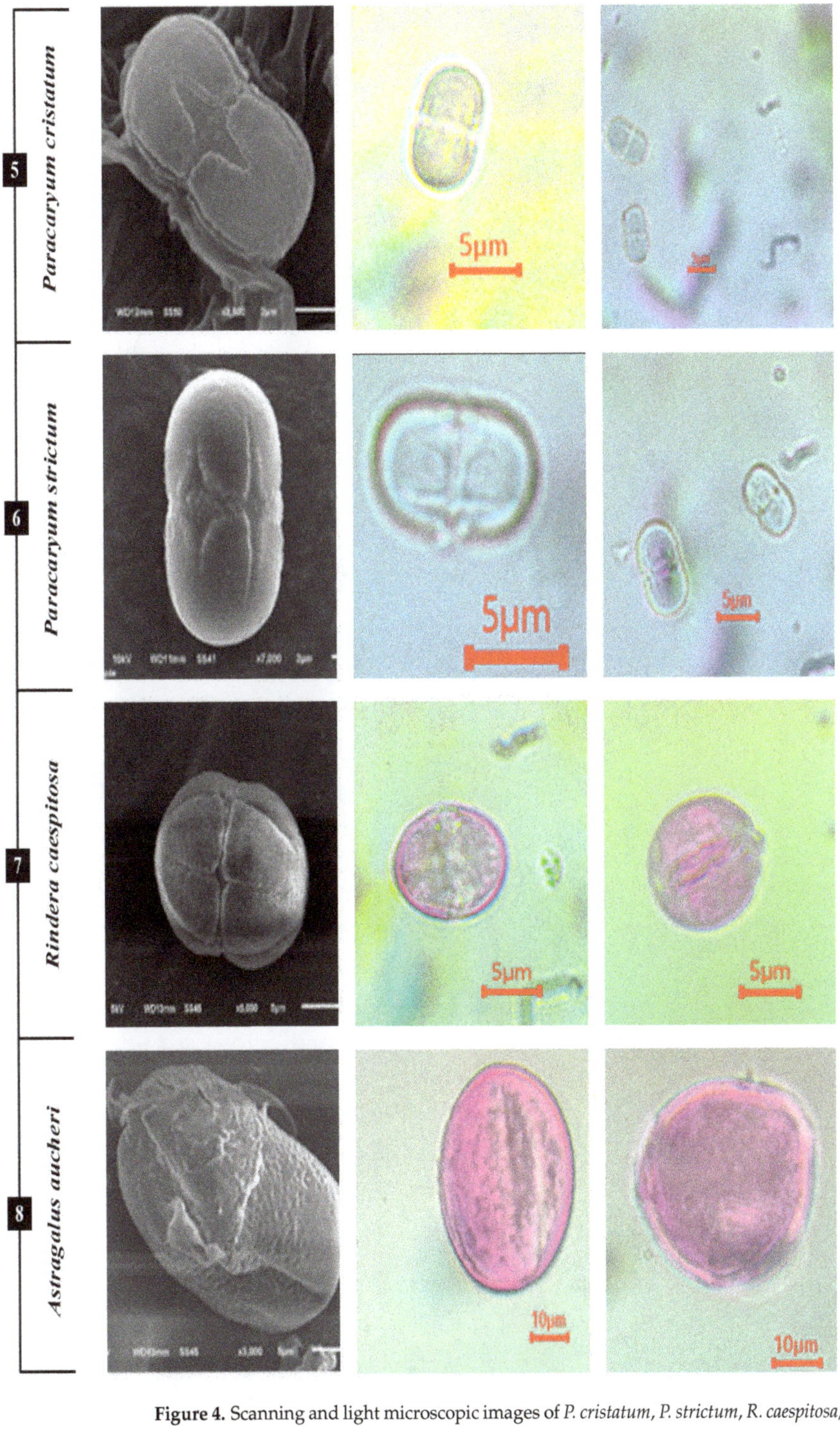

Figure 4. Scanning and light microscopic images of *P. cristatum*, *P. strictum*, *R. caespitosa*, and *A. aucheri*.

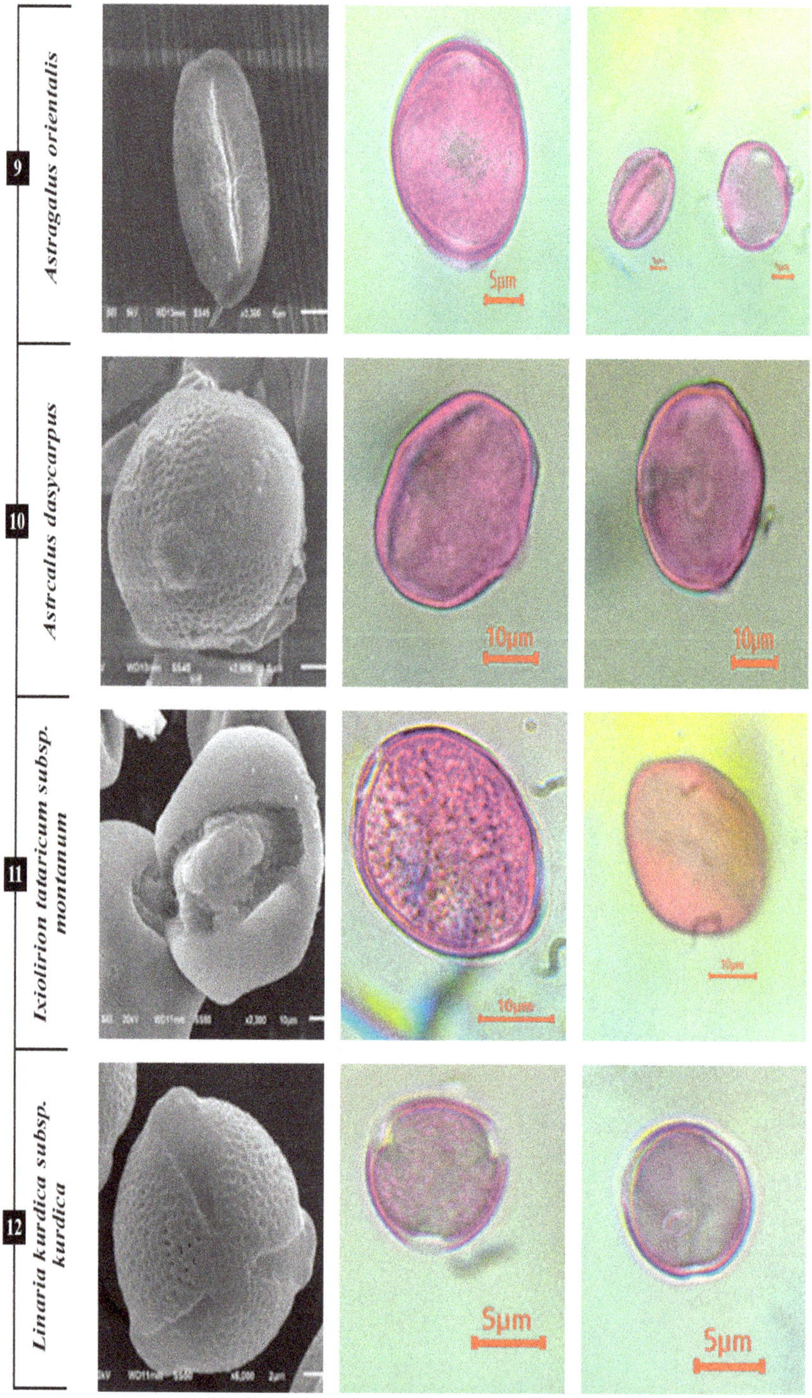

Figure 5. Scanning and light microscopic images of *A. orientalis*, *A. dasycarpus*, *I. tataricum* subsp. *montanum*, and *L. kurdica* subsp. *kurdica*.

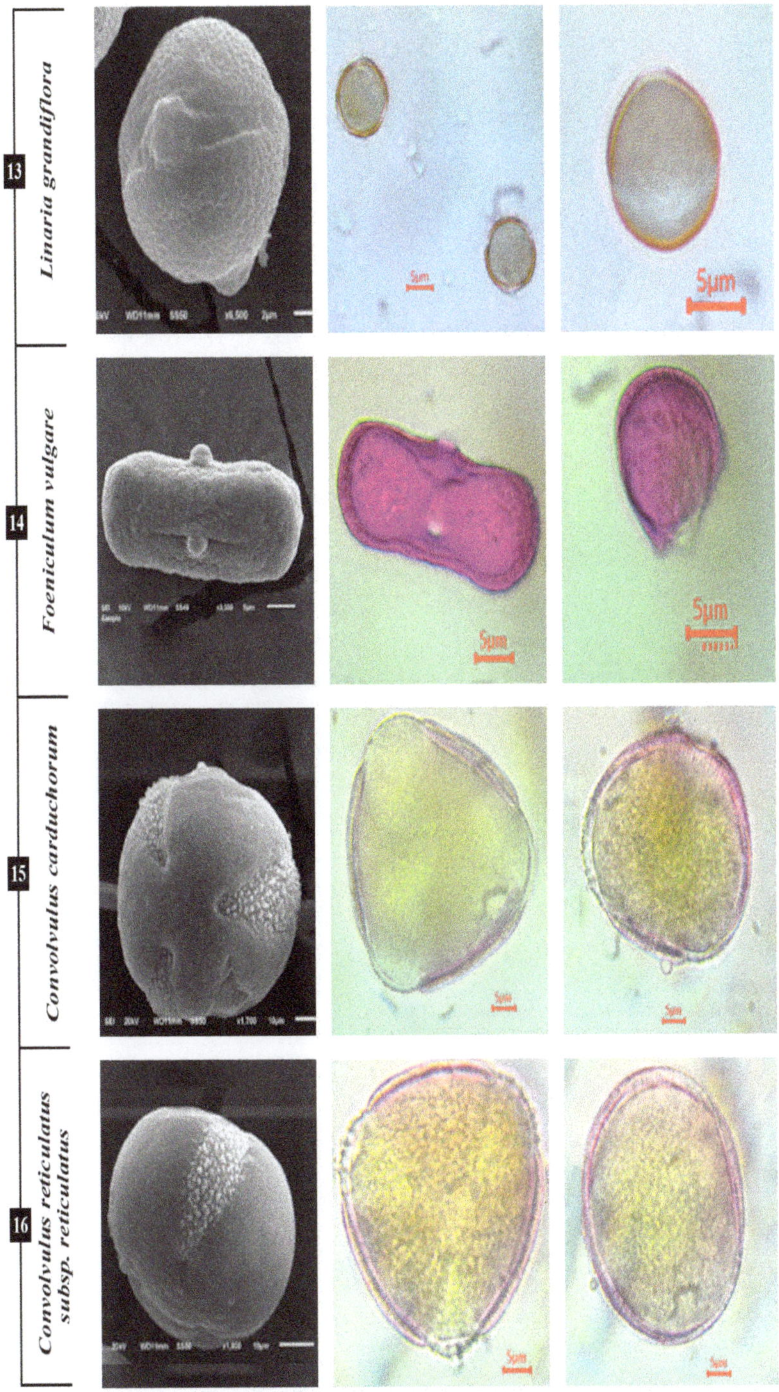

Figure 6. Scanning and light microscopic images of *L. grandiflora*, *F. vulgare*, *C. carduchorum* and *C. reticulatus* subsp. *reticulatus*.

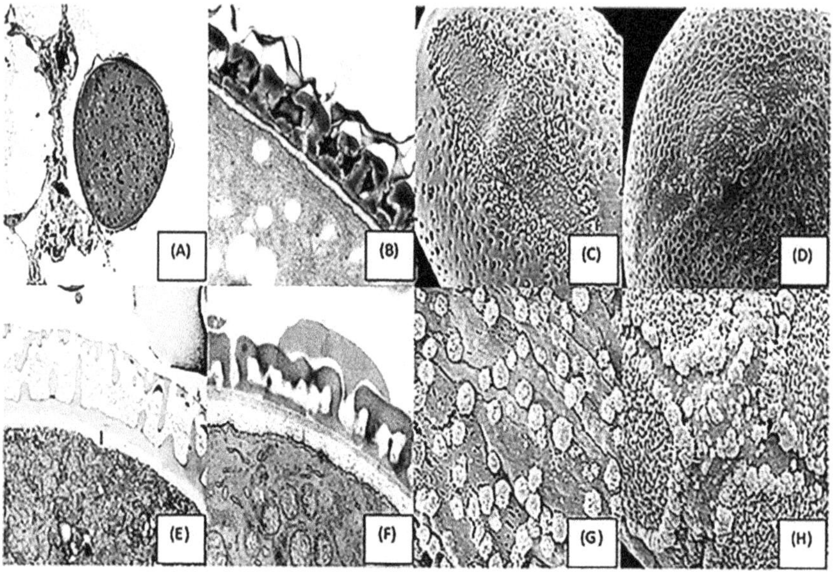

Figure 7. TEM micrographs: (**A**,**B**) *A. reptans*, (**C**) *A. aucheri*, (**D**) *A. orientalis*, (**E**) *C. reticulatus* subsp. *reticulatus*, (**F**) *L. kurdica* subsp. *kurdica*, and (**G**,**H**) *R. caespitosa*.

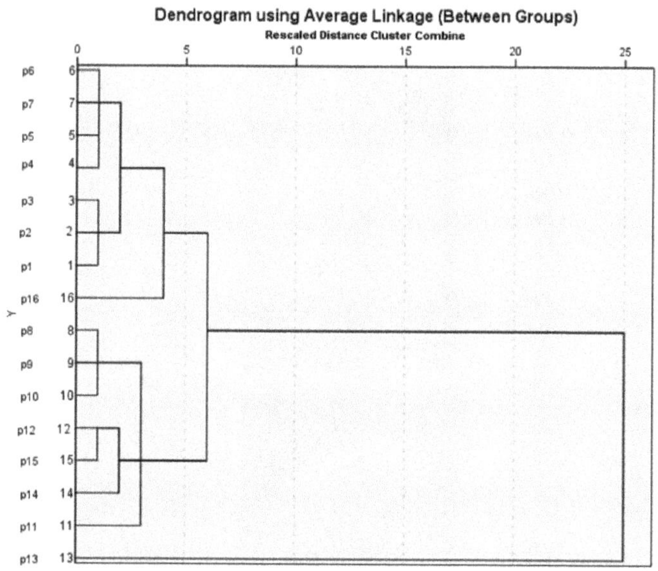

Figure 8. Clustering analysis of palynological data of taxa: Plant name p1—*Ajuga reptans;* p2—*Micromeria cremnophila* ssp. *anatolica;* p3—*Onosma bornmuelleri;* p4—*Onosma sericea;* p5—*Rindera caespitosa;* p6—*Paracaryum cristatum* ssp. *cristatum;* p7—*Paracaryum strictum;* p8—*Astragalus aucheri;* p9—*Astragalus aureus;* p10—*Astragalus dactylocarpus;* p11—*Ixiolirion tataricum* subsp. *montanum;* p12—*Linaria kurdica* subsp. *kurdica;* p13—*Linaria grandiflora;* p14—*Foeniculum vulgare;* p15—*Convolvulus carduchorum;* p16—*Convolvulus reticulatus* subsp. *reticulatus*.

Multivariate analysis was conducted to identify the structure of variability and to measure the distances between groups. These analyses were performed on complete data sets. The UPGMA (unweighted pair-group average linkage) clustering method based on Pearson distances was used to measure the similarities between each measured unit (Figure 9). Principle component analysis (PCA) and cluster analysis (CA) were used to evaluate the pollen morphology data of 16 taxa. The raw data were standardized to the same weight as previously reported.

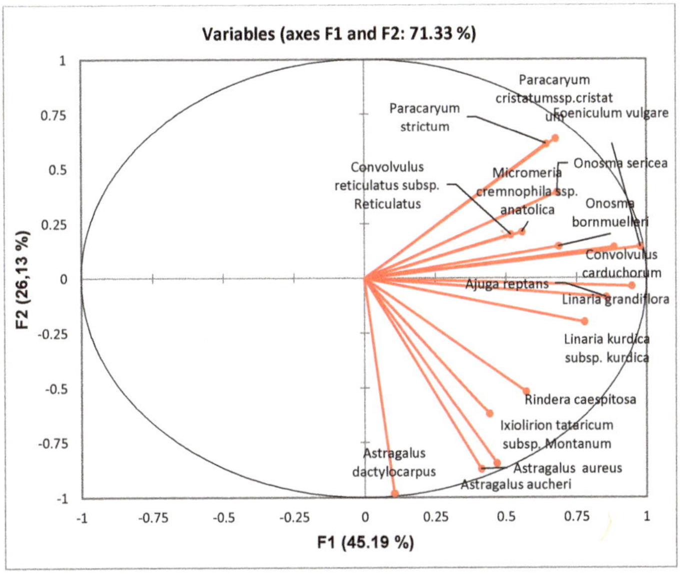

Figure 9. PCA correlation circle from the analysis of the palynological data of taxa.

3. Result

The micromorphological pollen features of selected 16 taxa from Turkey are summarized in Tables 2 and 3. The pollen's ultrastructure characteristic illustrations are shown in microscopic detail in Figures 3–7.

3.1. Pollen Descriptions

3.1.1. *A. reptans*

The results of light microscopy examinations in the pollen of the *Ajuga reptans* taxon revealed the tricolpate structure: polar axis of 22.4 ± (2.0) µm; equatorial axis of 22.7 ± (2.1) µm; pollen shape: oblate-spheroidal, (P/E = 0.98); polar aspect; colpus length of 16.7 ± (1.2) µm in equatorial aspect; and colpus width of 7.6 ± (1) µm. Exine thickness was 1.1 ± (0.1) µm, and their ornamentation showed a singletate granulate ornamentation according to SEM results (Figure 3).

3.1.2. *M. cremnophila* subsp. *anatolica*

Polar axis of 40 ± (2.3) µm, equatorial axis of 46.6 ± (1.3) µm, exine of 1.5–2.4 µm, and P/E: 0.85 sub-oblate. Pollen ornamentation showed reticulated ornamentation according to the SEM results.

3.1.3. *O. bornmuelleri* and *O. sericeum*

O. bornmuelleri have the following characteristics: polar axis 17.3 ± (1.0) µm; equatorial axis 14.7 ± (1) µm; pollen shape subprolate; (P/E = 1.17); polar aspect, colpus length 13.7 ± (0.1) µm in equatorial aspect; and pore width 3 × 3.29 µm. Exine thickness was 1.0 ± (0.5) µm, and their ornamentation showed a scabrad-granulate ornamentation according to SEM results. The polar axis of *O. sericeum* pollen grains was 17.8 ± (1.0) µm, equatorial axis was 14.9 ± (0.1) μm, pollen shape was subprolate, (P/E = 1.19), polar aspect, equatorial aspect, colpus length was 13.8 ± (0.1) μm, and pore width 2.83 μm × 2.29 μm. Exine thickness was 1.0 ± (0.5) μm, and the ornamentation showed a scabrad ornamentation according to the SEM results. Both *Onosma* species are similar in terms of pollen morphology (Figure 3).

3.1.4. *R. caespitosa*

Polar axis 13 ± (1.0) μm, equatorial axis 10 ± (0.1) μm, and pollen shape subprolate (P/E = 1.3). Exine thickness was 0.5 ± (0.3) μm, and ornamentation showed granulate ornamentation according to SEM results.

3.1.5. *P. cristatum* subsp. *cristatum, P. strictum*

The pollen grains have a 6-heterocolpate and are hexagonal from the polar view. The aperture margins are strongly thickened. Polar axis was 9.1 ± (0.3) μm, equatorial axis was 5.6 ± (0.1) μm, exine was 0.24 ± (0.2), P/E ratio was 1.62, and the pollen shape was prolate with respect to *P. cristatum* subsp. *cristatum*. The *P. strictum* pollen grains have a 6-heterocolpate and are hexagonal from the polar view. Polar axis was 10.1 ± (0.5) μm, equatorial axis was 5.2 ± (0.2) μm, exine was 0.23 ± (0.1), and P/E ratio was 1.94. The pollen was elliptical from the equatorial view, ±hexagonal from the polar view, and the pollen shape was prolate.

3.1.6. *A. aucheri, A. aureus, A. dasycarpum*

Polar axis 35.5 ± (1.9) μm, equatorial axis 23.6 ± 1.28 μm, P/E ratio 1.50, and pollen shape prolate in *A. aucheri*. According to the SEM results of the ornamentation, it was determined that they showed micro-reticular ornamentation. The *A. aureus* pollen grains have a polar axis of 35.8 ± 1.2 μm, equatorial axis of 23.5 ± 1.0 μm, P/E ratio of 1.52, and prolate pollen shape. Amb diameter average 20 ± 1.5. Exine thickness subtectate 0.5 ± 0.1 μm. According to the SEM results of the ornamentation, it was determined that they showed micro-reticulate ornamentation (Figure 5). Polar axis 38.7 ± 2.2 μm, equatorial axis 25.3 ± 1.5 μm, P/E ratio 1.52, and pollen shape prolate in *A. dasycarpum*. According to the SEM results of the ornamentation, it was determined that they showed micro-reticular ornamentation (Figures 4 and 5).

3.1.7. *I. tataricum* subsp. *montanum*

Pollen grains heteropolar, monosulcate, and polar axis 51.2 ± (4.3) mm, equatorial axis 41.6 ± (3.3) μm, P/E: 1.23 sub-prolate. Exine 1–1.8 μm, and the ornamentation showed a reticulate ornamentation according to SEM results.

3.1.8. *L. kurdica* subsp. *kurdica*

Pollens with radial symmetry, isopolar, tricolporate; polar axis 10.9 ± (1.2) μm, equatorial axis 11.2 ± (1.1) μm. Exine thickness 0.9 ± 0.2 μm. P/E: 0.97. Exine is reticulate. Colpus length 8.01 ± (2.3) μm; colpus width 2.4 ± (2.3) μm.

3.1.9. *L. grandiflora*

Pollens with radial symmetry, iso polar, tricolporate; polar axis 10.0 ± (1.2) μm, equatorial axis 10.3 ± (1.3) μm. Exine thickness 0.82 ± 0.2 μm. P/E: 0.97. Exine is reticulate. Colpus length 7.85 ± (1.2) μm; colpus width 3.8 ± (0.9) μm.

3.1.10. *F. vulgare*

Polar axis 26.86 ± (1.9) μm, equatorial axis 10.07 ± (1.8) μm, exine 1.8–2.1 μm, pollen shape 2.68 per-prolate, colpus length 14.90 μm, colpus width 4.9 μm, por çapı 3.8 μm. The ornamentation showed a rugulate-striate ornamentation according to the SEM results.

3.1.11. *C. carduchorum* and *C. reticulatus* subsp. *reticulatus*

Convolvulus species have an isopolar pollen shape. Pollen grains are radial and the pollen type is three zonocolpate in *C. carduchorum*. The pollen grains were symmetrical. The polar axis was 42.3 ± (1.3) m, equatorial axis was 43.5 ± (1.2) mm, exine was 2.1 ± (0.4), P/E: 0.97, and pollen shape was oblate-spheroidal in *C. carduchorum*. The polar axis was 42.3 ± (1.3) μm, equatorial axis was 43.5 ± (1.2) μm, exine was 2.1 ± (0.4) μm, P/E: 0.97, and pollen shape was oblate-spheroidal in *C. reticulatus* subsp. *reticulatus*.

Principle component analysis (PCA) and cluster analysis (CA) were performed to identify the palynological data of taxa. The PCA was then performed with Varimax rotation using the matrix correlation configuration. The main components of the principal component analysis were PC1 with 45.19% and with PC2 26.13%, respectively. The total load of PC1 and PC2 was 71.33%. The Kaiser–Meyer–Olkin (KMO) method was conducted to examine the correlation of the variables. KMO was at 0.798, which is considered acceptable. Barlett's test of sphericity also showed a statistical significance at alpha 0.04 for the data set. PCA analysis, which clarified the relationship between the taxa and pollen morphology, was explained with two possible groups (PC1 and PC2). The results are presented in Figure 9.

4. Discussion

The morphological parameters of pollen provide additional information in plant taxonomy and systematics. Pollen analyzed by LM showed high variability in equatorial diameter, polar diameter, width, and length of the colpi and exine. Variations were observed in the pollen morphology of a wide range of species selected from a particular area (Figure 2). This suggests that palynological data play a vital role in evolutionary studies. Pollen size and exine ornamentation are two of the most prominent diagnostic features of plant species in systematic studies. This study differs from previous studies on species collected from the study area in terms of qualitative and quantitative microscopic characteristics.

Kose et al. [28] studied the pollen grains of eight different *Ajuga* species and measured the pollen morphology of *A. reptans* suboblata-subprolata and tricolpatae, with P/E = 0.89 and the orientation is tectatae-granulate; they determined that the exine thickness was 1.46 μm. According to these results, *A. reptans* pollen grains collected from the Bingöl region and the pollen findings studied by Kose et al. show differences, albeit partially. In another study, *Onosma bornmuelleri* pollen grains have an equatorial appearance, a polar axis of 25.08 μm, and an equatorial axis of 21.47 μm [29]. Ornamentation in the polar region was determined as granulate. The exine is 1.05 μm thick [29]. In our study, the polar axis was 22.4 ± (2.0) μm and the equatorial axis was 22.7 ± (2.1) μm, while in the other study, the polar axis was 19.02 ± 2.15 μm, and the equatorial axis was 21.32 ± 2.1 μm. This situation caused different pollen shape results.

Bigazzi et al. [30] reported that *Paracaryum cristatum* pollen grains were small and had a polar axis of 10.7–14.4 μm, equatorial axis of 10.2–13.4 μm, were elliptic in the equatorial view and hexagonal in the polar view, and ranged from prolate-spheroidal to subprolate in shape (P/E: 1.06–1.25), with (3-)6-hetero-colpate and ectocingulate. The composed apertures are spindle-shaped and rarely rhombic, with margins not thickened or only slightly thickened; the endoapertures are elongate, about 1.5 × 3 mm, situated at the equator, and exhibited a granular membrane; the simple colpi are narrow and generally shorter than colporate apertures. According to the results of this study, the pollen's shape was found to be prolate with respect to *P. cristatum* subsp. *cristatum* and *P. strictum*. The pollen morphologies of *P. cristatum* subsp. *cristatum* found in this study were close to each other. *Rindera caespitosa*'s pollen shape was subprolate and ornamentation showed granulate ornamentation according to this study's SEM results. Our results show similarities with the studies conducted.

In a study of two sections of this genus, with respect to *Astragalus falcatus*, *A. odoratus*, *A. ornithopodioides*, *A. stevenianus* var. *stevenianus*, *A. stevenianus* var. *kochianus*, and

A. jodostachys species with a polar axis (P) of 24.53–40.76 μm and equatorial axis (E) of 21.82–31.48 μm, pollen shapes were determined as prolate-spheroidal, prolate, or subprolate. Ekici et al. studied the revision of the Hololeuce Bunge (*Astragalus*) section of Turkey. They found that the pollens of all species were tricolporate types and had reticulate ornamentation. As a result of pollen measurements, they stated that the pollen shapes showed slight differences, but pollen characteristics did not show systematically significant differences between species [31]. As a result of light and electron microscopy studies of the specimens in the Onobrychoidei section of the genus *Astragalus* L., they reported that the pollen was generally tricolporate, prolate, subprolate, or prolate-spheroidal. The polar axis of the pollen varied between 23.4 and 42.6 μm, and the equatorial axis varied between 14.3 and 36.4 μm; their external appearance was elliptic or flattened oval in the meridional optical region and trilobat and sometimes tetrabulate in the polar optical region [32]. The shape of pollen belonging to *A. aucheri* and *A. orientalis* species that we determined in this study is prolate. In this respect, our results show parallelism with the studies carried out. Akan et al. (2005) studied the pollen morphology of taxa belonging to the Alopecias Bunge section of the genus *Astragalus* in Turkey, employing light and electron microscopy. They found that the pollen shape of the taxa was subprolate or prolate-spheroidal, the pollen type was tricolpate, and ornamentation was microreticulate [33]. The ornamentation of the *Astragalus* species in our study was determined as reticulate. Thus, pollen, especially polar ornamentation in the region, differed from each other.

Oybak Dönmez and Işık [34] determined that the *Ixiolirion* sp.'s pollen ornamentation was micro-reticulate, and it has a polar axis of 50 ± (1.34), equatorial axis of 43 ± (1.26), and exine of 1–1.5 μm. In another study of *Linaria kurdica* subsp. *kurdica*, the pollen was tricolporate and prolate-spheroid; the polar axis was 16.95–15.37 μm, equatorial axis was 14.88–15.86 μm, and exine thickness was 0.94 μm (exine embellishment reticulate). The intin was 0.61 μm thick, the colpus length was 12.5 μm, and colpus width was determined to be 3.4 μm [35]. According to the results of this study, *L. kurdica* subsp. *kurdica* pollens are isopolar and tricolporate. However, the polar and equatorial lengths were found to be smaller in this taxon from Turkey.

The pollen morphology studies of the *Convolvulus* genus taxa in this study were presented for the first time in this study. Qualitative and quantitative observations of the pollen of *Convolvulus carduchorum* and *C. reticulatus* subsp. *reticulatus* were observed to be compatible within the genus [36].

Beekeeping is an agricultural activity in which honey bee colonies are used for the purpose of maximizing the worker bee population during the periods when the nectar flow is most abundant and for the production of honey, pollen, royal jelly, and pollination of plants [37]. Baydar and Gürel found that Fabaceae-type pollens are among the most preferred for honey bees and that pollen has an important role in the Fabaceae pollen of species belonging to the family compared to species from both families because they are much richer in protein and minerals [38]. Pollen quality is also important in pollen preferences. Fabaceae is the most important species in the flora. In our study, there are three specimens from the Fabaceae family. However, all pollens analyzed in this study were important species for beekeeping.

Qualitative and quantitative observations obtained as a result of palynological data were grouped by cluster analysis (Figure 8). According to this result, the taxa at the family and genus levels gave results that were consistent with their morphological and systematics. Taxa in the same genus were also found in close clades. Pollen morphology data were examined from all aspects with both statistical methods and data visualization techniques. All methods supported each other by making the study more comprehensive. Pollen morphologies studied for the first time have made a great contribution to the science of systematic taxonomy.

5. Conclusions

The analyzed plant taxa have 3-colporate, zono-colporate, mono-sulcate, and 6-colpate pollen with reticulate and micro-reticulate exine-sculpturing patterns. The pollen grains' qualitative data were utilized to evaluate pollen types and differentiate species at the genus level. It was concluded that individuals within the genus, which are morphologically similar to each other, can be easily distinguished by using the data obtained as a result of palynological studies. For example, the genus *Onosma*, *Paracaryum* in the family Boraginaceae in this study can easily be distinguished from each other thanks to the data provided by this and other studies in the study of pollen analysis in honey and beekeeping. The type and number of apertures, exine ornamentation, and pollen shapes of the species in these genera are different from each other. Furthermore, quantitative data concerning pollen are important tools for finding the similarities and dissimilarities between taxa. The morphological features of pollen that are provided in this study might be useful for systematic and phylogenetic analysis.

In this study, the comprehensive palynological significance of the reported species was examined. The available information emphasizes that pollen's morphology characteristics of the studied species can be used to accurately identify the species. Significant variation among palynological characteristics was studied during both quantitative and non-quantitative analyses. The traits examined in this study were reviewed and compared with previous literature. When comparing the current results of the reported species with previous results, a degree of similarity emerged. Taxonomic data based on qualitative traits were established for the investigated species for precise recognition and identification. Pollen morphology, as well as photographic and statistical data of the taxa, were determined in this study. While the pollen morphology of 16 taxa collected from the study area was studied for the first, the pollen morphologies of some taxa were also introduced to the literature for the first time in this study. The species analyzed with TEM were visualized in taxa collected from this region for the first time. This study contributed to bee plant research, melissopalynological, and systematic botanical flora studies.

Author Contributions: Writing—original draft, A.D.; Writing—review & editing, P.Y.S. All authors have read and agreed to the published version of the manuscript.

Funding: This research received no external funding.

Data Availability Statement: Not applicable.

Conflicts of Interest: The authors declare no conflict of interest.

References

1. Munsif, S.; Khan, M.A.; Ahmad, M.; Zafar, M.; Shah, G.M.; Yasmin, G. Comparative pollen studies of the genera Lantana, Verbena and Vitex of family Verbenaceae from Pakistan. *Int. J. Agric. Biol.* **2007**, *42*, 545–549.
2. Hyde, H.A.; Williams, D.A. The right word. *Pollen Anal. Circ.* **1944**, *8*, 6.
3. Lopes, A.C.V.; Souza, C.N.D.; Saba, M.D.; Gasparino, E.C. Pollen morphology of Malvaceae from Cerrado forest fragments: Details of aperture and ornamentation in the pollen types definition. *Palynology* **2022**, *46*, 1–15. [CrossRef]
4. Soares, E.L.; Landi, L.A.D.C.; Gasparino, E.C. Additions to the knowledge of the pollen morphology of some Fabaceae from Cerrado forest patches of Brazil. *Palynology* **2021**, *45*, 269–281. [CrossRef]
5. Anar, M.; Ahmad, M.; Zafar, S.; Elnaggar, A.Y.; Zafar, M.; Sultana, S.; Tariq, A.; Anjum, F.; Hussein, E.E.; Kiliç, Ö.; et al. Palynomorphological diversity of Asteraceous and Poaceous allergenic plant using microscopic techniques in lesser Himalaya-Pakistan. *Microsc. Res. Tech.* **2022**, *85*, 2061–2075. [CrossRef]
6. Ahmad, S.; Ahmad, M.; Fawzy Ramadan, M.; Sultana, S.; Papini, A.; Ullah, F.; Saqib, S.; Ayaz, A.; Ahmed Bazai, M.; Zaman, W.; et al. Palynological Study of Fossil Plants from Miocene Murree Formation of Pakistan: Clues to Investigate Palaeoclimate and Palaeoenvironment. *Agronomy* **2023**, *13*, 269. [CrossRef]
7. Usma, A.; Ahmad, M.; Zafar, M.; Sultana, S.; Ullah, F.; Saqib, S.; Ayaz, A.; Zaman, W. Palynological Study of Weed Flora from Potohar Plateau. *Agronomy* **2022**, *12*, 2500. [CrossRef]

8. Karaismailoğlu, M.C. Taxonomical, morphological, palynological, anatomical and ecological investigations on monotypic genus *Pachyphragma* from Turkey. *Pak. J. Bot.* **2019**, *51*, 1021–1026. [CrossRef]
9. Doğan, C.; Çakır, N. Pollen Morphology of the Genus *Arenaria* L. (Subgenus *Arenaria*) (Caryophyllaceae) in Turkey. *Pak. J. Bot.* **2019**, *51*, 6.
10. Khan, S.A.; Tahir, S.S.; Mangi, J.U. Colpate, colporate and porate pollen isolated from the shale of bara formation, Lakhra, Sindh, Pakistan. *Pak. J. Bot.* **2022**, *52*, 2095–2099. [CrossRef]
11. Raza, J.; Ahmad, M.; Zafar, M.; Athar, M.; Sultana, S.; Majeed, S.; Yaseen, G.; Imran, M.; Nazish, M.; Hussain, A. Comparative foliar anatomical and pollen morphological studies of Acanthaceae using light microscope and scanning electron microscope for effective microteaching in community. *Microsc. Res. Tech.* **2020**, *83*, 1103–1117. [CrossRef] [PubMed]
12. Asgari Nematian, M.; Ranjbar, M. Comparative pollen morphology and its systematic implications in Resedaceae in Iran. *Nord. J. Bot.* **2022**, *2*, 03394. [CrossRef]
13. Carlquist, S. Tribal interrelationships and phylogeny of the Asteraceae. *Aliso A J. Syst. Florist. Bot.* **1976**, *8*, 465–492. [CrossRef]
14. Erdtman, G. *Pollen Morphology and Plant Taxonomy: Angiosperms*; Brill Archive: Leiden, The Netherlands, 1986; Volume 1.
15. Gentry, A.H.; Tomb, A.S. Taxonomic implications of Bignoniaceae palynology. *Ann. Mo. Bot. Gard.* **1979**, *66*, 756–777. [CrossRef]
16. Ulukuş, D.; Tugay, O. Türkiye için endemik bir bitki olan *Cousinia foliosa* Boiss. & balansa (asteraceae) üzerine anatomik ve palinolojik çalışmalar. *Selçuk Üniversitesi Fen Fakültesi Fen Derg.* **2020**, *46*, 118–125.
17. Bahadur, S.; Taj, S.; Long, W.; Ahmad, M. Pollen morphology and its implication in the taxonomy of some selected tribes of the Asteraceae of Hainan Island South China. *Bot. Rev.* **2022**, *88*, 271–298. [CrossRef]
18. Bahadur, S.; Taj, S.; Long, W.; Hanif, U. Pollen Morphological Peculiarities of Selected Mimosoideae Taxa of Hainan Island and Their Taxonomic Relevance. *Agronomy* **2022**, *12*, 1122. [CrossRef]
19. Umber, F.; Zafar, M.; Ullah, R.; Bari, A.; Khan, M.Y.; Ahmad, M.; Sultana, S. Implication of light and scanning electron microscopy for pollen morphology of selected taxa of family Asteraceae and Brassicaceae. *Microsc. Res. Tech.* **2022**, *85*, 373–384. [CrossRef] [PubMed]
20. Majeed, S.; Zafar, M.; Ahmad, M.; Zafar, S.; Ghufran, A.; Ayoub, M.; Sultana, S.; Yaseen, G.; Raza, J. Morpho-palynological and anatomical studies in desert cacti (*Opuntia dillenii* and *Opuntia monacantha*) using light and scanning electron microscopy. *Microsc. Res. Tech.* **2022**, *85*, 2801–2812. [CrossRef]
21. Zafar, A.; Zafar, M.; Ahmad, M.; Khan, A.M.; Mahmood, T.; Kilic, O.; Fatima, A.; Habib, D.; Sultana, S.; Majeed, S.; et al. Microscopic (LM and SEM) visualization of pollen ultrastructure among honeybee flora from lower Margalla Hills and allied areas. *Microsc. Res. Tech.* **2022**, *85*, 3325–3338. [CrossRef]
22. Demirpolat, A.; Dogan, G.; Bagci, E. Pollen Morphology and Fruit Micromorphology of the Genus *Scandix* (Apiaceae). *Feb-Fresenius Environ. Bull.* **2021**, *30*, 12413.
23. Erdtman, G. Pollen morphology and plant taxonomy 4. Labiatae, Verbenaceae and Avicenniaceae. *Sven. Bot. Tidskr.* **1945**, *39*, 279–285.
24. Wodehouse, R.P. *Pollen grains. Their Structure, Identification and Significance in Science and Medicine*; McGraw-Hill Book Company: New York, NY, USA, 1935.
25. Faegri, K.; Iversen, J. *Textbook of Pollen Analysis*; Hafner Press: New York, NY, USA, 1975.
26. Ertman, G. *Pollen Morphology and Plant Taxonomy Angiosperms*; Ronald Press: New York, NY, USA, 1952.
27. Kılıç, N.; Yılmaz Dağdeviren, R.; Caner, H.; Akkemik, Ü. Türkiye'de Kullanılmakta Olan Palinoloji ve Polen Terimleri Üzerine Bir Değerlendirme ve Öneriler. *Avrasya Terim Derg.* **2020**, *8*, 98–108.
28. Kose, Y.B.; Erkara, I.; Alan, S. Pollen Morphology of Some Turkish *Ajuga* L. (Lamiaceae) and its Taxonomic Value. *Bangladesh J. Bot.* **2011**, *40*, 29–33. [CrossRef]
29. Turkmen, Z.; Coşkunçelebi, K.; Makbul, S.; Beyazoğlu, O.; Doğan, C. Pollen Morphology of *Onosma* L. (Boraginaceae) Taxa Distributed In Anatolia Kuzey-Doğu. *Fen Bilim. Enstitüsü Derg.* **2011**, *4*, 71–82.
30. Bigazzi, M.; Nardi, E.; Selvi, F. Palynological Contribution To The Systematics Of *Rindera* And The Allied Genera *Paracaryum* and *Solenanthus* (Boraginaceae-Cynoglosseae). *Willdenowia* **2006**, *36*, 37–46. [CrossRef]
31. Ekici, M.; Ekim, T. Revision of the Section Hololeuce Bunge of the Genus *Astragalus* L. *(Leguminosae) in Turkey, Turk. J. Bot.* **2004**, *28*, 307–347.
32. Pınar, M.N.; Ekici, M.; Aytaç, Z.; Akan, H.; Çeter, T. Pollen morphology of *Astragalus* L. sect. Onobrychoidei DC. (Fabaceae) in Turkey. *Turk. J. Bot.* **2009**, *33*, 291–303. [CrossRef]
33. Akan, H.; Tatlıdil, S.; Bıçakcı, A. Pollen morphology of *Astragalus* L. Section Alopecuroidei DC. (Fabaceae) in Turkey. *Int. J. Bot.* **2005**, *1*, 50–58. [CrossRef]
34. Oybak Dönmez, E.; Işık, S. Pollen morphology of Turkish Amaryllidaceae, Ixioliriaceae and Iridaceae. *Grana* **2008**, *47*, 15–38. [CrossRef]
35. Çelenk, S. Afyon Ve Çevresinde Yayılış Gösteren Linaria Miller Türlerinin Polen Morfolojileri. *Fırat Üniversitesi Mühendislik Bilim. Derg.* **2004**, *16*, 183–194.
36. Demirpolat, A.; Kılıç, Ö. A Scanning Electron Microscope Study: Investigation of Some *Convolvulus* L. Taxa (Convolvulaceae) as Morphological and Palynological. *J. Glob. Health Nat. Sci.* **2021**, *4*, 94–101.

37. Güler, A. *Bal Arısı (Apis Mellifera)*; Ondokuz Mayıs Üniversitesi, Ziraat Fakültesi: Samsun, Turkey, 2006; p. 574.
38. Baydar, H.; Gürel, F. Antalya doğal florasında bal arısı (*Apis mellifera*)' nın polen toplama aktivitesi, polen tercihi ve farklı polen tiplerinin morfolojik ve kalite özellikleri. *Turk. J. Agric. For.* **1998**, *22*, 475–482.

agronomy

MDPI

Article

Palynological Study of Fossil Plants from Miocene Murree Formation of Pakistan: Clues to Investigate Palaeoclimate and Palaeoenvironment

Shabir Ahmad [1], Mushtaq Ahmad [1,*], Mohamed Fawzy Ramadan [2], Shazia Sultana [1], Alessio Papini [3], Fazal Ullah [4], Saddam Saqib [5,6], Asma Ayaz [7], Mansoor Ahmed Bazai [8], Wajid Zaman [9,*] and Muhammad Zafar [1]

1 Department of Plant Sciences, Quaid-i-Azam University, Islamabad 45320, Pakistan
2 Department of Clinical Nutrition, Faculty of Applied Medical Sciences, Umm Al-Qura University, Makkah 21955, Saudi Arabia
3 Department of Biology, University of Florence, 50121 Florence, Italy
4 State Key Laboratory of Grassland Agro-Ecosystems, School of Life Sciences, Lanzhou University, Lanzhou 730020, China
5 State Key Laboratory of Systematic and Evolutionary Botany, Institute of Botany, Chinese Academy of Sciences, Beijing 100093, China
6 University of Chinese Academy of Sciences, Beijing 100049, China
7 State Key Laboratory of Biocatalysis and Enzyme Engineering, School of Life Sciences, Hubei University, Wuhan 434200, China
8 Livestock and Dairy Development Department Baluchistan, Quetta 87300, Pakistan
9 Department of Life Sciences, Yeungnam University, Gyeongsan 38541, Republic of Korea
* Correspondence: mushtaqflora@hotmail.com (M.A.); shangla123@gmail.com (W.Z.)

Citation: Ahmad, S.; Ahmad, M.; Fawzy Ramadan, M.; Sultana, S.; Papini, A.; Ullah, F.; Saqib, S.; Ayaz, A.; Ahmed Bazai, M.; Zaman, W.; et al. Palynological Study of Fossil Plants from Miocene Murree Formation of Pakistan: Clues to Investigate Palaeoclimate and Palaeoenvironment. *Agronomy* **2023**, *13*, 269. https://doi.org/10.3390/agronomy13010269

Academic Editors: Carmine Guarino and Rodolfo Gentili

Received: 8 November 2022
Revised: 29 December 2022
Accepted: 13 January 2023
Published: 16 January 2023

Abstract: Palaeoflora in Pakistan in the Miocene is characterized by its high biodiversity. The present study investigated the pollen of fossil plants from the Murree Formation of Pakistan. Shales and mudstones were collected from the Murree section located at the foothills of the Margalla hills and analyzed by palynofacies and palynostratigraphic analyses. In this paleopalynological study of the Miocene Murree Formations of Pakistan, 31 samples were analyzed using microscopic techniques containing 48 pollen types from 12 families. The quantitative and qualitative morphological features of pollen were determined using light and scanning electron microscopy to help identify the pollen grains. Exine ornamentations and spines were the most important diagnostic features for distinguishing one pollen grain from another. The maximum exine thickness was observed in *Ocimum basilicum* of 4.25 μm, whereas the maximum pollen diameter was recorded for *Pinus* of 69.5 μm. Lamiaceae, Asteraceae, and Poaceae were the dominant families. The results showed that the preservation of floral records was not optimal. The presented results provide data on the dominant fossil plant taxa that existed in Pakistan (23.03–5.33 Ma). The evolution and phytogeographical histories of fossil plants can be unraveled using rock sediments to preserve biodiversity.

Keywords: palynoflora; taxonomic; microscopy; systematics; vegetation's origin

1. Introduction

In the tertiary period, the Himalayas formed as a result of a collision between the Eurasian and Indian plates [1]. The Himalayas have been further classified into higher Himalayas, lesser Himalayas, and sub-Himalayas and are considered the main source for the Murree Formation. This formation has a strong tectonic affinity regarding the structural evolution with the Himalaya orogeny. It has been named Murree owing to the locality of Murree hills in the district of Rawalpindi, Pakistan. The Rawalpindi formation lies in the early Miocene period [2]. Therefore, it provides important information on how the fossil plant pollen record in the core–mantle boundary relates to the Indo–Asian tectonic collision and the environmental change preceding the Miocene–Eocene Transition [3]. In Pakistan, the Murree Formation comprises the area of Islamabad, Kohat, Murree, and Attock districts. Regarding the current climate (temperature and precipitation), four seasons are recorded in the study area: summer, autumn, winter, and spring. The summer season is longer than others; it begins in April and ends in September. Geographically, the country comprises mountains, plateaus, rivers, deserts, and forests. The regions with lower altitudes exhibit tropical and subtropical climates, while higher altitudes possess temperate environments [4,5].

Pakistan has Gondwanan, Eurasian, and Tethyan ancestors of the vegetation components. The Indus basin of Pakistan, which is in the center, south, and east of the country, represents the Gondwanan heritage [6]. The sedimentary strata of the Murree Formation consist of purple, grey, and dark red sandstone formed from a monotonous sequence. The lower part consists of calcareous sandstone containing numerous foraminifera, generally known as Fatehjang. In the Attock district, one of the types of sections is exposed at the north of the Dhok Maiki (Lat. 33°25′ N: Long. 72°35′). The main body of the formation consists of fossiliferous, silicified wood, plant remains, and fish [7–9]. The current study analyzed some fossil plant pollen records from different families. Pollen and spores are essential sources to determine the climate variations of the study area. They are important investigative tools in several fields, including paleobotany, archaeology, and geography, regarding the climate and other environmental factors affecting flora in any area [10,11]. The Stratigraphic committee of Pakistan has worked on the stratigraphic nomenclature in the country [6]. Paleo-environmental changes in Asia occurred mainly in the Miocene period. Therefore, numerous relevant fossils of the period can be obtained from the corresponding stratum [12]. The sedimentary strata of the Murree Formation consist of various coarse to fine-grained units. In the Attock district, the Murree Formation comprises lamellated claystone, sandstone, and conglomerates. In Kohat, the formation is characterized by medium to coarse-grained sandstones [13]. In northern Potwar, the formation reaches its maximum thickness of approximately 3030 m, but in western Kohat, it thins out and is approximately 9 m thick [4]. The main deposition of the Murree Formation occurred in a fluviatile environment that was affected by meandering rivers and turbidity currents, as evidenced by cyclic deposition of shale, siltstones, and sandstones, as sedimentary structures, such as calcite concretions, ripple marks, cross-bedding, worm barrows, and lithofacies (Figure 1). The sediments of the Murree Formation contain fossil materials of plant and animal remains. For example, in Fathehjang, numerous mammal bones have been discovered, including remains from different even-toed ungulates, rhinoceros, and hippopotamus-like animals [7]. Based on the mammalian fossil record, the age of the Murree Formation has been constrained to the Miocene.

A study was conducted to estimate the effects of climate change on Pakistan's natural forest ecosystems, using 1990 as the base year and assuming a 0.3 °C rise in temperature and changes in precipitation of 0, +1, and −1% decade^{-1}. The study predicted that the atmospheric CO_2 concentration of 350 ppmv will rise to 500 ppmv in 2050 and 575 ppmv in 2080. The current climate of the country was reported to vary. Some major biomes were noted, such as croplands, mosaic vegetation, evergreen forest, mosaic grassland, open grassland, barren area, and permanent snow. Owing to various climatic conditions, the variations were investigated in different kinds of forests in the study area, i.e., alpine tundra, mixed woodland, cold conifer, warm conifer, xerophytic wood, grassland, steppe, and desert (Figure 1). Extreme biomes on either side of the spectrum are more vulnerable to climate change than others. The alpine tundra loses approximately 32% of its area to other biomes, particularly those directly beneath it (such as cold coniferous/mixed forest). On the other hand, warm conifers expand their range significantly when temperatures and precipitation rise. Climatic variations led to significant shifts in forest areas from one biome to another in northern Pakistan. The overall climate has a significant negative impact on the natural and forest ecosystem of Pakistan [13–15].

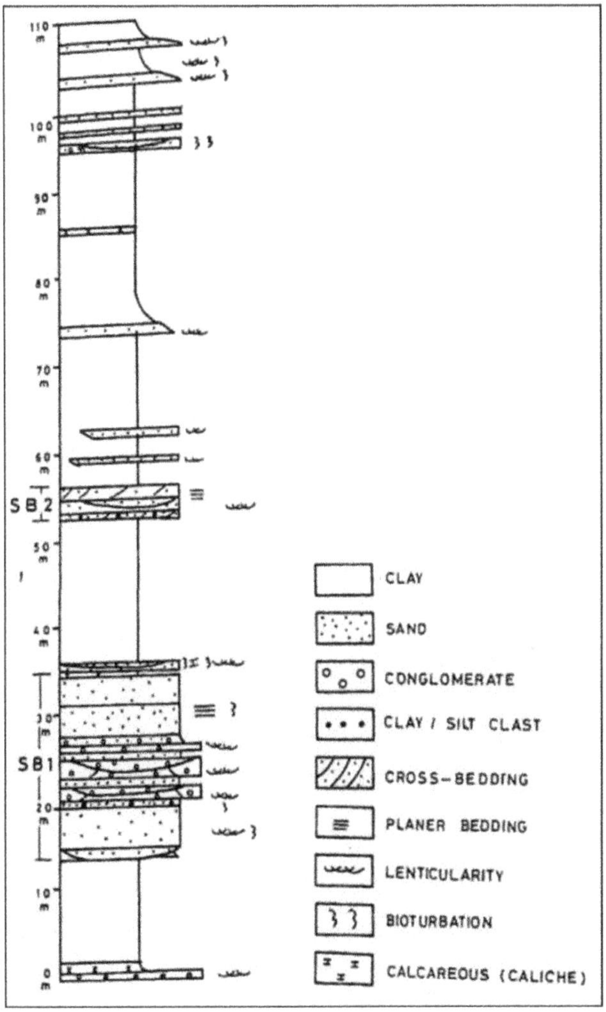

Figure 1. Stratigraphic column of the Murree Formation, Pakistan [13].

The multiple scenarios support the study area's rich and diverse vegetation. This diversity is reflected in the current flora of Pakistan (see http://legacy.tropicos.org/Project/Pakistan), where more than 6000 higher vascular plant species have been described, along with approximately 190 pteridophytes. Approximately 80% of the flowering plants were recorded. Important for this study is that the pollen morphology of plants occurring in various vegetation units of Pakistan is well documented. Optical and scanning electron microscopy-based pollen morphologies of over 70 plant families in Pakistan have already been published in a series of reputable papers [14]. These data make it practical to compare the fossil pollen grains to pollen from extant plants whose ancestors most likely occurred in the study area during the Miocene.

Palaeopalynology is the study of ancient spores and pollen occurring in sedimentary rocks. Pollen grains are the male reproductive parts (male gametophyte surrounded by a sporal wall) found in seed plants and show a very high degree of diversity in shapes and size [16,17]. The outer wall layer of the pollen grain is called the exine, while the inner layer is the intine; each is made up of several compounds, among which sporopollenin is the most important of the exine layer. Exine can be divided into a sexine (more external) and a more internal nexine. The wall of the pollen grain is resistant to the action of the external environmental conditions. In different families and genera, pollen varies in sculpturing, walls, size, shape diameters, colpi, and pores [18]. Spores are the long-distance dispersal units in seedless terrestrial plants, and monilophytes can also be used in paleopalynology. Pollen and the spore morphology also play a vital role in species identifications [19–21].

Palynological investigations in mountainous areas of Pakistan in the Miocene age are few and limited to some regions [22]. The primary goal of this investigation is to study the palynomorph assemblages found in the tidal and marine environments of the study area. The assemblages of palynoflora recovered from the rocks are relatively diverse and well-preserved. The floral characters were compared to those found in previous studies. In addition, several changes were made to the original matrix of morphological characters based on a previous study [23]. The following questions are addressed in the current study. What were the dominant types of vegetation in the investigated age? What were the biogeographic origins of Pakistan's modern flora? How did these pollen types compare with morphotaxa from Miocene records of adjacent regions?

The following approaches were made to answer all these questions. Palynomorphs' morphological features were investigated using qualitative and quantitative pollen characters to determine the paleo-climate of the past marine environment and further described along with the section near the Murree Formation. Paleopalynology knowledge is essential for knowing the past climate of the Miocene reconstruction of vegetation and paleo-environment. The purpose of this study was to investigate fossil plant pollen records preserved in the sedimentary strata of the Miocene and its correlation with the paleo-environment [12]. This research was very useful in investigating fossil pollen of the Miocene Murree Formation for the first time. The study helps reconstruct Miocene palaeovegetation and climatic changes occurring in the depositional environment. It also helps to estimate the regional palaeoclimate, paleophytogeographic history of the paleoflora and its relation to the modern flora of Pakistan.

2. Materials and Methods

The field surveys were conducted in November and December 2021 on the numerous different sequential horizons in the foothills of Margalla Islamabad, Pakistan, during which the latitude, longitude, rainfall, and temperature were recorded (Figure 2). The main deposition of the Murree Formation occurred in a fluviatile environment affected by meandering rivers and turbidity currents, as evidenced by the cyclic deposition of shales, siltstones, and sandstones, as well as sedimentary structures, such as calcite concretions, ripple marks, cross-bedding, worm barrows, and lithofacies.

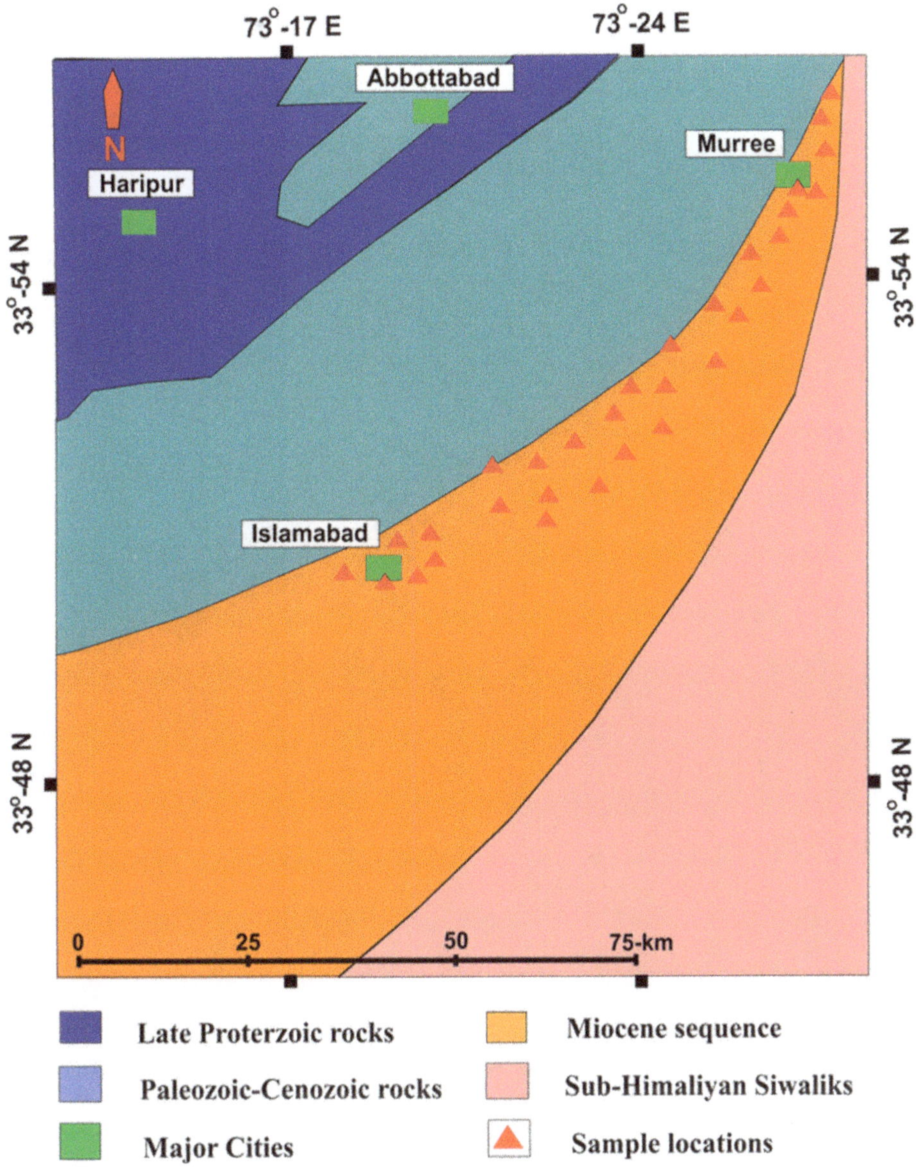

Figure 2. Map of the study area showing the sample localities [8].

2.1. Samples Collections

The Miocene Murree Formation has distinguished features, such as lithofacies, bed thickness, sedimentary structures, grain size, and color. Sandstones of the formation are generally fine-grained with a dominant reddish color followed by greenish and greyish color grains [13]. The study area is considered the Miocene Murree Formation based on these unique features and the discovery of the mammalian fossil record [7,13].

The samples were collected after digging 30 cm of sedimentary rock carefully to ensure that it contained palynomorphs (Figure 3). For paleopalynological purposes, 31 samples were collected from shales and mudstones along a 50 m thick Murree section and processed at the Plant Systematics and Biodiversity Laboratory, where all necessary instrumentations and chemicals are available. The samples were photographed, collected in polyethylene bags, and then brought to the laboratory for further processing. Palynological samples were also used to identify palynomorphs in plant taxonomy and systematics [24,25].

Figure 3. Field photography: (**A**) Hammer placed on the rock for cutting the rock sediments. (**B**) Sample collection from the study area. (**C**) Samples collected and placed in a polyethylene bag for palynological purposes. (**D**) View of the study area.

2.2. Sample Preparation

The preparation of sedimentary rock samples was undertaken using the standard protocol of palynology with some modifications [26,27]. The rock sediments were treated with 10% HCl for 1 day to remove carbonates and then treated with distilled water to neutralize them. The samples were again reacted with 40% HF for at least one day to remove silicate particles and treated with distilled water for neutralization. For further palynological processing, the samples were centrifuged at 2500 rpm to separate the heavy and light particles. Heavy liquid with a specific gravity of 1.9–2.0 was treated with samples and centrifuged again at 2500 rpm for five minutes. With a 150 μm mesh, the samples were sieved to remove the plant fragments and coarse debris. The resulting pollen residue was filtered through a 10 μm sieve and kept in distilled water until mounted into glycerol jelly on slides for pollen analysis. The macerated sample was washed and mounted in glycerin jelly. The pollen was classified into different classes based on the pollen size, shape, numbers, exine thickness, pores, and colpi. The pollen taxa were identified by comparing them with the regional palynolora reported elsewhere [25,26]. The Tortonian age of the Murree Formation was assigned to the late Miocene, based on the correlation of plant fossils, fossilized fish remains, and mammalian records and its geological age was confirmed [7]. The main morphological characteristics of the pollen grains to be investigated were as follows: dimension, aperture types, sculpturing, pollen diameter, and shape [1,25]. In the 31 samples analyzed, 15 were barren and 16 contained different types of pollen. With the help of optical and scanning microscopy, we described each sample's morphological features of the pollen grains (Figures 4–8).

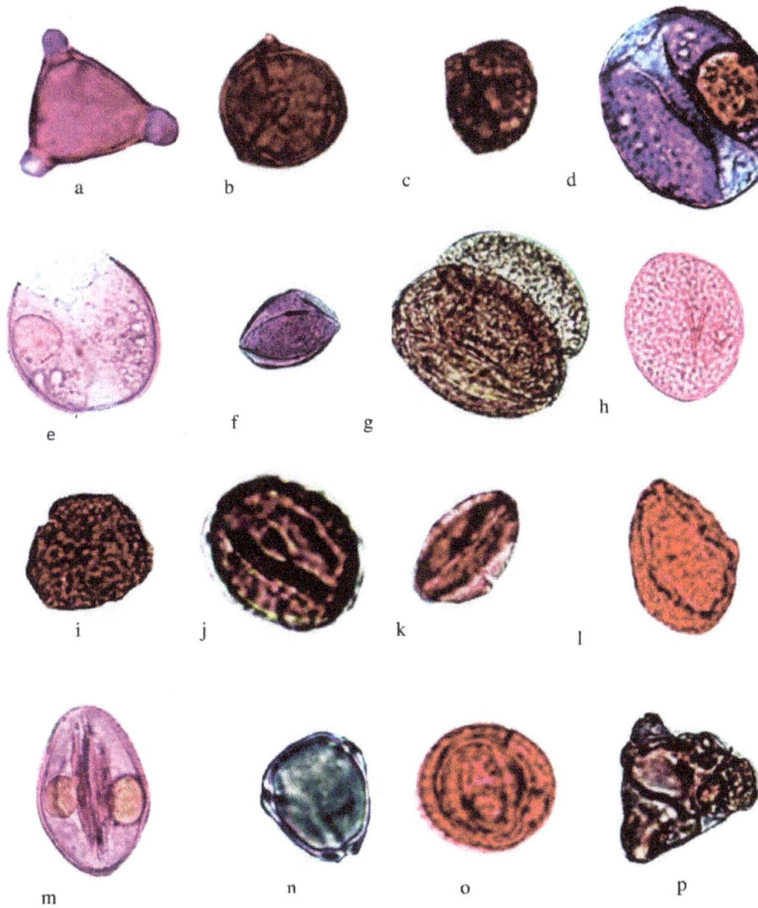

Figure 4. Optical microscopy images of pollen. (**a**) *Gevuina Avellana.* (**b**,**c**) *Sanguisorba minor.* (**d**–**g**) *Pinus* sp. (**h**) *Sparganium* sp. (**i**–**l**) *Quercus* sp. (**m**) *Juniper* sp. (**n**) Myrtaceae. (**o**) *Artemisia* sp. (**p**) *Interporopollenites* sp.

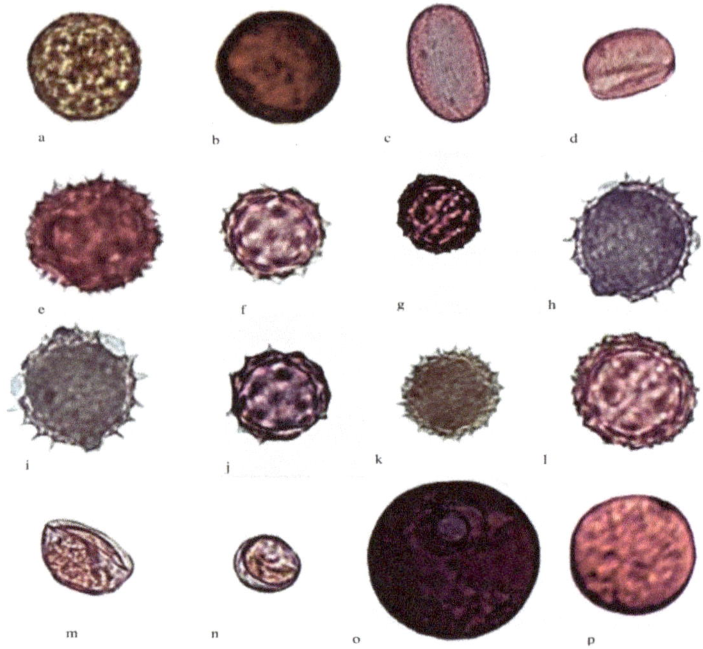

Figure 5. Optical microscopy images of pollen. (**a**) Chenopodiaceae. (**b**) Cupracaceae. (**c**,**d**) Liliaceae. (**e**–**l**) Asteraceae. (**m**–**p**) Poaceae.

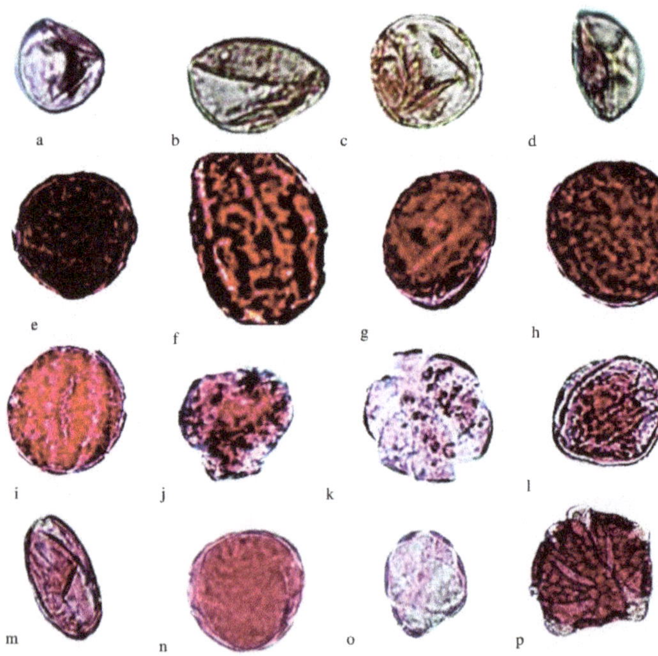

Figure 6. Optical microscopy images of pollen. (**a**–**d**) Poaceae. (**e**–**p**) Lamiaceae.

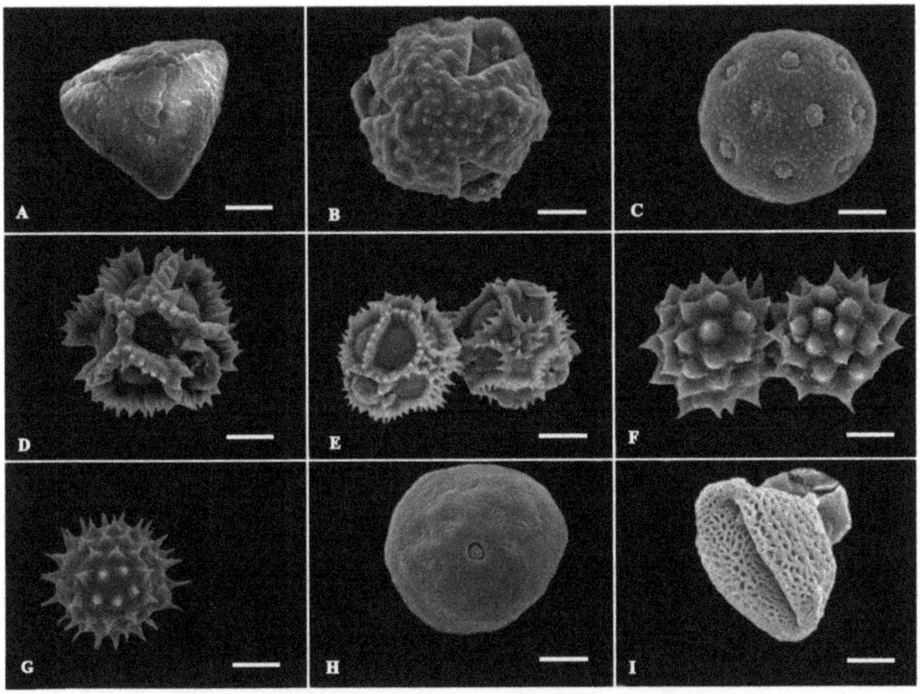

Figure 7. Scanning electron microscopy images of pollen. (**A**) The scale bar represents 5 µm for Myrtaceae, (**B**) 10 µm for *Artemisia* spp. (Asteraceae), (**C**) 5 µm for Chenopodiaceae, (**D**) 5 µm for Cichoriodeae (Asteraceae), (**E–G**) 10 µm for other Asteraceae, (**H**) 5 µm for Poaceae, and (**I**) 5 µm for Lamiaceae.

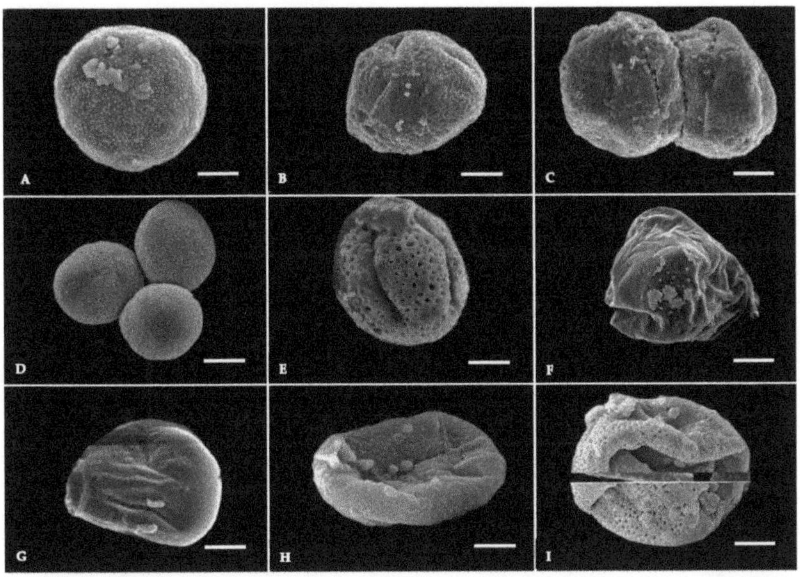

Figure 8. Scanning electron microscopy images of pollen. (**A–I**) The scale bar represents 10 µm for Lamiaceae.

2.3. Samples' Identifications

From the late 1970s, the use of standard microscopic techniques has been established as a standard for investigating pollen of fossil plants. Since then, a vast amount of data on the pollen morphology of numerous plant systematic lineages has been gathered. These include books or articles in series that focus on specific genera, families, or species groups. Many smaller publications focused on a single species or multiple species. As a result, prehistoric pollen can be compared using informational descriptions and high-resolution pictures of actual plant material. This makes it possible to identify many fossil pollen grains confidently and gives us the context we need to place the fossil pollen record in a palaeo-botanical and palaeo-biogeographic context. The diagnostic features from the literature under light and scanning electron microscopy were compiled to identify the palynological record of fossil plants. One of the most important criteria is the sculpturing pattern of pollen grains. Different kinds of sculpturing patterns, i.e., psilate, reticulate, and echinate, have been noted in the pollen grains. In addition, the above feature, number of pores, colpi, and spines are the critical features for species identification [19]. The fossil plant source of the pollen grain identification was also made by referencing a collection of modern pollen grains and the published literature. The identification was carried out up to the family, genera, and sometimes to the species level [25].

3. Results

In the present study, 89 pollen types were recorded from 31 samples, of which 48 pollen grains were identified as belonging to 12 different families. Qualitative and quantitative pollen characters are shown in tabular form (Tables 1 and 2) and micrographs are shown in Figures 4–8. The results of the present study showed that the morphological features of pollen were sufficiently variable to be considered helpful for the classification at the species level. The fossil palynomorphs were investigated using optical and scanning electron microscopy.

Table 1. Micromorphological characteristics of pollen—qualitative.

S. No	Species/Taxon/Family	Pollen Shape	Pollen Type	Colpi/Pore	Ornamentations	Spines
1	*Gevuina Avellana* Molina	Triangular	Tricolporate	P	Psilate	A
2	*Sanguisorba minor* Scop	Circular	-	A	Psilate	A
3		Spheroidal	Tricolpaorte	P	Psilate	A
4	*Pinus* spp.	Circular	Monoporate	A	Regulate	A
5		Subprolate	Dicolporate	P	Psilate	A
6		Peroblate	-	P	Psilate	A
7		Angular	Monoporate	A	Regulate	A
8	*Sparganium* spp.	Circular	Monoporate	P	Psilate	A
9	*Quercus* spp.	Inter-angular	Tricolpate	P	Regulate	A
10		Prolate-spheroidal	-	A	Psilate	A
11		Prolate	-	A	Psilate	A
12		Prolate-spheroidal	-	A	Regulate	A
13	*Juniperous* spp.	Prolate	-	A	Psilate	A
14	Myrtaceae	Angular	Tricolpate	P	Psilate	A
15	*Artemisia* spp.	Spheroidal	-	A	Regulate	A

Table 1. *Cont.*

S. No	Species/Taxon/Family	Pollen Shape	Pollen Type	Colpi/Pore	Ornamentations	Spines
16	*Interporopollenites* spp.	Inter-semilobate		A	Psilate	A
17	Chenopodiaceae	Spheroidal	Pentaporate	P	Regulate	A
18	Cupressaceae	Spheroidal	-	A	Psilate	A
19	Liliaceae	Prolate	-	A	Psilate	A
20		Peroblate	-	A	Psilate	A
21	Cichoriodeae	Spheroidal	Polyporate	P	Echinate	A
22	Asteraceae	Spheroidal	-	A	Echinate	P
23		Spheroidal	-	A	Echinate	P
24		Spheroidal	-	A	Echinate	P
25		Prolate-spheroidal	Tricolporate	P	Echinate	P
26		Circular	-	A	Echinate	P
27		Spheroidal	-	A	Echinate	P
28		Circular	Tricolporate	P	Echinate	P
29	Poaceae	Elliptic	Monoporate	P	Psilate	A
30		Spheroidal	-	A	Psilate	A
31		Prolate	-	A	Psilate	A
32		Subprolate	-	A	Psilate	A
33		Spheroidal	-	A	Psilate	A
34		Inter-subangular	-	A	Psilate	A
35		Circular	-	A	Psilate	A
36		Prolate-spheroidal	-	A	Psilate	A
37	Lmiaceae	Prolate-spheroidal	-	A	-	A
38		Semi-angular	-	A	Psilate	A
39		Circular	-	A	-	A
40		Circular	-	A	-	A
41		Angular	-	A	Psilate	A
42		Circular	-	A	-	A
43		Angular	-	A	-	A
44		Circular-lobed	-	A	Psilate	A
45		Prolate	-	A	-	A
46		Circular-lobed	-	A	Psilate	A
47		Circular	-	A	Psilate	A
48	*Ocimum basilicum* L.	Oblate-spheroidal	Hexacolporate	P	Reticulate	A

Note: P = Present; A = Absent.

Table 2. Quantitative features of fossil plants pollen from the Murree Formation, Pakistan.

S. No	Species Name	Exine Thickness (μm)	Pollen Diameter (μm)	Colpi Length (μm)	Colpi Width (μm)
1	*Gevuina Avellana* Molina	2.25	25.5	7.0	6.25
2	*Sanguisorba minor* Scop	3.25	21.75	10.75	2.25
3		4.00	38.5	-	-
4	*Pinus* spp.	2.5	55.75	-	-
5		3.75	59.00	13.00	17.25
6		3.5	69.5	-	-
7		1.25	53.5	-	-
8	*Sparganium* spp.	2.75	38.00	-	-
9	*Quercus* spp.	1.6	32.25	2	
10		3.25	21.00	-	-
11		1.25	29.25	-	-
12		2.00	51.25	-	-
13	*Juniperous* spp.	3.5	56.5	-	-
14	Myrtaceae	4.07	31.8	-	-
15	*Artemisia* spp.	1.5	22.5	-	-
16	*Interporopollenites* spp.	1.25	40.25	2.5	3
17	Chenopodiaceae	3.5	25.25	-	-
18	Cupressaceae	3.25	26.00	-	-
19	Liliaceae	1.00	42.5	-	-
20		2.5	32.25	-	-
21	Cichoriodeae	3.5	22.75	-	-
22	Asteraceae	4.00	17.5	-	-
23		3.75	17.25	-	-
24		3.75	22.00	-	-
25		3.5	38.5	-	-
26		3.5	36.75	-	-
27		3.25	32.00	-	-
28		3.75	21.00	-	-
29	Poaceae	2.25	31.00	-	-
32		2.00	27.5	-	-
31		1.75	23.00	-	-
32		2.00	26.25	-	-
33		1.25	40.00	-	-
34		2.25	25.25	-	-
35		1.75	31.75	-	-
36		1.25	16.5	-	-

Table 2. *Cont.*

S. No	Species Name	Exine Thickness (µm)	Pollen Diameter(µm)	Colpi Length(µm)	Colpi Width(µm)
37	Lamiaceae	1.25	31.00	-	-
38		3.00	30.25	-	-
39		3.25	39.00	-	-
40		2.5	38.25	-	-
41		2.5	40.25	-	-
42		3.00	38.25	-	-
43		2.25	31.25	-	-
44		2.25	37.5	4.75	5.25
45		2.25	38.5	-	-
46		2.75	32.75	-	-
47		3.00	43.00	-	-
48	*Ocimum basilicum* L.	4.25	57.23	22.6	4.22

Approximately 60% of the pollen grains from the study area were identified. An investigation of pollen in thin sections was difficult because the identifications can be made complicated by unfavorable cuts (and damages) through the specimens. The specimens were compared with Pakistan's extinct and modern plants in shape, size, sculpturing pattern, exine thickness, and aperture morphology. The species were identified using microscopic techniques based on morphological features.

3.1. *Proteaceae*

Gevuina avellana pollen was tricolporate, triangular, isopolar, and radially symmetrical with psilate ornamentations; the exine thickness was 2.25 µm; the pollen diameter was 25.5 µm; the colpi length was 7.00 µm; the colpi width was 6.25 µm.

3.2. *Rosaceae*

An unidentified genus and species belonging to Rosaceae is listed. The pollen was spheroidal, suboblate, striate micro-perforate, and tricolporate; the exine thickness was 4.00 µm, and the pollen diameter was 38.5 µm.

3.3. *Pinaceae*

Pinus pollen was circular, subprolate, peroblate, monoporate, angular, dicolporate, psilate, and regulate. The exine thickness was 1.25–3.5 µm; the colpi length was 17.25 µm; the colpi width was 17.25 µm; the pollen diameter was 55.75–69.5 µm.

3.4. *Typhaceae*

Sparganium pollen was circular, monoporate, and psilate; exine thickness was 2.75 µm and the pollen diameter was 38.00 µm.

3.5. *Fagaceae*

An unidentified genus and species belonging to Fagaceae is listed. The pollen was inter-angular, prolate-spheroidal, prolate, tricolpate, psilate, and regulate. The exine thickness ranged from 1.25 to 3.25 µm; the pollen diameter was 21.00–32.25 µm; the colpi length was 7.00 µm; the colpi width was 6.25 µm.

3.6. Cupressaceae

The *Interporopollenites* pollen is inter-semilobate and psilate; the exine thickness was 1.25 µm and the pollen diameter was 40.25 µm. *Juniperus* pollen was prolate and psilate; the exine thickness was 3.5 µm and the pollen diameter was 56.5 µm. *Cupressaceae* sp. (unidentified species) pollen was spheroidal and psilate; exine thickness was 3.25 µm and pollen diameter was 26.00 µm.

3.7. Myrtaceae

Unidentified genus and species belonging to Myrtaceae. Pollen angular, tricolpate, and psilate. The exine thickness was 3.00 µm and the pollen diameter was 23.00 µm.

3.8. Asteraceae

Artemisia pollen was spheroidal and regulate; the exine thickness was 1.5 µm and the pollen diameter was 22.5 µm. An unidentified genus and species belonging to the subfamily Cichorioideae were found. The pollen was spheroidal and echinate; the exine thickness was 3.5 µm and the pollen diameter was 22.75 µm. While the rest of the pollen is spheroidal, prolate-spheroidal, circular, and tricolporate; the exine thickness ranged from 3.25 to 4.00 µm and the pollen diameter ranged from 17.25 to 38.5 µm.

3.9. Chenopodiaceae

An unidentified genus and species belonging to Chenopodiaceae were found. The pollen was pheroidal, pentaporate, and regulate; the exine thickness was 3.5 µm and the pollen diameter was 25.25 µm.

3.10. Liliaceae

An unidentified genus and species belonging to Liliaceae was found. The pollen was prolate, peroblate, and psilate; the exine thickness was 1.00–2.5 µm and the pollen diameter ranged from 32.25 to 42.5 µm.

3.11. Poaceae

Unidentified genus and species belonging to Poaceae. The pollen was spheroidal, circular, prolate, subprolate, inter-subangular, monoporate, and psilate; the exine thickness ranged from 1.75 to 2.25 µm. The pollen diameter was 16.50–40.00 µm. The colpi length was 7.00 µm and the colpi width was 6.25 µm.

3.12. Lamiaceae

An unidentified genus and species belonging to Lamiaceae was found. The pollen was prolate-spheroidal, semi-angular, angular, circular, and psilate. The exine thickness was 1.25–3.25 µm; the pollen diameter was 30.25–43.00 µm; the colpi length was 7.00 µm; the colpi width was 6.25 µm.

4. Discussion

In the present study, 48 species were reported belonging to 12 different families. In this research, most of the pollen belongs to families of angiosperms: Lamiaceae, Poaceae, Asteraceae, Liliaceae, Cupressaceae, Pinaceae, Fagaceae, Typhaceae, Myrtaceae, Rosaceae, Chenopodiaceae, and Proteaceae (Table 3). Optical and scanning electron microscopy revealed the qualitative and quantitative characteristics of pollen, including the exine thickness, pollen diameter, colpus length, and width. Different pollen shapes were examined, i.e., triangular, circular, spheroidal, prolate, suboblate, subprolate, inter-subangular, and semi-angular. Three types of exine ornamentations were investigated, i.e., psilate, reticulate, and echinate. Similarly, different pollen types were noted, such as tricolporate, monoporate, dicolpate, and pentaporate. In this study, the majority of the species investigated were previously reported in many other regions, i.e., South Asia, Africa, Europe, and America. The contribution of the present work can be considered very important in the understanding of the paleoflora of the study area [28,29]. The existence of pollen in the study area and its identifications is a very reliable tool for the reconstruction of the palaeo-environment and palaeoecology of the area [30]. The floral record of Miocene strata provides a better source as an indicator of the environment.

Table 3. Pollen category, families, taxa, and their counts.

S. No	Category	Family	Taxa/Tribe	Counts
1	Angiosperm	Proteaceae	*Gevuina Avellana* Molina	1
2	Angiosperm	Rosaceae	*Sanguisorba minor* Scop	2
3	Gymnosperm	Pinaceae	*Pinus* spp.	4
4	Angiosperm	Typhaceae	*Sparganium* spp.	1
5	Angiosperm	Fagaceae	*Quercus* spp.	4
6	Angiosperm	Asteraceae	*Artemisia* spp.	1
7	Gymnosperm	Cupressaceae	*Juniperous* spp. *Interporopollenites* spp. -	1 1 1
8	Angiosperm	Lamiaceae	-	12
9	Angiosperm	Chenopodiaceae	-	1
10	Angiosperm	Asteraceae	-	7
11	Angiosperm	Poaceae	-	8
12	Angiosperm	Liliaceae	-	2
13	Angiosperm	Asteraceae	Cichoriodeae	1
14	Angiosperm	Myrtaceae	-	1

The study of fossil pollen records from the Miocene Epoch sedimentary strata was discussed comprehensively using palynological analysis from micro-fossils [22]. The *Quercus* pollen of Miocene strata was compared with that of the deciduous forest of China based on scanning electron microscopy showing uniformly fine granule, regulate, scabrate-verrucate, and rod-like elements of exine sculpturing [30]. The presence of *Quercus* in the study area revealed a warm temperate climate, and *Pinus* indicates a temperate climate. Cupressaceae, Lamiaceae, Fagaceae, and Rosaceae were investigated in other coeval excavations showing accordance with the present results. In the late Miocene–Pleistocene, mixed forests were described for the Yunnan Plateau of China, particularly coniferous plants belonging to the genus *Abies* and *Picea* pollen [31]. Asteraceae, Poaceae, Fabaceae, Cupressaceae, Liliaceae, Myrtaceae, and Pinaceae were studied in the palynological study of the late Miocene from Siwalik sediments of Uttar Pradesh, which also supports the present findings for evidence in the fossil records of pollen and provides information on the presence of minimal rain with a warm, humid climate in Southeast Asia, such as India and China [32–37]. An early Miocene study on the Dulte Formation was carried out to determine the linkage of palynology and palaeoecology [23,38]. Using combined optical microscopy and SEM, the pollen morphology of Miocene flora for Bignoniaceae, Myrtales, and Lamiales has been reported and showed variations within different palynological features, which are useful for species classification and identification [39]. The formation has marine and tidal influences showing that the vegetation grew near coastal areas, such as salt marshes or mangroves [28]. The indication of the palaeo-environment appears to be in accordance with a previous study [12].

Pollen grains of Asteraceae today were recorded as having tricolporate and were spheroidal, echinated, and zonocolporated [40,41]. In the present study, the Asteraceae pollen was reported to be spheroidal, prolate-spheroidal, and with echinated sculpturing.

A palynological study of the current Poaceae members showed a monoporate-diporate aperture, aerolate and scarbate tectum, while in this study, they were mostly spheroidal, subprolate shapes and inter-subangular and psilate ornamentations [40]. Climatic variations occurred during the Miocene and were responsible for the development and spreading of grasses. Similarly, the past pollen record of Pinaceae in the area was monad, oblate, and nearly circular with narrow attachment alveolate structuring and vesiculate-bisaccate, while in this study, it had a bisaccate, circular-angular shape, and with psilate-regulate exine sculpturing [41]. *Quercus* pollen grains were eutectate, prolate, sculpturing perforate, microverrucate, microregulate (scanning electron microscopy), and scabrate (optical microscopy), while basic sculpturing units were rod-shaped [42]. Several species of *Ocimum* (Lamiaceae) originated from India and then migrated towards the east and west. The palynological record of fossil pollen noted in the literature was mainly hexacolpate, as in the present study with reticulate exine ornamentations. The biogeographical history and phylogeny of other Lamiaceae originating from China and spreading towards the Southwest are closely related to the present results [43,44]. A palaeoecological record of *Sanguisobra minor* (Rosaceae) from the late Eocene-late Miocene was reported from southern central Anatolia [45]. The present study provides data about a larger distribution area with respect to what was previously thought [45]. The Miocene epoch of Pakistan is very important because of the significant environmental changes, but the fossil vegetation records were investigated. The Miocene Murree Formation palynoflora in relation to the palaeoclimate of Pakistan were compared with the modern flora with the fossil record in the depositional environments in China [46,47]. The pollen record for the late Miocene epoch of the Tibetan Plateau was recorded, showing the climatic changes and vegetation distribution briefly as with the present results [48]. *Sanguisorba minor* was dominant in the present investigation showing two different pollen types, while data about the genus *Pinus* showed four distinct pollen grains. Lamiaceae can also be considered dominant, with 12 species attributed on the base of the pollen grains, followed by nine Asteraceae, four Poaceae, and two Liliaceae, while the remaining families contain only one pollen grain. *Ocimum basilicum* had the highest exine thickness (4.25 μm), corresponding to the other literature data. Regarding the pollen diameter, *Pinus* had a maximum value of 69.5 μm, whereas the pollen grains attributable to Asteraceae had a minimum value of 17.25 μm on average. The fossil record of Asteraceae from the Miocene of Patagonia has been reported as being echinated and tricolporated in nature, whereas the current Asteraceae analyzed pollen grains were circular, tricolpate, and echinate exine ornamentations having high thickness [27]. *Gevuina Avellana* (Proteaceae) is considered the oldest species, having psilate exine ornamentations, and radially symmetrical and tricolpated pollen.

This study showed that the paleoflora from Murree formation in the foothills of Margalla provide important information on the paleo-environmental changes in the study area. Fossil-based floral studies were neglected in Pakistan, despite being ecologically and stratigraphically very important, and have never been explored in the proposed study area [29]. Additional research to combine findings from many proxies, including plants and isotope data, is required to comprehend the palaeoclimate history of this region. This type of study will be very important for the paleobotany, paleo-environment, and biogeography of Pakistan, being the first in the investigated area. In addition, such a study helps understand the vegetation origin, the effects of climatic changes with time, and in solving taxonomic problems [49,50].

5. Conclusions

Palynological studies of fossil plants from the Miocene Murree Formation of Pakistan were dominated by angiospermic pollen. This study examined the past vegetation's record, its correlations with the past climate, and its reconstructions. Using optical and scanning electron microscopy, various morphological features were examined, which helped in species identifications. Pollen morphological features, such as pollen size, shape, exine ornamentations, and symmetry, are important for species identifications in plant taxonomy and systematics. The overall results of the Murree Formation are similar to what was previously reported, but some variations in the qualitative and quantitative characteristics of pollen were observed. Further studies to identify the species and genus using advanced microscopic techniques, such as transmission electron microscopy (TEM), are recommended.

Author Contributions: S.A., M.A., M.Z. and S.S. (Shazia Sultana) designed the research, performed the experiments, and analyzed the data., W.Z., A.P., M.F.R., F.U., A.A., M.A., M.A.B. and S.S. (Saddam Saqib) helped in visualization, methodology, writing—review and editing, investigation, resources, and data curation. All authors have read and agreed to the published version of the manuscript.

Funding: Deanship of Scientific Research at Umm Al-Qura University, Grant Code: 22UQU4430043DSR05.

Data Availability Statement: Not applicable.

Acknowledgments: The authors would like to thank the Deanship of Scientific Research at Umm Al-Qura University for supporting this work by Grant Code: 22UQU4430043DSR05. The authors also offer thanks to Carlos Jaramillo from the Smithsonian Institute of Panama and Reza Safaierad from University of Liege, Belgium for his continuous guidance and support thoroughly.

Conflicts of Interest: The authors declare no conflict of interest.

References

1. Meigs, A.J.; Douglas, W.B.; Richard, A.B. Middle-late Miocene (>10 Ma) formation of the Main Boundary thrust in the western Himalaya. *Geology* **1995**, *23*, 423–426. [CrossRef]
2. Zaheer, M.; Khan, M.S.; Mughal, M.S.; Abbasi, N. Petrography, provenance, diaganesis and depositional environment of Murree formation in Jhelum Valley, Sub Himalayas, Azad Jammu and Kashmir, Pakistan. *Arab. J. Geosci.* **2017**, *10*, 514. [CrossRef]
3. Strother, S.L.; Salzmann, U.; Sangiorgi, F.; Bijl, P.K.; Pross, J.; Escutia, C.; Salabarnada, A.; Pound, M.J.; Voss, J.; Woodward, J. A new quantitative approach to identify reworking in Eocene to Miocene pollen records from offshore Antarctica using red fluorescence and digital imaging. *Biogeosciences* **2017**, *14*, 2089–2100. [CrossRef]
4. Iqbal, J.; Shaikh, N.; Ahmed, M.; Zaman, W.; Khan, A.; Ayaz, A.; El-Ansary, D.O.; Sharma, H.; Elansary, H.O.; Park, S. Floristic Association of Moist Temperate Forests of Shangla District, Delineated by a Multivariate Approach. *Agronomy* **2022**, *12*, 1723. [CrossRef]
5. Gilani, H.; Goheer, M.A.; Ahmad, H.; Hussain, K. Under predicted climate change: Distribution and ecological niche modelling of six native tree species in Gilgit-Baltistan, Pakistan. *Ecol. Indic.* **2020**, *111*, 106049. [CrossRef]
6. Liu, F.; Peng, H.; Bomfleur, B.; Kerp, H.; Zhu, H.; Shen, S. Palynology and vegetation dynamics across the Permian–Triassic boundary in southern Tibet. *Earth-Sci. Rev.* **2020**, *209*, 103278. [CrossRef]
7. Shah, S.I. Stratigraphy of Pakistan. *Mem. Geol. Surv. Pak.* **1977**, *12*, 138.
8. Mughal, M.S.; Zhang, C.; Du, D.; Zhang, L.; Mustafa, S.; Hameed, F.; Khan, M.R.; Zaheer, M.; Blaise, D. Petrography and provenance of the Early Miocene Murree Formation, Himalayan Foreland Basin, Muzaffarabad, Pakistan. *J. Asian Earth Sci.* **2018**, *162*, 25–40. [CrossRef]
9. Boyd, M. Phytoliths as paleoenvironmental indicators in a dune field on the northern Great Plains. *J Arid Environ.* **2005**, *61*, 357–375. [CrossRef]
10. Edwards, K.J. Pollen, women, war and other things: Reflections on the history of palynology. *Veg. Hist. Archaeobot.* **2018**, *27*, 319–335. [CrossRef]
11. Malkani, M.S.; Mahmood, Z.; Somro, N.; Shaikh, S. Revised Stratigraphy and Mineral Resources of Kirthar Basin, Pakistan. *Mem. Geol. Surv. Pak.* **2017**, *24*, 1–134.
12. Liu, Y.; Song, C.; Meng, Q.; He, P.; Yang, R.; Huang, R.; Chen, S.; Wang, D.; Xing, Z. Paleoclimate change since the Miocene inferred from clay-mineral records of the Jiuquan Basin, NW China. *Palaeogeogr. Palaeoclimatol. Palaeoecol.* **2020**, *550*, 109730. [CrossRef]
13. Yar, M.; Hanif, M.; Sajid, M. Lithofacies and petrography of Miocene Murree Formation, Peshawar basin, NW Pakistan: Implications for provenance and paleoclimate. *Arab. J. Geosci.* **2021**, *14*, 714. [CrossRef]
14. Perveen, A.; Qaiser, M. Pollen flora of Pakistan—LXXII. Ericaceae. *Pak. J. Bot.* **2013**, *45*, 977–979.
15. Siddiqui, K.M.; Mohammad, I.; Ayaz, M. Forest ecosystem climate change impact assessment and adaptation strategies for Pakistan. *Clim. Res.* **1999**, *12*, 195–203. [CrossRef]
16. Williams, C.G. The Male Gametophyte Enclosed in a Pollen Wall. In *Conifer Reproductive Biology*; Springer: Berlin/Heidelberg, Germany, 2009; pp. 69–89.
17. Usma, A.; Ahmad, M.; Zafar, M.; Sultana, S.; Ullah, F.; Saqib, S.; Ayaz, A.; Zaman, W. Palynological Study of Weed Flora from Potohar Plateau. *Agronomy* **2022**, *12*, 2500. [CrossRef]

18. Mander, L.; Punyasena, S.W. Fossil pollen and spores in paleoecology. In *Methods in Paleoecology*; Springer: Cham, Switzerland, 2018; pp. 215–234.
19. Jaramillo, C.A.; Moreno, E.; Ramírez, V.; da Silva-Caminha, S.A.; de la Barrera, A.; de la Barrera, A.; Sanchez, C.R.; Morón, S.; Herrer, F.; Escobar, J. Palynological record of the last 20 million years in Panama. In *Paleobotany and Biogeography: A Festschrift for Alan Graham in His 80th Year*; Stevens, W.D., Montiel, O.M., Raven, P., Eds.; Missouri Botanical Garden Press: St. Louis, MO, USA, 2014; pp. 134–253.
20. Shah, S.N.; Ahmad, M.; Zafar, M.; Ullah, F.; Zaman, W.; Malik, K.; Rashid, N.; Gul, S. Taxonomic importance of spore morphology in Thelypteridaceae from Northern Pakistan. *Microsc. Res. Tech.* **2019**, *82*, 1326–1333. [CrossRef]
21. Punt, W.; Hoen, P.P.; Blackmore, S.; Nilsson, S.; Le Thomas, A. Glossary of pollen and spore terminology. *Rev. Palaeobot Palynol.* **2007**, *143*, 1–81. [CrossRef]
22. De Franceschi, D.; Hoorn, C.; Antoine, P.O.; Cheema, I.U.; Flynn, L.J.; Lindsay, E.H.; Marivaux, L.; Métais, G.; Rajpar, A.R.; Welcomme, J.L. Floral data from the mid-Cenozoic of central Pakistan. *Rev. Palaeobot. Palynol.* **2008**, *150*, 115–129. [CrossRef]
23. Faegri, K.; Kaland, P.E.; Krzywinski, K. *Textbook of Pollen Analysis*; John Wiley & Sons Ltd.: Hoboken, NJ, USA, 1989.
24. Svetlana, P.; Torsten, U.; Anna, A.; Valentina, T.; Polina, T.; Yaowu, X. Early Miocene flora of central Kazakhstan (Turgai Plateau) and its paleoenvironmental implications. *Plant Divers.* **2019**, *41*, 183–197. [CrossRef]
25. Ahmad, M.; Ali, M.; Zafar, M.; Sultana, S.; Majeed, S.; Yaseen, G.; Ahmad, S. Palynological diversity of Melliferous flora around apiaries from district Mardan Khyber Pakhtunkhwa-Pakistan. *Bot. Rev.* **2022**, *88*, 299–332.
26. Flynn, L.J.; Morgan, M.E. An unusual diatomyid rodent from an infrequently sampled Late Miocene interval in the Siwaliks of Pakistan. *Palaeontol. Electron.* **2005**, *8*, 17A.
27. Palazzesi, L.; Barreda, V.; Tellería, M.C. Fossil pollen grains of Asteraceae from the Miocene of Patagonia: Barnadesioideae affinity. *Rev. Palaeobot. Palynol.* **2009**, *155*, 83–88. [CrossRef]
28. Zetter, R.; Ferguson, D.K.; Mohr, B.A. Discriminating fossil evergreen and deciduous Quercus pollen: A case study from the Miocene of eastern China. *Rev. Palaeobot. Palynol.* **2007**, *145*, 289–303.
29. Khan, A.N.; Collins, A.E.; Qazi, F. Causes and extent of environmental impacts of landslide hazard in the Himalayan region: A case study of Murree, Pakistan. *Nat. Hazards* **2011**, *57*, 413–434.
30. Denk, T.; Bouchal, J.M. New Fagaceous pollen taxa from the Miocene Søby flora of Denmark and their biogeographic implications. *Am. J. Bot.* **2021**, *108*, 1500–1524. [CrossRef]
31. Wu, F.; Gao, S.; Tang, F.; Meng, Q.; An, C. A late Miocene-early Pleistocene palynological record and its climatic and tectonic implications for the Yunnan Plateau, China. *Palaeogeogr. Paleoclimatol. Palaeoecol.* **2019**, *530*, 190–199. [CrossRef]
32. Sarkar, S.; Bhattacharya, A.P.; Singh, H.P. Palynology of Middle Siwalik sediments (Late Miocene) from Bagh Rao, Uttar Pradesh. *Palaeobotanist* **1994**, *42*, 199–209. [CrossRef]
33. Mandaokar, B.D. Palynoflora from the Keifang Formation (early Miocene) Mizoram, India and its environmental significance. *J. Palaeontol. Soc. India* **2002**, *47*, 77–83.
34. Mehrotra, R.C.; Bera, S.K.; Basumatary, S.K.; Srivastava, G. Study of fossil wood from the Middle–Late Miocene sediments of Dhemaji and Lakhimpur districts of Assam, India and its palaeoecological and palaeophytogeographical implications. *J. Earth Syst. Sci.* **2011**, *120*, 681–701. [CrossRef]
35. Paruya, D.K.; Ghosh, R.; Bismas, O.; Bera, M.; Bera, S. Dispersed fungal remains from the Neogene Siwalik forest of sub Himalayan Arunachal Pradesh, India and their palaeoenvironmental indicative values. *J. Mycopathol. Res.* **2017**, *55*, 303–307.
36. Wang, W.M.; Harley, M.M. The Miocene genus Fupingopollenites: Comparisons with ultrastructure and pseudocolpi in modern pollen. *Rev. Palaeobot. Palynol.* **2004**, *131*, 117–145. [CrossRef]
37. Xu, J.X.; Ferguson, D.K.; Li, C.S.; Wang, Y.F. Late Miocene vegetation and climate of the Lühe region in Yunnan, southwestern China. *Rev. Palaeobot. Palynol.* **2008**, *148*, 36–59. [CrossRef]
38. Quamar, M.F.; Singh, P.; Garg, A.; Tripathi, S.; Farooqui, A.; Shukla, A.N.; Prasad, N. Pollen characters and their evolutionary and taxonomic significance: Using light and confocal laser scanning microscope to study diverse plant pollen taxa from central India. *Palynology* **2022**, *46*, 1–13. [CrossRef]
39. Ahmad, M.; Bano, A.; Zafar, M.; Khan, M.A.; Chaudhry, M.J.I.; Sultana, S. Pollen morphology of some species of the family asteraceae from the alpine zone, Deosai Plateau, northern Pakistan. *Palynology* **2013**, *37*, 189–195. [CrossRef]
40. Usma, A.; Ahmad, M.; Zafar, M.; Sultana, S.; Lubna; Kalsoom, N.; Zaman, W.; Ullah, F. Micromorphological variations and taxonomic implications of caryopses of some grasses from Pakistan. *Wulfenia* **2020**, *27*, 86–96.
41. Khan, R.; Ul Abidin, S.Z.; Ahmad, M.; Zafar, M.; Liu, J.; Amina, H. Palyno-morphological characteristics of gymnosperm flora of Pakistan and its taxonomic implications with LM and SEM methods. *Microsc. Res. Tech.* **2018**, *81*, 74–87. [CrossRef]
42. Bouchal, J.M.; Zetter, R.; Grímsson, F.; Denk, T. The middle Miocene palynoflora and palaeoenvironments of Eskihisar (Yatağan basin, south-western Anatolia): A combined LM and SEM investigation. *Bot. J. Linn. Soc.* **2016**, *182*, 14–79. [CrossRef]
43. Kar, R.K. On the Indian origin of *Ocimum* (Lamiaceae): A palynological approach. *Palaeobotanist* **1996**, *43*, 43–50. [CrossRef]
44. Yu, X.Q.; Maki, M.; Drew, B.T.; Paton, A.J.; Li, H.W.; Zhao, J.L.; Conran, J.G.; Li, J. Phylogeny and historical biogeography of *Isodon* (Lamiaceae): Rapid radiation in south-west China and Miocene overland dispersal into Africa. *Mol. Phylogenet. Evol.* **2014**, *77*, 183–194. [CrossRef]

45. Akgün, F.; Kayseri-Özer, M.S.; Tekin, E.; Varol, B.; Şen, Ş.; Herece, E.; Gündoğan, İ.; Sözeri, K.; Us, M.S. Late Eocene to Late Miocene palaeoecological and palaeoenvironmental dynamics of the Ereğli-Ulukışla Basin (Southern Central Anatolia). *Geol. J.* **2021**, *56*, 673–703. [CrossRef]
46. Guo, Z.T.; Sun, B.; Zhang, Z.S.; Peng, S.Z.; Xiao, G.Q.; Ge, J.Y.; Hao, Q.Z.; Qiao, Y.S.; Liang, M.Y.; Liu, J.F.; et al. A major reorganization of Asian climate by the early Miocene. *Clim. Past* **2008**, *4*, 153–174. [CrossRef]
47. Yang, Y.; Wang, W.M.; Shu, J.W.; Chen, W. Miocene palynoflora from Shengxian Formation, Zhejiang Province, southeast China and its palaeovegetational and palaeoenvironmental implications. *Rev. Palaeobot. Palynol.* **2018**, *259*, 185–197. [CrossRef]
48. Liu, J.; Li, J.J.; Song, C.H.; Yu, H.; Peng, T.J.; Hui, Z.C.; Ye, X.Y. Palynological evidence for late Miocene stepwise aridification on the northeastern Tibetan Plateau. *Clim. Past* **2016**, *12*, 1473–1484. [CrossRef]
49. Florenzano, A. The history of pastoral activities in S Italy inferred from palynology: A long-term perspective to support biodiversity awareness. *Sustainability* **2019**, *11*, 404. [CrossRef]
50. Camuera, J.; Jiménez-Moreno, G.; Ramos-Román, M.J.; García-Alix, A.; Toney, J.L.; Anderson, R.S.; Jiménez-Espejo, F.; Bright, J.; Webster, C.; Yanes, Y.; et al. Vegetation and climate changes during the last two glacial-interglacial cycles in the western Mediterranean: A new long pollen record from Padul (southern Iberian Peninsula). *Quat. Sci. Rev.* **2019**, *205*, 86–105. [CrossRef]

agronomy

MDPI

Article

Ophioglossum lusitanicum L.: New Records of Plant Community and 92/43/EEC Habitat in Italy

Enrico Vito Perrino [1,*], Valeria Tomaselli [2,*], Robert Philipp Wagensommer [3], Giuseppe Nicola Silletti [4], Assunta Esposito [5] and Adriano Stinca [5,*]

1 CIHEAM, Mediterranean Agronomic Institute of Bari, Via Ceglie 9, 70010 Valenzano, Italy
2 Department of Biosciences, Biotechnologies and Environment, University of Bari "Aldo Moro", Via Orabona 4, 70125 Bari, Italy
3 Faculty of Education, Free University of Bozen-Bolzano, Viale Ratisbona 16, 39042 Bressanone, Italy
4 General Div. (r), 70100 Bari, Italy
5 Department of Environmental, Biological and Pharmaceutical Sciences and Technologies, University of Campania Luigi Vanvitelli, 81100 Caserta, Italy
* Correspondence: perrino@iamb.it (E.V.P.); valeria.tomaselli@uniba.it (V.T.); adriano.stinca@unicampania.it (A.S.)

Abstract: In this paper, integrating field surveys and literature data, an analysis of *Ophioglossum lusitanicum* plant communities and related 92/43/EEC habitats are reported for Italy. Two new syntaxa, *Euphorbio exiguae-Ophioglossetum lusitanici* ass. nova hoc loco and *trifolietosum scabri* subass. nova hoc loco of the *Rumici bucephalophori-Ophioglossetum lusitanici* were described in the Apulia and Campania regions.Both types of vegetation identified in Apulia, Campania, and Sicily regions represent two different aspects of the same priority habitat: "pseudo-steppe with grasses and annuals of the *Thero-Brachypodietea*" (habitat code 6220*). A phytosociological and ecological dataset of the literature and new field surveys highlighting the soil type as parameters affecting the vegetation cover of this small fernlike plant, with the *Trachynion distachyae* Rivas–Martínez, 1978 alliance on calcareous soils and *Helianthemion guttati* Br.-Bl. in Br.-Bl. et al., 1940 alliance on volcanic soils. Many species of other types of annual meadows have been identified within *Ophioglossum* communities due to the very small patches of land, where they have been found, and ecological conditions that facilitate this phenomenon of the transgression of other therophytes species.

Keywords: fernlike plant; new syntaxa; Ophioglossaceae; phytosociological study

Citation: Perrino, E.V.; Tomaselli, V.; Wagensommer, R.P.; Silletti, G.N.; Esposito, A.; Stinca, A. *Ophioglossum lusitanicum* L.: New Records of Plant Community and 92/43/EEC Habitat in Italy. *Agronomy* **2022**, *12*, 3188. https://doi.org/10.3390/agronomy12123188

Academic Editors: Alessio Papini, Mushtaq Ahmad, Fazal Ullah and Wajid Zaman

Received: 9 November 2022
Accepted: 10 December 2022
Published: 15 December 2022

1. Introduction

Ophioglossum L. is a cosmopolitan monophyletic genus, belonging to the oldest existing evolutionary line of the vascular cryptogams megaphylls [1,2], and includes about 25–30 species [1–3]. The number of chromosomes in the genus ranges from 2n = 240 (*O. lusitanicum* L.) to 2n = 1440 (*O. reticulatum* L.), the latter being the highest chromosomal number known in plants [4]. *Ophioglossum lusitanicum* is a small fernlike plant a few centimeters high (from 2 to 7 cm) (Figure 1), difficult to observe in the field due to its size and ecology, it is generally found mixed and undercover with the biggest plants [5]. Its distribution is very fragmented and includes the temperate zones of the Atlantic coast and extends from the Macaronesian region eastwards up to India, including the Mediterranean basin [6] and Iran [7]. Three species of the genus *Ophioglossum* have been reported in Italy: *O. azoricum* C. Presl is only present in Latium and Veneto, while it is doubtful in Tuscany; *O. vulgatum* L. is present in all Italian regions, except for Valle d'Aosta and Apulia; and *O. lusitanicum* L. is present in all Tyrrhenian regions, Apulia, Sicily, and Sardinia [8]. In Italy, all the species of *Ophioglossum* are of conservation interest, as reported in the Regional Red Lists [9]. Indeed, *O. lusitanicum* (Figure 1) was evaluated as lower risk (LR) in Tuscany, Latium, and Apulia, and endangered (EN) in Calabria [9]. Regarding the Apulia region, it has been found in some sites of Salento [10], and only one in Polignano a Mare [11,12] and on Gargano promontory [13]. This small fernlike plant was assessed as a characteristic species of several syntaxa: in the Mediterranean Basin with the alliance of *Cicendio-Solenopsion laurentiae* Brullo and Minissale 1998 [14–16], in Spain with the *Ophioglosso-Cicendietum filiformis* Rivas Goday, 1970 [17], in western France with the *Ophioglosso lusitanici-Isoëtion histricis* de Foucault 1988 [18], and all aspects referable to the class of *Isoëto-Nanojuncetea* Br.-Bl. et Tx. in Br.-Bl. et al., 1952. In addition, the *Scillo-Ophioglossetum lusitanici* Ballesteros and Sagarra, 1984 association was described in Spain [19], which differ from the previous vegetation, as it falls under the *Helianthemion guttati* Br.-Bl. in Br.-Bl. et al., 1940 alliance. Both types of vegetation fall within the priority habitat of Directive 92/43 EEC [20]. *Helianthemion guttati* belongs to the habitat "pseudo-steppe with grasses and annuals of the *Thero-Brachypodietea*" (code 6220*), while the aspects of *Isoëto-Nanojuncetea* belongs to "Mediterranean temporary ponds" (code 3170*). In Croatia, unlike other countries, it grows mainly in disturbed habitats (e.g., paths and edges of paths) in garrigue vegetation [3].Given the considerable conservation interest of this species and, in some cases, in the habitat in which it grows, a more in-depth research can provide an important contribution to the knowledge of its distribution and ecology. In this study, through the integration of new field data with literature data, a phytosociological analysis of *O. lusitanicum* plant communities occurs in some Italian regions. The field investigations were carried out both on volcanic and calcareous soils, in order to assess whether the hypothesis of different types of a geological substrate can determine a difference in the floristic composition between plant communities.

(a)

(b)

Figure 1. (**a**) Small patches of soil with *O. lusitanicum* community; (**b**) Individuals of *O. lusitanicum* in late winter. Locality: Polignano a Mare (Bari). Date: 4 March 2012. Photos by Enrico Vito Perrino.

2. Materials and Methods

Study sites are located in Italy and are representative of soils whose parent material is volcanic (Elba Island in Tuscany: T1–T2; Mt. Vesuvius in Campania: CV1–CV8; Vulcano Island in Sicily: S1–S5;) and calcareous (Tifatini Mts. in Campania: CT1–CT3; Santeramo in Colle in loc. Alessandrelli in Apulia: P1–P5) in origin (Figure 2).

The investigated sites of Tuscany, Apulia, and Sicily are part of the Special Area of Conservation (SAC) of the Natura 2000 network; that is, SAC "Elba orientale" (code IT5160102), SAC "Murgia Alta" (code 9120007), and SAC "Isola di Vulcano" (code ITA030027). In Campania, Mt. Vesuvius study site falls into SAC "Vesuvio e M. Somma" (code IT8030037), SAC "Vesuvio" (code IT8030036), and Vesuvius National Park.

Figure 2. The sampling sites evaluated for the phytosociological classification.

Field surveys, performed by applying the Braun–Blanquet method [21], were carried out in Campania and Apulia in 2016 and 2021, between the late winter and spring periods. Plant species cover was visually assessed based on the following abundance scores: (r) <1% and rare; (+) <1%; (1) 1–5%; (2) 6–25%; (3) 26–50%; (5) 51–75%; and (5) 76–100%. The collected plant material is preserved in the *Herbarium Austroitalicum* of the University of Campania Luigi Vanvitelli (19 March 2016, *Stinca A.*, IT 3702) and *Herbarium Horti Botanici Barensis* of the University of Bari (3 March 2012, *Perrino E.V. et Signorile G.*, BI 35965) (IT and BI, acronyms follow Thiers 2022) [22]. As far as Tuscany and Sicily sites, vegetation data were taken from specialistic literature [23,24]. In total, 21 relevés were selected and analyzed. For the taxa identification, Flora Europaea [25] and

Flora d'Italia [26] were used, while for the nomenclature we followed Bartolucci et al. [8] and Galasso et al. [27].

A multivariate analysis using XLSTAT software [28] was applied to the data matrix to test the degree of floristic similarity between vegetation relevés. To this aim, the plant sampling cover scale was transformed into the ordinal scale according to Van der Maarel [29] and the obtained matrix was processed by agglomerative hierarchical clustering (AHC), using Spearman's correlation coefficient. In addition, the principal component analysis (PCA) was performed with the application of the Euclidean biplot, using Spearman's correlation coefficient, to the graphical exploration of the distances among relevés.

3. Results and Discussion

Many communities of *O. lusitanicum* usually have early spring phenology and a typically Mediterranean character, with an Atlantic–Mediterranean climate, moist acid soils submerged or waterlogged in winter and dry in late spring [14], and the presence of higher microstationary humidity compared to the surrounding soil, as observed in Tuscany [30]. These ecological features are confirmed with new data in the present study and in other non-Mediterranean countries, such as Iran, where a dense population of this species has been identified in the forest clearings of the submontane belt, inside small sandy patches [7], and in Algeria in the clearing of cork oaks, in small areas of moist sandy soil [31].

In Italy, the vegetation with this interesting fern is characterized by small pools ranging from a few centimeters to a few square meters, often within other vegetation types, such as shrublands As a result, these plant communities are rich in transgressive taxa of other grassland syntaxa.

The classification and ordination analysis of the vegetation relevés (a data matrix of 145 taxa $\times$ 21 phytosociological relevés) are depicted in Figures 3 and 4, respectively. The dendrogram (Figure 3) shows two clearly distinct groups: cluster I includes all the plant communities of volcanic soils divided in turn into three subgroups (IA: Tuscany, IB: Sicily, IC: Campania–Mt. Vesuvius) while cluster II includes all the relevés carried out on limestone soils and grouping two subclusters (IIA: Campania–Tifatini Mts., IIB: Apulia). These results are consistent with those of the biplot showing the separation of the same groups and subgroups according to the first and second principal components. In detail, the clusters of relevés are highlighted along the first axis (F1), which accounted for 20.74% of the variation (to improve readability, species names are not reported) and clearly reflect the separation of the soil types. However, the phytosociological relevés showed a separation according to the second axis (F2), which accounted for 14.28% of the remaining variation and clearly reflected mainly their geographical location and the characteristics of the plant communities (Figure 4).

These results highlight significant differences between calciphilous (*Brachypodietalia distachyae* Rivas–Martínez, 1978) and acidophilous communities (*Helianthemetalia guttati* Br.-Bl. in Br.-Bl. et al., 1940) and fully confirm our initial hypothesis regarding the effect of soil type on the floristic composition of the studied plant cenoses.

3.1. Calciphilous Communities (Brachypodietalia distachyae)

For the surveys carried out in loc. Alessandrelli (Apulia) and on the Tifatini Mts. (Campania) on calcareous substrates, at relatively low altitudes (i.e., 325 to 460 m.a.s.l.) in the hinterland, *O. lusitanicum* and *Euphorbia exigua* can be considered characteristic taxa of a new phytosociological association: *Euphorbio exiguae-Ophioglossetum lusitanici* ass. nova hoc loco (*holotypus*: relevé P4 in Table 1, Figure 5). In Apulia, the high coverage of *Brachypodium distachyon* suggests that this plant community in the annual meadows of *Trachynion distachyae*, the alliance of *Brachypodietalia distachyae* and *Stipo-Trachynietea distachyae* (habitat 6220*), is identified by the presence of *Hypochaeris achyrophorus, Ononis reclinata, Linum strictum, Trifolium scabrum, Romulea bulbocodium, Hippocrepis biflora,* and other species (Table 1). Due to the small extension of this community, we find many transgressive species of other grassland communities well represented in the surrounding territory, especially those of *Helianthemetea guttati* (habitat 6220*) and *Festuco valesiacae-Brometea erecti* (habitat 62A0). The plant cenoses found in Campania and Apulia are very similar, even if slightly different in some aspects of their floristic composition. It is interesting to note that the Apulian sites are slightly richer in characteristic species of the order *Brachypodietalia distachyae* and *Stipo-Trachynietea distachyae*, though it is above all the presence and high coverage of *Brachypodium distachyon*, a species characteristic of the *Trachynion distachyae* which differentiates, from the vegetational point of view, Campania (subgroup IIA, CT1-CT3) from Apulia (subgroup IIB, P1-P5) (Figure 3), even spatially (Figure 4).

3.2. Acidophilous Communities (*Helianthemetalia guttati*)

On the volcanic soils of Vulcano Island and Mt. Vesuvius, where this species has recently been rediscovered [32], a peculiar aspect of *Helianthemetalia guttati* communities has been attributed to this fern, the *Rumici bucephalophori-Ophioglossetum lusitanici* Médail, Pavon, Lo Cascio and Pasta 2016 association [24], due to the presence of *Rumex bucephalophorus*, *Ornithopus compressus*, *Trifolium arvense* subsp. *arvense*, *Lagurus ovatus*, *Silene mutabilis*, *Aira caryophyllea*, *Tolpis virgata* subsp. *virgata*, *Silene gallica*, and other species. This consideration is not statistically supported, because the Campania and Sicilian sites belong to different subgroups (IB, IC) (Figure 3) spatially distant (Figure 4). It seems there are good relationships between communities and environments at Mt. Vesuvius (Table 2), in which the main environmental aspect, compared to other communities, is the higher altitude while, from the vegetational point of view, a crucial role is played by the high coverage of *Trifolium scabrum* and the presence of some transgressive species of the *Festuco valesiacae-Brometea erecti*. For these reasons, the communities occurring at Mt. Vesuvius have been described as a new endemic subassociation of the *Rumici bucephalophori-Ophioglossetum lusitanici*: *trifolietosum scabri* subass. nova hoc loco (*holotypus*: relevé CV4, Table 2, Figure 6). The vegetation of Mt. Vesuvius could be referred to as the priority habitat 6220*; however, the characteristics of the volcanic soil do not exclude in the future, through further studies, its attribution to the habitat 8320 "*Fields of lava and natural excavations*", which is very complex with little known and characterized by pioneer vegetation with few species, often endemics.

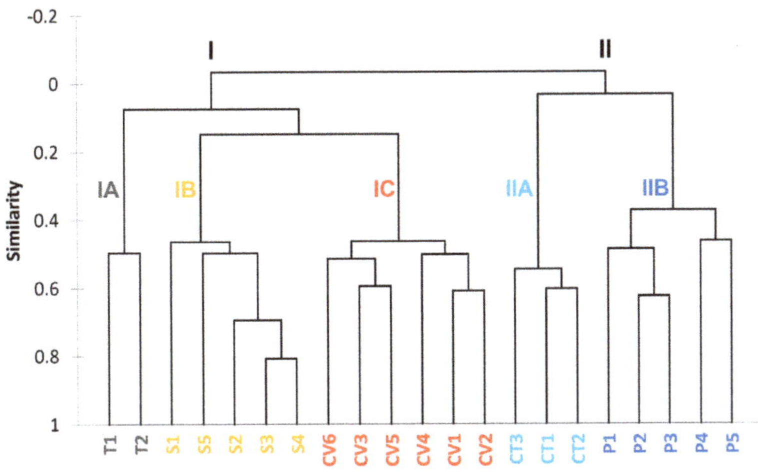

Figure 3. Dendrogram of the agglomerative hierarchical clustering of the vegetation relevés (T1–T2: Tuscany; S1–S5: Sicily; CV1–CV6: Campania–Mt. Vesuvius; CT1–CT3: Campania–Tifatini Mts.; P1–P5: Apulia).

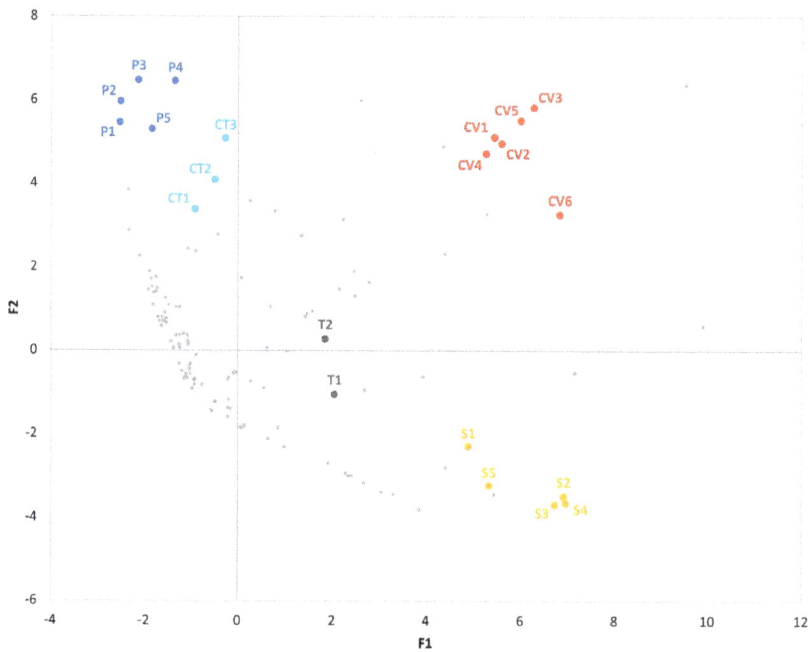

Figure 4. Biplot of the principal component analysis in the first two principal components space of the vegetation relevés (T1–T2: Tuscany; S1–S5: Sicily; CV1–CV6: Campania–Mt. Vesuvius; CT1–CT3: Campania–Tifatini Mts.; P1–P5: Apulia); light gray circles indicate taxa whose names are not reported to improve the readability of the figure.

Table 1. *Euphorbio exiguae-Ophioglossetum lusitanici* ass. nova hoc loco. Further details of relevés and sporadic species are reported in Appendices A and B, respectively.

Sampling Number	**1**	**2**	**3**	**4**	**5**	**6**	**7**	**8**
Identification Relevé Code	**P1**	**P2**	**P3**	**P4**	**P5**	**CT1**	**CT2**	**CT3**
Altitude (m a.s.l.)	460	458	457	450	450	325	325	325
Aspect	W	W	W	W	W	W	WSW	SW
Slope (°)	10	7	7	8	8	2	3	3
Area of relevé (m^2)	1	1	1	1	1	1	1	1
Stoniness (%)	20	15	10	10	10	20	5	5
Rockiness (%)	2	5	2	5	5	-	-	-
Total cover (%)	90	90	85	85	80	-	-	-
Vascular plant layer cover (%)	80	80	75	75	70	50	80	60
Moss layer cover (%)	45	40	45	40	40	70	60	90
Number of taxa	29	33	28	29	28	34	29	28
Holotypus (≠)				≠				
Charact. *Euphorbio exiguae-Ophioglossetum lusitanici* ass. nova hoc loco								
Ophioglossum lusitanicum L.	2	1	1	1	+	2	3	3
Euphorbia exigua L. subsp. *exigua*	1	1	+	1	+	+	+	+
Charact. *Trachynion distachyae* Rivas-Martínez, 1978								
Brachypodium distachyon (L.) P.Beauv.	2	3	3	3	1	-	-	-
Bupleurum baldense Turra	-	-	-	-	-	-	+	-
Charact. *Brachypodietalia distachyae* Rivas-Martínez, 1978 and *Stipo-Trachynietea distachyae* Brullo in Brullo et al., 2001								
Hypochaeris achyrophorus L.	1	+	+	1	+	-	-	-
Ononis reclinata L.	1	+	1	-	-	+	-	+

Table 1. *Cont.*

Sampling Number	1	2	3	4	5	6	7	8
Identification Relevé Code	**P1**	**P2**	**P3**	**P4**	**P5**	**CT1**	**CT2**	**CT3**
Linum strictum L.	-	1	-	+	+	1	+	-
Trifolium scabrum L.	-	+	2	1	2	-	-	-
Romulea bulbocodium (L.) Sebast. & Mauri	-	-	+	-	-	1	+	1
Polygala monspeliaca L.	1	+	-	+	-	-	-	-
Valantia muralis L.	+	-	+	+	-	-	-	-
Hippocrepis biflora Spreng.	-	-	-	-	-	+	1	2
Stipellula capensis (Thunb.) Röser & H.R.Hamasha	-	-	-	2	1	-	-	-
Transg. *Helianthemetea guttati* Rivas Goday & Rivas–Martínez, 1963								
Briza maxima L.	-	+	+	+	+	-	+	-
Medicago minima (L.) L.	+	+	+	1	-	-	-	-
Trifolium stellatum L.	+	+	1	-	+	-	-	-
Helianthemum salicifolium (L.) Mill.	-	1	1	+	+	-	-	-
Trifolium campestre Schreb.	1	+	+	-	-	-	-	-
Aira elegans Willd. subsp. *elegans*	+	-	-	-	-	-	+	+
Lysimachia linum-stellatum L.	-	-	-	+	-	+	-	+
Cerastium brachypetalum Desp. ex Pers. subsp. *tenoreanum* (Ser.) Soó	-	-	-	-	-	+	+	+
Tuberaria guttata (L.) Fourr.	-	-	-	-	-	+	1	+
Onobrychis caput-galli (L.) Lam.	-	+	1	-	-	-	-	-
Hedypnois rhagadioloides (L.) F.W.Schmidt	-	-	-	+	+	-	-	-
Trifolium cherleri L.	-	-	-	+	+	-	-	-
Filago pygmaea L.	-	-	+	-	-	-	-	-
Alkanna tinctoria Tausch subsp. *tinctoria*	-	-	-	+	-	-	-	-
Trifolium arvense L. subsp. *arvense*	-	-	-	+	-	-	-	-
Cynosurus echinatus L.	-	-	-	-	+	-	-	-
Hippocrepis ciliata Willd.	-	-	-	-	+	-	-	-
Transg. *Festuco valesiacae-Brometea erecti* Br.-Bl. & Tüxen ex Br.-Bl., 1949								
Anthyllis vulneraria L. subsp. *rubriflora* (DC.) Arcang.	+	+	-	+	-	+	+	-
Bellardia trixago (L.) All.	+	-	-	+	+	-	-	+
Anacamptis morio (L.) R.M.Bateman, Pridgeon & M.W.Chase	-	+	+	+	+	-	-	-
Poterium sanguisorba L. s.l.	-	-	-	+	-	2	1	+
Colchicum cupanii Guss. subsp. *cupanii*	1	1	1	-	-	-	-	-
Scorzonera villosa Scop. subsp. *columnae* (Guss.) Nyman	+	+	-	-	+	-	-	-
Stipa austroitalica Martinovský s.l.	+	+	-	-	+	-	-	-
Thymus spinulosus Ten.	+	-	-	+	+	-	-	-
Petrorhagia saxifraga (L.) Link subsp. *gasparrinii* (Guss.) Pignatti ex Greuter & Burdet	-	-	-	-	-	+	+	+
Teucrium chamaedrys L. subsp. *chamaedrys*	-	-	-	-	-	+	1	-
Muscari comosum (L.) Mill.	-	+	-	-	-	-	-	-
Ophrys bertolonii Moretti	-	+	-	-	-	-	-	-
Eryngium campestre L.	-	-	-	-	+	-	-	-
Helianthemum nummularium (L.) Mill. s.l.	-	-	-	-	-	+	-	-
Centaurium erythraea Rafn subsp. *erythraea*	-	-	-	-	-	-	+	-
Brachypodium rupestre (Host) Roem. Schult.	-	-	-	-	-	-	-	+
Transg. *Chenopodietea* Br.-Bl. in Br.-Bl. et al., 1952								
Sherardia arvensis L.	1	+	+	+	-	+	+	+
Lysimachia arvensis (L.) U.Manns & Anderb. subsp. *arvensis*	1	-	+	+	+	-	+	+
Triticum vagans (Jord. & Fourr.) Greuter	+	-	+	1	-	-	-	-
Hypericum triquetrifolium Turra	-	-	-	+	-	-	-	-
Erodium cicutarium (L.) L'Hér.	-	-	-	-	-	-	+	-
Transg. *Lygeo sparti-Stipetea tenacissimae* Rivas–Martínez, 1978								
Reichardia picroides (L.) Roth	+	-	+	-	+	+	-	+
Anacamptis papilionacea (L.) R.M.Bateman, Pridgeon & M.W.Chase	+	-	+	-	-	-	+	+
Daucus carota L. subsp. *carota*	-	+	-	-	+	+	-	-
Carlina corymbosa L.	+	-	-	-	-	+	-	-
Convolvulus elegantissimus Mill.	+	-	-	-	-	-	-	-
Thapsia asclepium L.	-	1	-	-	-	-	-	-
Transg. *Ononido-Rosmarinetea* Br.-Bl. in A.Bolos & Vayreda, 1950								
Micromeria graeca (L.) Benth. ex Rchb. s.l.	2	+	+	-	-	-	-	2
Osyris alba L.	-	1	+	-	-	-	-	-
Teucrium capitatum L. subsp. *capitatum*	-	-	-	-	-	+	-	-

Figure 5. *Euphorbio exiguae-Ophioglossetum lusitanici* ass. nova hoc loco. Date: 3 May 2021, Santeramo in Colle, in loc. Alessandrelli (Bari). Photo by Enrico Vito Perrino.

Table 2. *Rumici bucephalophori-Ophioglossetum lusitanici* and *trifolietosum scabri* subass. nova hoc loco. Further details of relevés and sporadic species are reported in Appendices A and B, respectively.

Sampling Number **Identification Relevé Code**	**9** **S1 ***	**10** **S2 ***	**11** **S3 ***	**12** **S4 ***	**13** **S5 ***	**14** **CV1**	**15** **CV2**	**16** **CV3**	**17** **CV4**	**18** **CV5**	**19** **CV6**	**20** **T1 ****	**21** **T2 ****
Altitude (m a.s.l.)	25	35	35	35	35	479	479	479	479	479	479	84	140
Aspect	-	NW	NW	NW	NW	-	WNW	N	-	-	-	-	-
Slope (°)	-	20	20	20	25	-	1	2	-	-	-	-	-
Area of relevé (m^2)	1	1	1	1	1	1.2	0.05	0.06	0.5	0.2	0.06	0.3	0.25
Stoniness (%)	-	-	-	-	-	-	-	-	-	-	-	-	-
Rockiness (%)	-	-	-	-	-	10	10	-	5	10	10	-	-
Total cover (%)	-	-	-	-	-	-	-	-	-	-	-	40	30
Vascular plant layer cover (%)	-	-	-	-	-	70	70	70	95	70	90	-	-
Moss layer cover (%)	-	-	-	-	-	90	90	100	10	90	70	-	-
Number of taxa	19	16	14	13	13	23	15	10	23	19	14	15	8
Holotypus (≠)									≠				
Charact. *trifolietosum scabri* subass. nova hoc loco													
Trifolium scabrum L.	-	-	-	-	-	1	1	1	+	1	-	-	-
Charact. *Rumici bucephalophori-Ophioglossetum lusitanici* Médail, Pavon, Lo Cascio & Pasta, 2016													
Ophioglossum lusitanicum L.	3	2	3	3	2	4	3	4	5	4	4	3	3
Rumex bucephalophorus L. s.l.	1	2	2	3	2	1	+	1	1	1	+	-	-
Charact. *Helianthemion guttati* Br.-Bl. in Br.-Bl. et al., 1940, *Helianthemetalia guttati* Br.-Bl. in Br.-Bl. et al., 1940 *Helianthemetea guttati* Rivas Goday & Rivas–Martínez, 1963													
Ornithopus compressus L.	-	1	2	+	+	+	-	1	-	+	1	-	-
Trifolium arvense L. subsp. *arvense*	-	+	-	-	-	+	+	+	+	+	2	-	-
Lagurus ovatus L.	+	2	3	+	-	-	+	-	-	-	-	+	-
Silene mutabilis L.	+	+	1	+	+	-	-	-	-	-	-	-	-
Aira caryophyllea L.	+	1	-	-	-	-	-	-	+	+	-	+	-
Tolpis virgata (Desf.) Bertol. subsp. *virgata*	+	+	+	-	+	-	-	-	-	-	-	-	-
Silene gallica L.	-	-	-	-	-	+	2	-	+	-	+	-	-
Cerastium semidecandrum L.	-	-	-	-	-	+	-	+	-	+	+	-	-
Briza maxima L.	+	-	-	-	-	-	-	+	-	+	-	-	-
Ornithopus pinnatus (Mill.) Druce	-	+	2	+	-	-	-	-	-	-	-	-	-

Table 2. *Cont.*

Sampling Number Identification Relevé Code	9 S1 *	10 S2 *	11 S3 *	12 S4 *	13 S5 *	14 CV1	15 CV2	16 CV3	17 CV4	18 CV5	19 CV6	20 T1 **	21 T2 **
Galium parisiense L.	-	+	+	+	-	-	-	-	-	-	-	-	-
Phedimus stellatus (L.) Raf.	-	-	-	-	-	1	+	-	+	-	-	-	-
Lupinus angustifolius L.	-	-	-	-	-	-	+	-	+	1	-	-	-
Logfia gallica (L.) Cosson & Germ.	-	+	-	-	-	-	-	-	-	-	-	-	-
Sedum rubens L.	-	-	-	-	1	-	-	-	-	-	-	-	-
Trifolium cherleri L.	-	-	-	-	-	-	-	-	-	-	-	1	-
Cynosurus echinatus L.	-	-	-	-	-	-	-	-	-	-	-	+	-
Festuca lachenalii (C.C.Gmel.) Spenn.	-	-	-	-	-	-	-	-	-	-	-	-	+
Transg. *Echio-Galactition tomentosae* O.De Bolòs et Molinier, 1969													
Echium plantagineum L.	2	3	2	1	+	-	-	-	-	-	2	-	-
Galactites tomentosus Moench	2	-	1	-	-	-	-	-	-	-	-	-	-
Transg. *Festuco valesiacae-Brometea erecti* Br.-Bl. & Tüxen ex Br.-Bl., 1949													
Petrorhagia saxifraga (L.) Link subsp. *gasparrinii* (Guss.) Pignatti ex Greuter & Burdet	-	-	-	-	-	+	+	+	+	-	-	-	-
Allium vineale L.	-	-	-	-	-	+	+	-	-	+	-	-	-
Sixalix atropurpurea (L.) Greuter & Burdet	-	-	-	-	-	-	+	-	-	-	+	-	-
Muscari comosum (L.) Mill.	-	-	-	-	-	-	-	1	+	-	-	-	-
Transg. *Lygeo sparti-Stipetea tenacissimae* Rivas–Martínez, 1978													
Hyparrhenia hirta (L.) Stapf subsp. *hirta*	2	-	1	2	+	-	-	-	-	-	-	-	-
Daucus carota L. subsp. *carota*	-	-	-	-	-	+	+	-	-	-	-	-	-
Lathyrus clymenum L.	-	-	-	-	-	-	-	-	+	-	-	-	-
Reichardia picroides (L.) Roth	-	-	-	-	-	-	-	-	-	+	-	-	-

* from [24], ** from [23].

Figure 6. *Rumici bucephalophori-Ophioglossetum lusitanici* subass. *trifolietosum scabri* subass. nova hoc loco. Date: 19 March 2016, Mt. Vesuvius, Ercolano, between Casa Cantoniera and Osservatorio Vesuviano (Naples). Photo by Adriano Stinca.

In Elba Island (Tuscany), *O. lusitanicum* identifies microcoenoses on a small riparian plateau along a stream near Vallone di Madonna del Monserrato, consisting of sandy sediments where it grows with other species of different ecological contingents, such as *Petrosedum rupestre*, *Centranthus calcitrapa*, *Senecio vulgaris*, and *Geranium purpureum*, testifying that we are dealing with transitional aspects in a progressive release from the wet temporary ponds. For this reason, it was not possible to assign the Elba Island plant community with *O. lusitanicum* to its own syntaxon (Table 2), even if the authors in the syntaxonomic scheme report *Isoëtion durieui* Br.-Bl., 1935 [23], confirming the vegetational versatility of this species. This variability is confirmed in other sites of Elba Island by other authors [30] who highlight, even if without phytosociological relevés, the presence of hygrophilous stations of *Isoëtion durieui* due to the presence of *Isoëtes durieui* Bory with *Juncus bufonius* L. and *J. capitatus* Weigel. In the same locality, at higher altitudes between 250 and 750 m a.s.l., were reported the ephemeral meadows of late spring on granitic soils, referable to the *Helianthemetalia guttati* class, the latter interpreted as a secondary series with respect to the communities of the *Isoëtion*, which would be of primary role. Statistically, although falling within the same group (I) of the Sicilian and Campania–Mt. Vesuvius sites, it must refer to another subgroup (IA) (Figure 3) and, spatially, it has a distinct position (Figure 4). Therefore, further field studies are necessary to clarify the syntaxonomic position of the plant populations of Tuscany.

3.3. *Syntaxonomical Scheme of Surveyed Vegetation*

Stipo-Trachynietea distachyae Brullo in Brullo et al., 2001
Brachypodietalia distachyae Rivas–Martínez, 1978
Trachynion distachyae Rivas–Martínez, 1978
Euphorbio exiguae-Ophioglossetum lusitanici Perrino, Stinca & Tomaselli, 2022 ass. nova hoc loco
Helianthemetea guttati Rivas Goday and Rivas–Martínez, 1963
Helianthemetalia guttati Br.-Bl. in Br.-Bl. et al., 1940
Helianthemion guttati Br.-Bl. in Br.-Bl. et al., 1940
Rumici bucephalophori-Ophioglossetum lusitanici Médail, Pavon, Lo Cascio & Pasta, 2016
trifolietosum scabri Stinca, Perrino & Tomaselli, 2022 subass. nova hoc loco

4. Conclusions

The *Ophioglossum* L. species, such as *O. azoricum* and *O. vulgatum*, although reported in the Italian Regional Red Lists [9], as stated by other authors [33], have become increasingly rare due to: (1) their particular biological cycle; and (2) human impact on their growth environments. The new *O. lusitanicum* communities identified in Apulia, Campania, and Sicily regions represent two different aspects of the same priority habitat (i.e., "pseudo-steppe with grasses and annuals of the *Thero-Brachypodietea*" (habitat code 6220*), as well as described in Spain (*Scillo-Ophioglossetum lusitanici* Ballesteros & Sagarra, 1984), even if doubts persist for the plant populations of Mt. Vesuvius. This phytosociological framework differs from the vegetation described for this species in other European territories, often attributed to *Isoëto-Nanojuncetea* Br.-Bl. et Tx. in Br.-Bl. et al., 1952 (e.g., *Ophioglosso lusitanici-Isoëtion histricis* de Foucault, 1988) [14], which constitute a different habitat (habitat code 3170*).

The type of soil, together with the altimetric parameter, with some exceptions, is confirmed to be the key that discriminates the different types of vegetation and habitat detected. *Isoëtion durieui* (habitat 3170*) at low altitudes, often near the coast, prefers calcareous soils, while the annual meadows of *Brachypodietalia distachyae* (habitat 6220*) are found at relatively higher altitudes and inland. Nevertheless, given the limited number of investigated sites, the matter deserves further investigation. Finally, the aspects of *Helianthemetalia guttati* (habitat 6220*) have been reported on soils of a different nature, volcanic in Sicily and Campania, and granitic in Tuscany (Elba Island). These environmental and ecological assessments deserve to be deepened with further investigations in other Mediterranean territories to verify their reliability on a wide scale, including the primary and secondary roles. In any case, we emphasize that the description of new syntaxa, such as the description of new taxa, is the first step towards local and worldwide biodiversity conservation.

Author Contributions: Conceptualization, E.V.P. and V.T.; methodology, E.V.P., V.T. and A.S.; software, E.V.P., V.T. and A.S.; validation, E.V.P., V.T., G.N.S., R.P.W., A.E. and A.S.; formal analysis, E.V.P., V.T. and A.S.; investigation, E.V.P., V.T., A.E. and A.S.; data curation, E.V.P., V.T., G.N.S. and A.S.; writing—original draft preparation, E.V.P.; writing—review and editing, E.V.P., V.T., G.N.S., R.P.W., A.E. and A.S.; visualization, E.V.P., V.T., G.N.S., R.P.W., A.E. and A.S.; supervision, E.V.P., V.T., A.S. and R.P.W.; pictures, E.V.P. and A.S. All authors have read and agreed to the published version of the manuscript.

Funding: This research did not receive funding.

Data Availability Statement: The data that support the findings of this study are available from the corresponding authors upon reasonable request.

Conflicts of Interest: The authors declare no conflict of interest.

Appendix A Locality, municipality, region, date, identification relevé code, and geographic coordinates (UTM-WGS84) of the surveys analyzed (if available)

Table 1: **Rel. 1**: Santeramo in Colle in loc. Alessandrelli (Bari), Apulia, 3 May 2021, P1, 33T 644123 E—4514234 N; **Rel. 2**: Santeramo in Colle in loc. Alessandrelli (Bari), Apulia, 3 May 2021, P2, 33T 644091 E—4514226 N; **Rel. 3**: Santeramo in Colle in loc. Alessandrelli (Bari), Apulia, 3 May 2021, P3, 33T 644053 E—4514228 N; **Rel. 4**: Santeramo in Colle in loc. Alessandrelli (Bari), Apulia, 3 May 2021, P4, 33T 644012 E—4514227 N; **Rel. 5**: Santeramo in Colle in loc. Alessandrelli (Bari), Apulia, 3 May 2021, P5, 33T 643832 E—4514539 N; **Rel. 6**: Tifatini Mts., Caserta, Campania, 2 April 2016, CT1, 33T 444068 E—4550816 N; **Rel. 7**: Tifatini Mts., Caserta, Campania, 2 April 2016, CT2, 33T 444085 E—4550830 N; **Rel. 8**: Tifatini Mts., Caserta, Campania, 2 April 2016, CT3, 33T 444108 E—4550818 N.

Table 2: **Rel. 9**: Vulcano Island, Lipari (Messina), Sicily, 5 November 2016, S1* (from [24]), no geographic coordinates; **Rel. 10**: Vulcano Island, Lipari (Messina), Sicily, 5 November 2016, S2*, no geographic coordinates; **Rel. 11**: Vulcano Island, Lipari (Messina), Sicily, 5 November 2016, S3*, no geographic coordinates; **Rel. 12**: Vulcano Island, Lipari (Messina), Sicily, 5 November 2016, S4*, no geographic coordinates; **Rel. 13**: Vulcano Island, Lipari (Messina), Sicily, 5 November 2016, S5*, no geographic coordinates; **Rel. 14**: Mt. Vesuvius, Ercolano (between locality Casa Cantoniera and Osservatorio Vesuviano) (Naples), Campania, 19 March 2016, CV1, 33T 448579 E—4519275 N; **Rel. 15**: Mt. Vesuvius, Ercolano (between Casa Cantoniera and Osservatorio Vesuviano) (Naples), Campania, 19 March 2016, CV2, 33T 448574 E—4519281 N; **Rel. 16**: Mt. Vesuvius, Ercolano (between Casa Cantoniera and Osservatorio Vesuviano) (Naples), Campania, 19 March 2016, CV3, 33T 448573 E—4519291 N; **Rel. 17**: Mt. Vesuvius, Ercolano (between Casa Cantoniera and Osservatorio Vesuviano) (Naples), Campania, 19 March 2016, CV4, 33T 448558 E—4519301 N; **Rel. 18**: Mt. Vesuvius, Ercolano (between Casa Cantoniera and Osservatorio Vesuviano) (Naples), Campania, 19 March 2016, CV5, 33T 448549 E—4519307 N; **Rel. 19**: Mt. Vesuvius, Ercolano (between Casa Cantoniera and Osservatorio Vesuviano) (Naples), Campania, 19 March 2016, CV6, 33T 448531 E—4519312 N; **Rel. 20**: Madonna di Monserrato (Elba Island) (Livorno), Tuscany, in the period 1999–2005, T1** (from [23]), 32T 613859 E—4737753 N; **Rel. 21**: Madonna di Monserrato (Elba Island) (Livorno), Tuscany, in the period 1999–2005, T2**, 32T 613878—E 4737806 N.

Appendix B Other sporadic species found in the vegetation relevés

Table 1: **Rel. 1**: *Avena barbata* Pott ex Link (+), *Dasypyrum villosum* (L.) P.Candargy (+), *Linum usitatissimum* L. subsp. *angustifolium* (Huds.) Thell. (+), *Neotinea lactea* (Poir.) R.M.Bateman, Pridgeon & M.W.Chase (+). **Rel. 2**: *Avena barbata* Pott ex Link (+), *Dasypyrum villosum* (L.) P.Candargy (+), *Festuca danthonii* Asch. & Graebn. (+), *Lathyrus cicera* L. (+), *Linum usitatissimum* L. subsp. *angustifolium* (Huds.) Thell. (+), *Neotinea lactea* (Poir.) R.M.Bateman, Pridgeon & M.W.Chase (+), *Tordylium apulum* L. (+), *Valerianella* sp. (+), *Vicia* sp. (+). **Rel. 3**: *Bromus* sp. (+), *Dasypyrum villosum* (L.) P.Candargy (+), *Festuca danthonii* Asch. & Graebn. (1), *Lathyrus cicera* L. (+), *Tordylium apulum* L. (+), *Valerianella* sp. (+). **Rel. 4**:

Bromus sp. (+), *Crepis* sp. (+), *Dasypyrum villosum* (L.) P.Candargy (+), *Lathyrus cicera* L. (+). **Rel. 5**: *Avena barbata* Pott ex Link (+), *Echium vulgare* L. subsp. *pustulatum* (Sm.) Bonnier & Layens (+), *Galactites tomentosus* Moench (+), *Lathyrus cicera* L. (+), *Linum usitatissimum* L. subsp. *angustifolium* (Huds.) Thell. (+), *Plantago afra* L. (+), *Tordylium apulum* L. (+). **Rel. 6**: *Allium* sp. (+), *Anemone hortensis* L. subsp. *hortensis* (+), *Biscutella didyma* L. (+), *Crepis* sp. (+), *Euphorbia helioscopia* L. subsp. *helioscopia* (+), *Hypochaeris radicata* L. (+), *Lotus edulis* L. (+), *Parentucellia latifolia* (L.) Caruel (+), *Picris hieracioides* L. subsp. *hieracioides* (+), *Plantago lanceolata* L. (+), *Poa bulbosa* L. subsp. *bulbosa* (1), *Sonchus bulbosus* (L.) N.Kilian & Greuter subsp. *bulbosus* (+), *Trifolium* sp. (+). **Rel. 7**: *Anemone hortensis* L. subsp. *hortensis* (+), *Biscutella didyma* L. (+), *Crepis* sp. (+), *Euphorbia helioscopia* L. subsp. *helioscopia* (+), *Lotus edulis* L. (+), *Parentucellia latifolia* (L.) Caruel (1), *Poa bulbosa* L. subsp. *bulbosa* (1), *Ranunculus* sp. (2), *Trifolium* sp. (2). **Rel. 8**: *Allium* sp. (+), *Carex flacca* Schreb. subsp. *erythrostachys* (Hoppe) Holub (+), *Coronilla scorpioides* (L.) W.D.J.Koch (+), *Crepis* sp. (+), *Euphorbia helioscopia* L. subsp. *helioscopia* (+), *Lotus edulis* L. (+), *Parentucellia latifolia* (L.) Caruel (+), *Poa bulbosa* L. subsp. *bulbosa* (1), *Trigonella* sp. (+), *Trifolium* sp. (1).

Table 2: **Rel. 1**: *Erodium cicutarium* (L.) L'Hér. (1), *Geranium molle* L. (1), *Heliotropium europaeum* L. (+), *Lobularia maritima* (L.) Desv. (+), *Lysimachia arvensis* (L.) U. Manns and Anderb. subsp. *arvensis* (+), *Medicago* sp. (+), *Orobanche minor* Sm. (+), *Plantago coronopus* L. (+), *Sherardia arvensis* L. (+). **Rel. 2**: *Erodium botrys* (Cav.) Bertol. (2), *Fumaria capreolata* L. subsp. *capreolata* (+), *Geranium molle* L. (+), *Medicago* sp. (+). **Rel. 3**: *Galium aparine* L. (+), *Geranium molle* L. (+), *Serapias* sp. (2). **Rel. 4**: *Erodium botrys* (Cav.) Bertol. (1), *Geranium molle* L. (+), *Medicago* sp. (+), *Serapias* sp. (+). **Rel. 5**: *Anisantha madritensis* (L.) Nevski subsp. *madritensis* (2), *Citrullus colocynthis* (L.) Schrad. (1), *Erodium botrys* (Cav.) Bertol. (+), *Fumaria capreolata* L. subsp. *capreolata* (+), *Plantago coronopus* L. (+). **Rel. 6**: *Calendula arvensis* (Vaill.) L. (1), *Carduus pycnocephalus* L. subsp. *pycnocephalus* (+), *Crepis* sp. (+), *Euphorbia peplus* L. (+), *Medicago arabica* (L.) Huds. (+), *Micromeria graeca* (L.) Benth. ex Rchb. s.l. (+), *Petrosedum rupestre* (L.) P.V.Heath (1), *Poa bulbosa* L. subsp. *bulbosa* (1), *Raphanus raphanistrum* L. subsp. *landra* (Moretti ex DC.) Bonnier & Layens (1), *Silene vulgaris* (Moench) Garcke subsp. *tenoreana* (Colla) Soldano & F.Conti (+), *Trigonella* sp. (+), *Veronica cymbalaria* Bodard subsp. *cymbalaria* (+). **Rel. 7**: *Euphorbia peplus* L. (+), *Poa bulbosa* L. subsp. *bulbosa* (1), *Trigonella* sp. (+). **Rel. 8**: *Petrosedum rupestre* (L.) P.V.Heath (+), *Poa bulbosa* L. subsp. *bulbosa* (+). **Rel. 9**: *Campanula erinus* L. (+), *Catapodium rigidum* (L.) C.E.Hubb. subsp. *rigidum* (+), *Euphorbia peplus* L. (+), *Holcus lanatus* L. subsp. *lanatus* (+), *Lobularia maritima* (L.) Desv. (1), *Lysimachia arvensis* (L.) U.Manns and Anderb. subsp. *arvensis* (+), *Micromeria graeca* (L.) Benth. ex Rchb. s.l. (+), *Petrosedum rupestre* (L.) P.V.Heath (1), *Poa bulbosa* L. subsp. *bulbosa* (+), *Serapias* sp. (+), *Sonchus asper* (L.) Hill subsp. *asper*, *Trigonella* sp. (1). **Rel. 10**: *Crepis* sp. (+), *Euphorbia peplus* L. (+), *Lysimachia arvensis* (L.) U.Manns & Anderb. subsp. *arvensis* (+), *Petrosedum rupestre* (L.) P.V.Heath (+), *Poa bulbosa* L. subsp. *bulbosa* (+), *Serapias* sp. (+), *Sonchus asper* (L.) Hill subsp. *asper* (+), *Spartium junceum* L. (+). **Rel. 11**: *Crepis* sp. (+), *Petrosedum rupestre* (L.) P.V.Heath (+), *Poa bulbosa* L. subsp. *bulbosa* (1), *Serapias* sp. (+), *Trigonella* sp. (+), *Vicia pseudocracca* Bertol. (1). **Rel. 12**: *Allium triquetrum* L. (+), *Centranthus calcitrapae* (L.) Dufr. subsp. *calcitrapae* (+), *Cerastium glomeratum* Thuill. (+), *Dactylis glomerata* L. s.l. (+), *Geranium molle* L. (+), *Geranium purpureum* Vill. (+), *Petrosedum rupestre* (L.) P.V.Heath (+), *Polycarpon tetraphyllum* L. (+), *Senecio vulgaris* L. (+), *Sherardia arvensis* L. (+), *Vicia disperma* DC. (+). **Rel. 13**: *Centranthus calcitrapae* (L.) Dufr. subsp. *calcitrapae* (+), *Geranium purpureum* Vill. (+), *Petrosedum rupestre* (L.) P.V.Heath (1), *Senecio vulgaris* L. (+), *Veronica arvensis* L. (+), *Vicia disperma* DC. (+).

References

1. Hauk, W.D.; Park, C.R.; Chase, M.W. Phylogenetic studies of Ophioglossaceae: Evidence from *rbcL* and *trnL-F* plastid DNA sequences and morphology. *Mol. Phylogenet. Evol.* **2003**, *28*, 131–151. [CrossRef] [PubMed]
2. Smith, A.R.; Pryer, K.M.; Schuettpelz, E.; Korall, P.; Schneider, H.; Wolf, P.G. A classification for extant ferns. *Taxon* **2006**, *55*, 705–731. [CrossRef]

3. Brana, S.; Vuković, N.; Kaligarič, M. Least adder's-tongue (*Ophioglossum lusitanicum* L.) in Croatia—Distribution, ecology and conservation. *Acta Bot. Croat.* **2014**, *73*, 471–480. [CrossRef]
4. Khandelwal, S. Chromosome evolution in the genus *Ophioglossum* L. *Bot. J. Linn. Soc.* **1990**, *102*, 205–217. [CrossRef]
5. Medjahdi, B.; Letreuch-Belarouci, A.; Prelly, R. Actualisation du catalogue des pteridophytes du Nord-ouest algerien (Region de Tlemcen). *Acta Bot. Malacitana.* **2013**, *38*, 33–39. [CrossRef]
6. Prelli, R. *Les Fougères et Plantes Alliées de France et'd'Europe occidentale*; Belin: Paris, France, 2001.
7. Naqinezhad, A.R.; Kavousi, K. The easternmost distribution of *Ophioglossum lusitanicum* L. (Ophioglossaceae), new to flora of Iran. *Iran J. Bot.* **2004**, *10*, 167–171.
8. Bartolucci, F.; Peruzzi, L.; Galasso, G.; Albano, A.; Alessandrini, A.; Ardenghi, N.M.G.; Astuti, G.; Bacchetta, G.; Ballelli, S.; Banfi, E.; et al. An updated checklist of the vascular flora native to Italy. *Plant Biosyst.* **2018**, *152*, 179–303. [CrossRef]
9. Conti, F.; Manzi, A.; Pedrotti, F. *Liste Rosse Regionali delle Piante d'Italia*; WWF Italia, Società Botanica Italiana, CIAS, Univ.: Camerino, Italy, 1997; p. 139.
10. Beccarisi, L.; Chiriacò, L.; Marchiori, S.; Medagli, P. Felci (Filicopsida) spontanee del Salento (Puglia, Italia). *Inform. Bot. Ital.* **2001**, *33*, 341–349.
11. Perrino, E.V.; Signorile, G.; Marvulli, M. A first checklist of the vascular flora of the Polignano a Mare coast (Apulia, southern Italy). *Nat. Croat.* **2013**, *22*, 295–318.
12. Perrino, E.V.; Wagensommer, R.P.; Silletti, G.N.; Signorile, G.; Angiulli, F. Nuovi dati distributivi e relazione con la Direttiva 92/43/CEE di taxa critici pugliesi dalla Provincia di Bari. *Inform. Bot. Ital.* **2013**, *45*, 53–62.
13. Licht, W.; Wagensommer, R.P. *Flora Vascolare del Gargano e delle Isole Tremiti. Chiavi Analitiche per la Determinazione*; Grenzi, C., Ed.; Parco Nazionale del Gargano: Foggia, Italy, 2020; pp. 955+LXIV.
14. Brullo, S.; Minissale, P. Considerazioni sintassonomiche sulla classe *Isoeto-Nanojuncetea*. *Itinera Geobot.* **1998**, *11*, 263–290.
15. Tomaselli, V.; Beccarisi, L.; Brullo, S.; Cambria, S.; Forte, L.; Minissale, P.; Veronico, G. Phytosociological research on temporary ponds in Apulia (southern Italy). *Mediterr. Bot.* **2020**, *41*, 15–41. [CrossRef]
16. Brullo, S.; Brullo, C.; Sciandrello, S.; Tavilla, G.; Cambria, S.; Tomaselli, V.; Ilardi, V.; Giusso del Galdo, G.; Minissale, P. The Plant Communities of the Class *Isoëto-Nanojuncetea* in Sicily. *Plants* **2022**, *11*, 1214. [CrossRef]
17. Rivas Goday, S. Revision de las comunidades hispanas de la clase *Isoeto-Nanojuncetea* Br.-Bl. et Tx. 1943. *Anal. Inst. Bot. Cavanilles* **1970**, *27*, 225–276.
18. De Foucault, B. Les végétations herbacées basses amphibies: Systémique, structuralisme, synsystématique. *Dissert Bot.* **1988**, *121*, 1–150.
19. Ballesteros, I.; Sagarra, E. Sobre l'estructura i la dinàmica de les comunitats terofítiques humides (classe *Isoeto-Nanojuncetea*) i els pradells amb *Ophioglossum lusitanicum* L. del Massís de Cadiretes (la Selva). *Collect Bot.* **1984**, *15*, 39–57.
20. European Commission. Council Directive 92/43/EEC of 21 May 1992 on the Conservation of Natural Habitats and of Wild Fauna and Flora, 1995–2007. Available online: http://eurlex.europa.eu/LexUriServ/LexUriServ.do?uri=CELEX:31992L0043:EN:HTML (accessed on 9 October 2022).
21. Braun-Blanquet, J. *Plant Sociology*; McGraw-Hill: New York, NY, USA, 1932; p. 439.
22. Thiers, B. Index Herbariorum: A Global Directory of Public Herbaria and Associated Staff [Continuously Updated]. Available online: http://sweetgum.nybg.org/ih/ (accessed on 3 November 2022).
23. Foggi, B.; Cartei, L.; Pignotti, L.; Signorini, M.A.; Viciani, D.; Dell'Olmo, L.; Menicagli, E. Il paesaggio vegetale dell'Isola d'Elba (Arcipelago Toscano). Studio fitosociologico e cartografico. *Fitosociologia* **2006**, *43*, 3–95.
24. Médail, F.; Pavon, D.; Lo Cascio, P.; Pasta, S. Sulla presenza di *Ophioglossum lusitanicum* L. (Psilotopsida) nell'isola di Vulcano e aggiunte alla flora vascolare delle Isole Eolie (Sicilia nord-orientale). *Nat. Sicil.* **2016**, *S. IV, XL*, 51–66.
25. Tutin, T.G.; Heywood, V.H.; Burges, N.A.; Moore, D.M.; Valentine, D.H.; Walters, S.M.; Webb, D.A. *Flora Europaea 1–5*, 1st ed.; Cambridge University Press: Cambridge, UK, 1968–1980.
26. Pignatti, S.; Guarino, R.; La Rosa, M. *Flora d'Italia 1–4*; Edagricole: Bologna, Italy, 2017–2019.
27. Galasso, G.; Conti, F.; Peruzzi, L.; Ardenghi, N.M.G.; Banfi, E.; Celesti-Grapow, L.; Albano, A.; Alessandrini, A.; Bacchetta, G.; Ballelli, S.; et al. An updated checklist of the vascular flora alien to Italy. *Plant Biosyst.* **2018**, *152*, 556–592. [CrossRef]
28. Addinsoft. XLSTAT Statistical and Data Analysis Solution; Version 2022.4.1. Available online: https://www.xlstat.com/en/ (accessed on 3 November 2022).
29. Van der Maarel, E. Transformation of cover-abundance values in phytosociology and its effects on community similarity. *Vegetatio* **1979**, *39*, 97–114.
30. Carta, A.; Pierini, B.; Peruzzi, L. Distribuzione di *Ophioglossum lusitanicum* L. (Psilotopsida) in Toscana. *Inform. Bot. Ital.* **2008**, *40*, 135–141.
31. Medjahdi, B.; Letreuch-Belrouci, A.; Babali, I. Une Nouvelle station d'*Ophioglossum lusitanicum* L. en Oranie (Algerie Occidentale). *Acta Bot. Malacit.* **2018**, *42*, 177–179. [CrossRef]
32. Stinca, A. Rediscovery of two presumably extinct plant vascular species in Mt. Vesuvius, Italy. *Plant Sociol. in press*.
33. Giovannini, A.; Pierini, B. *Ophioglossum* (Ophioglossaceae, Polypodiophyta) sul Monte Pisano. Due nuove stazioni di *O. azoricum* e di *O. vulgatum*. *Atti. Soc. Tosc. Sci. Nat. Mem. Ser. B* **2006**, *113*, 67–68.

agronomy

Article

Adaptive Responses of Four Medicinal Plants to High Altitude Oxidative Stresses through the Regulation of Antioxidants and Secondary Metabolites

Ibrahim A. Ibrahim [1], A. A. Jabbour [2], Awatif M. Abdulmajeed [3], Mohamed E. Elhady [4,5], Yaser A. Almaroai [2] and Ahmed M. Hashim [6,*]

1 Department of Pharmacology and Toxicology, Faculty of Medicine, Umm Al-Qura University, Makkah 21955, Saudi Arabia
2 Department of Biology, Faculty of Applied Sciences, Umm Al-Qura University, Makkah 21955, Saudi Arabia
3 Biology Department, Faculty of Science, University of Tabuk, Umluj 46429, Saudi Arabia
4 Department of Botany and Microbiology, Faculty of Science, Al Azhar University, Cairo 11865, Egypt
5 Biology Department, Faculty of Science and Arts, Al-Baha University, Al Mandaq 65581, Saudi Arabia
6 Botany Department, Faculty of Science, Ain Shams University, Cairo 11865, Egypt
* Correspondence: hashim-a-m@sci.asu.edu.eg

Abstract: The conservation of medicinal plants, particularly endangered or endemic species, is of the utmost importance, especially in light of inevitable climate change and its consequences. Species inhabiting high altitudes adopt exceptional defense mechanisms in response to abiotic stresses as a survival strategy. The objective of the current study was to investigate the effects of altitudinal variations on secondary metabolite accumulation and antioxidant enzyme capacity in four plants (*Cotoneaster orbicularis*, *Crataegus x sinaica*, *Echinops spinosissimus* subsp. *Spinosissimus*, and *Tanacetum sinaicum*) growing naturally on the Sinai Peninsula's high mountains. Plant leaves and soil samples were collected from three altitudes between 1500 and 2250 m a.s.l. to evaluate the adaptive responses of these species in relation to high-altitude oxidative stresses. The results showed that at higher altitudes, the electrical conductivity and the micronutrient contents of the soil decreased, which may be due to the prevalence of silt and clay decreasing at higher altitudes. Chlorophyll a, chlorophyll b, ascorbic acid, and total soluble protein showed similar results in relation to higher altitudes for all species. On the other hand, proline, total soluble sugars, carotenoids, phenols, tannins, and flavonoids increased in response to high altitudes. The activity levels of catalase and ascorbic acid peroxidase showed a significant increase aligned with higher altitudes, while a significant decrease in activity levels was obtained for polyphenol oxidase. In conclusion, the present findings showed that *Cotoneaster orbicularis* exhibited the maximum response for coping with high-altitude stresses, followed by the remaining three species regarding the level of biochemical and physiological responses. The present work will help formulate conservation plans for important medicinal species.

Keywords: antioxidant capacity; bioactive compounds; endangered medicinal species; high altitude environment; oxidative stresses; protectorate

Citation: Ibrahim, I.A.; Jabbour, A.A.; Abdulmajeed, A.M.; Elhady, M.E.; Almaroai, Y.A.; Hashim, A.M. Adaptive Responses of Four Medicinal Plants to High Altitude Oxidative Stresses through the Regulation of Antioxidants and Secondary Metabolites. *Agronomy* **2022**, *12*, 3032. https://doi.org/10.3390/agronomy12123032

Academic Editors: Alessio Papini, Mushtaq Ahmad, Fazal Ullah and Wajid Zaman

Received: 5 November 2022
Accepted: 26 November 2022
Published: 30 November 2022

1. Introduction

One of the main issues regarding the impact of climate change is the destruction of particular natural habitats, specifically in fragile environments, as these changes control the growth dynamics, distribution, and population structure of plants [1–3]. A species' survival in its ecosystem depends on its ability for habitat adaptation, and most plant species have evolved a diversified array of biochemical responses to deal with different environmental challenges [4,5]. To maintain a homeostatic level of photosynthetic efficiency, plants either adapt their physiological and morphological characters or adjust their photosynthetic capacity in response to the environment [6,7].

It is anticipated that the continuous consequences of climate change will increase the frequency of droughts, nutrient deficiency, and temperature increase, and the combined effects of these stresses on species' survival will be more harmful than their individual effects [8].

Although the Sinai Peninsula only comprises 6.1% of Egypt's total land area, it is home to about 1262 different plant species. Approximately 62 endemic plant species exist in Egypt, with the Sinai Mountains acting as home to half of them. According to Boulos [9], Saint Katherine Protectorate (SKP) is one of the Middle East's most floristically diverse locations, with 17 endemic taxa (28% of Egyptian endemic taxa), comprising a sizable fraction of the region's endemism [10–12]. Six microhabitats, including wadis, gorges, caves, basins, slopes, and terraces, can be differentiated in this area due to the variety of SKP's landforms and geologic features. The first two microhabitats include dykes that trap water, resulting in substantial plant cover [13].

Due to the altitude of the SKP mountains, plant growth, diversity, and distribution are limited by harsh and complicated environmental circumstances [14]. High illumination, strong winds, low levels of CO_2 and O_2, gravely thin soils with low water and nutrient contents, and low temperatures are only a few of these harsh environmental conditions [15,16]. Even though these stressors could retard plant growth, many plants have endured and progressed, developing a variety of adaptive defense mechanisms in response to these abiotic stresses [17,18].

In addition to being a valuable assortment of natural products, plant secondary metabolites (SMs) have also been linked to the activation and augmentation of plant defense mechanisms against biotic and abiotic environmental stresses [19,20]. Likewise, environmental variables such as soil microorganisms, water availability, soil pH, nutrients, temperature (high/low), high altitude, drought, and light conditions strongly affect the quality and quantity of SMs production among plant populations [21–23]. These SMs actually assist plants in maintaining physiological and morphological functions. For instance, they increase flavonoids, tannins, and phenols as antioxidants; accumulate soluble protein, soluble sugars, and proline to fend off drought via the regulation of the water-holding potential; and upregulate the xanthophyll cycle pigment pool by increasing carotenoids relative to chlorophylls, lessening the harmful effect of high light intensity [24–26].

On the other hand, reactive oxygen species (ROS) production, including singlet oxygen, hydroxyl radicals, and superoxide, increases as a result of high-altitude environmental stresses (light intensity, low temperature, and drought), amplifying the possibility of oxidative damage [27–29]. The combination of these stresses can seriously harm plants by limiting photosynthesis, closing stomata, and inhibiting various physiological processes, including antioxidant activities [30–32].

Cells achieve ROS homeostasis when the ratio of ROS synthesis to ROS scavenging is balanced. However, under typical circumstances, ROS production is minimal and typically balanced by antioxidant molecules [33]. Plants can scavenge ROS via two antioxidant systems: 1. the non-enzymatic antioxidant system, comprised of compounds of basic antioxidant characteristics, such as glutathione, flavonoids, phenolic compounds (water soluble), carotenoids, and a-tocopherols (lipid-soluble); and 2. the enzymatic antioxidant system, which comprises catalase (CAT), peroxidase (POD), superoxide dismutase (SOD), and ascorbate peroxidase (APX) [34,35]. Additionally, malondialdehyde (MDA) is considered an important indicator of abiotic stress (especially drought), and it is the end product of cell membrane lipid peroxidation. Frequently, MDA will increase with decreasing soil water content [36–38].

Four plant species of high value (medicinal, solid fuel, and edible) inhabiting the SKP mountains were investigated in this study (namely, *Cotoneaster orbicularis* Schltdl., *Crataegus x sinaica* Boiss., *Echinops spinosissimus* subsp. *Spinosissimus* Turra., and *Tanacetum sinaicum* Del.). Despite the medical and ecological importance of these species, little is known about their ecophysiology, particularly at high altitudes in the Sinai Peninsula. Consequently, understanding the different adaptive mechanisms of these species is crucial for predicting their responses to climate change in the future and assisting researchers in planning to conserve and sustain these species. In this study, the SMs, antioxidant capacity, and physiology of the four plants were analyzed in relation to altitudinal variation. Additionally, the eco-physiological response of these plants to exposure to different stresses at higher altitudes is investigated.

2. Materials and Methods

2.1. Study Area

The current study was carried out in the mountainous region of Saint Katherine protectorate (SKP) in the south part of Sinai Peninsula (Figure 1). SKP was declared as a protected area in 1996 by the (EEAA) [39]. It is located between 28°30′ and 28°35′ N and 33°55′ and 34°30′ E, covering an area of roughly 4350 km^2. The region is distinguished by the presence of Egypt's highest and steepest mountains, which reach elevations of up to 2624 m above sea level (a.s.l.) [40,41]. As a result of its distinctive geological, climatic, and morphological characteristics, this dry mountainous ecosystem is home to numerous rare, endemic, and medicinal plant species [10,40].

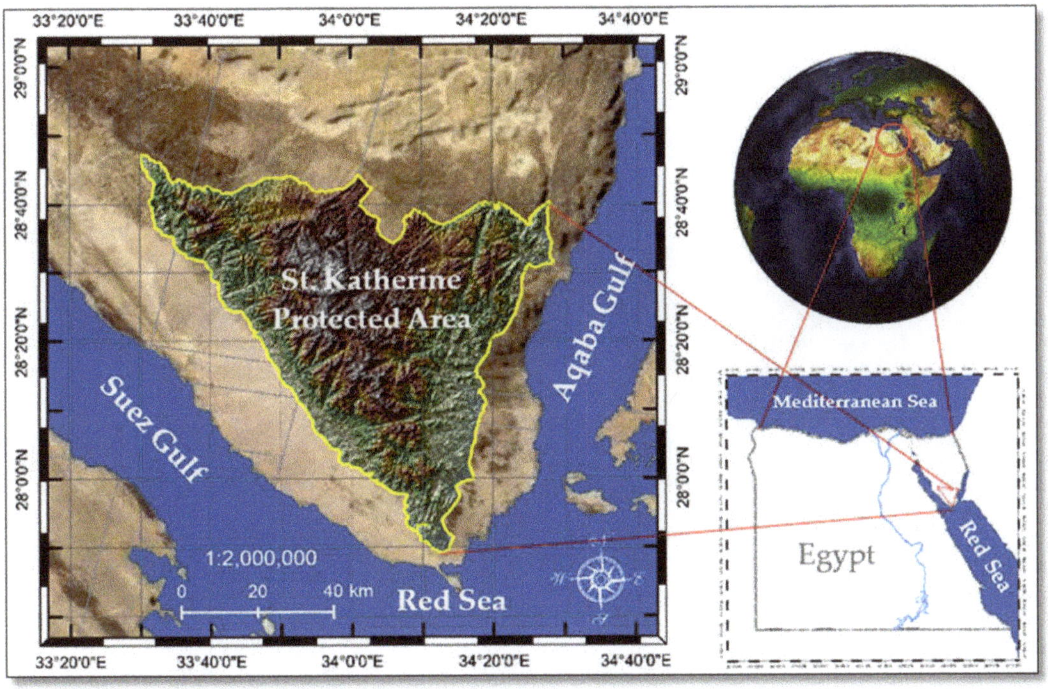

Figure 1. Map of the study area for Saint Katherine protectorate, Sinai Peninsula, Egypt (Base map: acquired using Landsat 8-2022-P/R: 178/39).

The plant materials were collected from El-Gebel Al-Ahmer (red mountain) at SKP in April/May 2022. The study area lies in a Saharan-Mediterranean hyper-arid zone, with extremely cold winters and hot, dry summers. Due to its great altitudes, SKP is the coldest place in Sinai [42,43]. The average temperature ranges from 5.4 °C to 25.2 °C, with the mean minimum temperature occurring in February (7.8 °C) and the mean highest temperature occurring in August (36 °C) [44].

2.2. Target Species

In Table 1, specifics concerning the target species are given, including their scientific names, their families, habitat, distribution, status, uses, distribution, altitude, and field photos. Four medicinal plant species (namely, *Cotoneaster orbicularis*, *Crataegus x sinaica*, *Echinops spinosissimus* subsp. *Spinosissimus*, and *Tanacetum sinaicum*) were collected from three different altitudes (1500–1750, 1750–2000, and 2000–2250 m a.s.l.) of the SKP mountains (Figure 2). Samples were taken from the shoot systems (4 species × 3 different individuals for each × 3 altitudes) of the four plant species under investigation, with a total of 36 plant samples, which were stored frozen at (−20 °C). Next, the samples were either extracted to determine the photosynthetic pigments, total antioxidant capacity, and enzyme, proline, ascorbic acid, and malondialdehyde levels, or air-dried to determine the quantity of tannins, carbohydrates, flavonoids, and phenolic compounds.

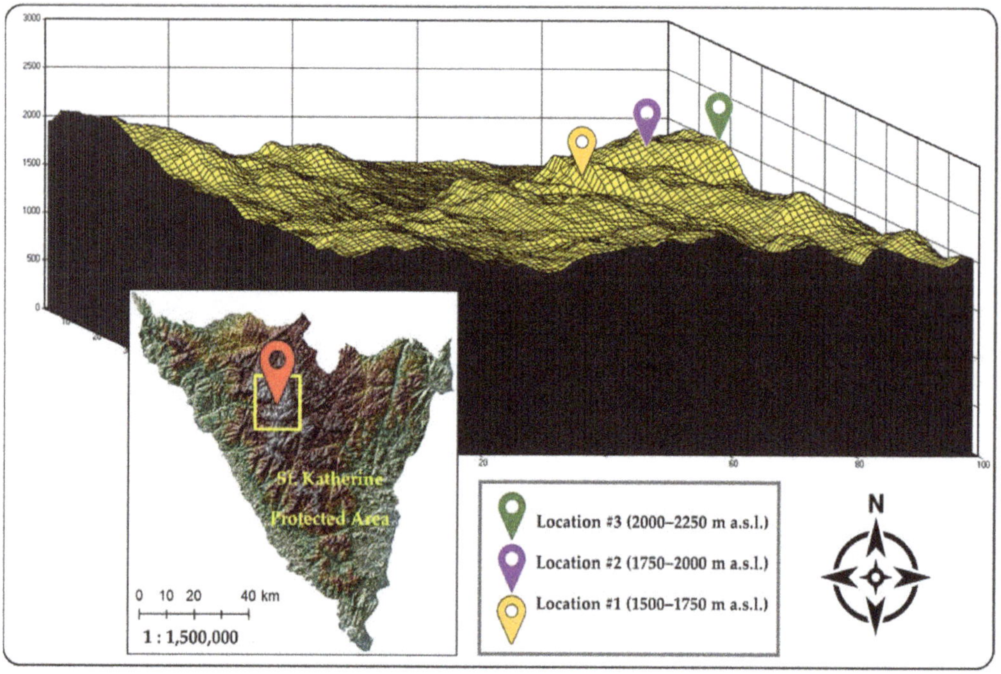

Figure 2. Surface profile showing the three locations of data collection at different altitudes in Saint Katherine protectorate.

Collected taxa were identified according to the work of Boulos, Täckholm, and Zohary [9,45,46]. Taxa names were clarified against Kew's Plants of the World Online [47], which was also used as a reference for family and genera classification.

Table 1. Family, scientific name, habitat, status, uses, distribution, and field photos of the four target species.

No.	Family	Plant Species	Habitat, Status, Uses, and Distribution	Field Photo
1	Rosaceae	*Cotoneaster orbicularis* Schltdl.	- Distribution: It is restricted to the rugged South Sinai and Eastern Arabia. - Status: Endangered - Habitat: It flourish in crevices and in soft dykes. - Uses: fuel, folk medicine, and animal grazing. - Refs. [48–50].	
2		*Crataegus x sinaica* Boiss.	- Distribution: It is confined to the South Sinai mountains. - Status: Endangered - Habitat: It grows in the SKP's elevated high wadis. - Uses: fuel, folk medicine, and food. - Refs. [51–53].	
3	Asteraceae	*Echinops spinosissimus* subsp. *Spinosissimus* Turra.	- Distribution: It grows well across the Sahara, notably near Sinai and the Red Sea. - Status: Common - Habitat: It grows on coastal calcareous dunes, wadi beds, and on gravelly to rocky surfaces. - Uses: fuel, folk medicine, and animal grazing. - Refs. [48,54,55].	
4		*Tanacetum sinaicum* Del.	- Distribution: It is confined to the South Sinai mountains. - Status: Endemic to Egypt - Habitat: It grows on the rocky slopes of SKP. - Uses: folk medicine. - Refs. [47,56,57]	

2.3. Soil Analysis

Twelve soil samples were taken from each studied altitude at a depth of 0–30 cm (one soil sample for each plant individual at each altitude). The samples were then air-dried to determine their physical and chemical characteristics. Mechanical analysis (texture) was determined according to the technique of [58] using the hydrometer method to separate sand, silt, and clay. Using the potentiometric method [59], pH value, total dissolved salts (TDS), and electric conductivity (E.C.) were determined in the soil–water extracts (1:5). Using the method designed by Johnson and Ulrich (versene titration method) [60], magnesium and calcium were measured volumetrically. According to the method of Shapiro and Brannock [61], potassium and sodium were determined by flame photometry. Bicarbonates and chlorides were determined by the titration method according to the work of [62,63]. Sulphates and calcium carbonates were determined according to the method of Bardsley and Lancaster [63]. Organic carbon (O.C.) and organic matter (O.M.) were determined using the titration method described by Piper [64].

2.4. Phytochemical Assay

All the chemicals used in this investigation were of high purity, purchased from Sigma Aldrich Chemical Co., Germany, and all organic solvents were of AR grade.

2.4.1. Extraction and Estimation of Photosynthetic Pigments

The photosynthetic pigments in terms of Chl a, Chl b and carotenoids were extracted in 80% acetone. Measurements were performed using a spectrophotometer (Spectronic 601, Milton Roy Company, Ivyland, PA, USA). Pigment contents were calculated according to the method of Metzner et al. [65].

2.4.2. Extraction and Estimation of Total Soluble Sugars

Following the method of Homme, et al. [66], sugars were extracted. The method of Loewus [67] was used to measure the total soluble sugar contents by reacting 0.1 mL of the ethanolic extract with 3 mL of newly made anthrone reagent. Utilizing the glucose standard curve, the soluble sugar concentration was quantified and represented as μg glucose equivalent/g dry weight.

2.4.3. Extraction and Estimation of Total Phenolic, Flavonoid, and Tannin Content

According to Malik and Singh's method [68], the total phenolic molecules were extracted, and their content was measured (concentrations were estimated based on the pyrogallol standard curve as gallic acid equivalents/g). According to the Harborne method [69], after being extracted, the total flavonoids were calorimetrically determined by reacting with $AlCl_3$ (the concentration of total flavonoids, given as μg/g dry weight, was determined using the quercetin standard curve). Finally, following the method outlined by Ejikeme et al. [70], the tannins were extracted and then quantified.

2.5. Biochemical Assay

2.5.1. Extraction and Estimation of Total Soluble Proteins and Proline Content

Total proteins were extracted by grinding 0.5 g of fresh leaf tissue in 1 mL of phosphate buffer (0.1 M, pH 7.0), before being preserved in ice. Then, the concentration of soluble proteins was assessed according to the method of Bradford [71]. The amount of free proline expressed as μg/g fresh weight was calculated according to the method Bates et al. [72].

2.5.2. Extraction and Estimation of Malondialdehyde

According to the method of Heath and Packer [73], the quantity of lipid peroxidation was assessed by counting the MDA amount generated by the thiobarbituric acid reaction (results were expressed as μ mol g^{-1} fresh weight).

2.6. Enzymes Extraction and Assays

The enzymes were extracted following the method of Mukheriee and Choudhuri, [74]. A total of 250 mg of fresh tissue were crushed coarsely using a pestle in a cooled mortar after being frozen in liquid nitrogen. Then, the ground powder was added to 10 mL of 100 mM phosphate buffer (KH_2PO_4/K_2HPO_4, pH 6.8), and centrifuged at 20,000× *g* for twenty minutes. Using the same buffer, the supernatant was diluted to a specified volume and used as an enzyme extract to determine the activity of various enzymes.

The super oxide dismutase (SOD) activity level was estimated following the method of Dhindsa et al. [75]. The SOD activity was measured as unit/mg protein. The CAT activity level was determined following the method of Hermans et al. [76]. The Catalase (CAT) activity was measured as mM of H_2O_2/g FW/min. The peroxidase (POD) activity was calculated according to the Kar and Mishra method [77], with slight modifications. Quinone/g f.w/min was used to express POD activity. Finally, ascorbate peroxidase (APX) activity was assessed according to the method of Koricheva et al., with slight modifications [78]. The APX activity was expressed as mM of ascorbate oxidized/g f.w/min.

2.7. Determination of Ascorbic Acid (AsA) and Total Antioxidant Capacity (TAC)

According to the method of Kampfenkel et al. [79], ascorbic acid extracted from fresh tissue weighing (0.1 g) was homogenized in one milliliter of trichloroacetic acid solution (6% (w/v)); the homogenate was then centrifuged at 12,000× g and 4 °C for ten minutes. Ascorbic acid estimation was performed using the supernatant. TAC was measured according to the method of Prieto et al. [80] using phosphomolybdenum.

2.8. Statistical Analysis

Data of four species (*Co. orbicularis, Cr. sinaica, Ec. spinosisimus,* and *Ta. sinaicum*) at three different altitudes (1500–1750, 1750–2000, and 2000–2250 m a.s.l) was analyzed to obtain the physiological and ecological parameters using Excel 365, Minitab 20, IBM SPSS 26, and PC-ORD 5. Three plant and soil samples were collected from each species at each altitude, with a total of 36 soil and plant samples (3 samples × 4 species × 3 altitudes). Data was cleansed before running any statistical analyses. Missing data and mistyping errors were checked. Descriptive statistics, including mean and standard error of mean, were calculated for the physiological parameters (CAT, SOD, APX, POX, total phenols, flavonoids, tannins, TAC, AsA, soluble sugar, total sol. protein, proline, MDA, chl. a, chl. b, and carotenoids) and soil parameters (pH, E.C., TDS, Cl^-(meq/L), $CaCO_3^-$ (%), SO_4^- (Mg/L), HCO_3^- (%), Ca^{++} (meq/L), Mg^{++} (meq/L), Na^+ (mlq/L), K^+ (mlq/L), O.C. (%), O.M. (%), sand %, Clay%, and silt%). Inferential statistics were used to compare the results of different groups. All variables parametric assumptions were tested, and the Box Cox transformation for non-normal dependent variables was applied using the optimal λ method, whenever needed.

Two-way ANOVA was used to evaluate the physiological and soil parameters for each altitudinal variation and different species, using the general fit linear model in Minitab 20. Bar charts of all variables, showing the mean and SE of each variable in relation to different factors, were produced using Excel 365. The results showed a good fit for different models, while normal residual probability plots showed a linear attitude for all analyses after data transformation. The p value was considered significant at $\alpha < 0.05$. Post hoc analyses of the interactions among all groups were conducted using the Tukey test for pairwise comparisons. The results of the post hoc analyses are represented with letters, where groups that share the same letters are non-significantly different, while different letters express significant differences among different groups.

Multivariate analyses of physiological and soil parameters were performed using PC-ORD 5. Two-way cluster analysis (TWCA) and principal component analysis (PCA) were conducted for all parameters regarding different species from different altitudes. Two-way cluster analysis was achieved using Euclidean distance as a distance measurement and Ward's method as a linkage method. PCA was plotted using PC1 and PC2, and correlations of all parameters and species from different altitudes were obtained. The TWCA dendrogram and the PCA joint plot were graphed using PC-ORD5.

3. Results and Discussion

3.1. Soil Analysis

In natural environments, such as mountains, the main determinant of plant species distribution and growth are moisture and temperature gradients, which combine the influence of numerous physical factors. The moisture content depends on the land form type, slope degree, nature of the soil surface, soil texture, and soil organic content, whereas the temperature is a net product of the elevation and the slope aspect [81–83]. Moreover, in the present work, there is a marked decrease in the distribution and intensity of species with increasing altitude. This may be correlated with changes in the physical and chemical characteristics of the soil collected from the three locations at different altitudes.

Performing a principal component analysis (PCA) showed that altitude is a major factor separating the 36 soil samples. PCA output showed that the first altitude (1500–1750) tends to correlate positively with axis 1, while the third altitude (2000–2250) showed a negative correlation with axis 1. The second altitude (1750–2000) showed no correlation with axis 1. On the contrary, axis 2 showed no correlation with any altitude. Generally, pH value, sand, and Ca^{+2} contents showed a correlation with the highest altitude (2000–2250), while lower altitudes were correlated with silt, clay, (OC), (OM), Cl^{-}, K^{+}, Mg^{+2}, and E.C. (Figure 3).

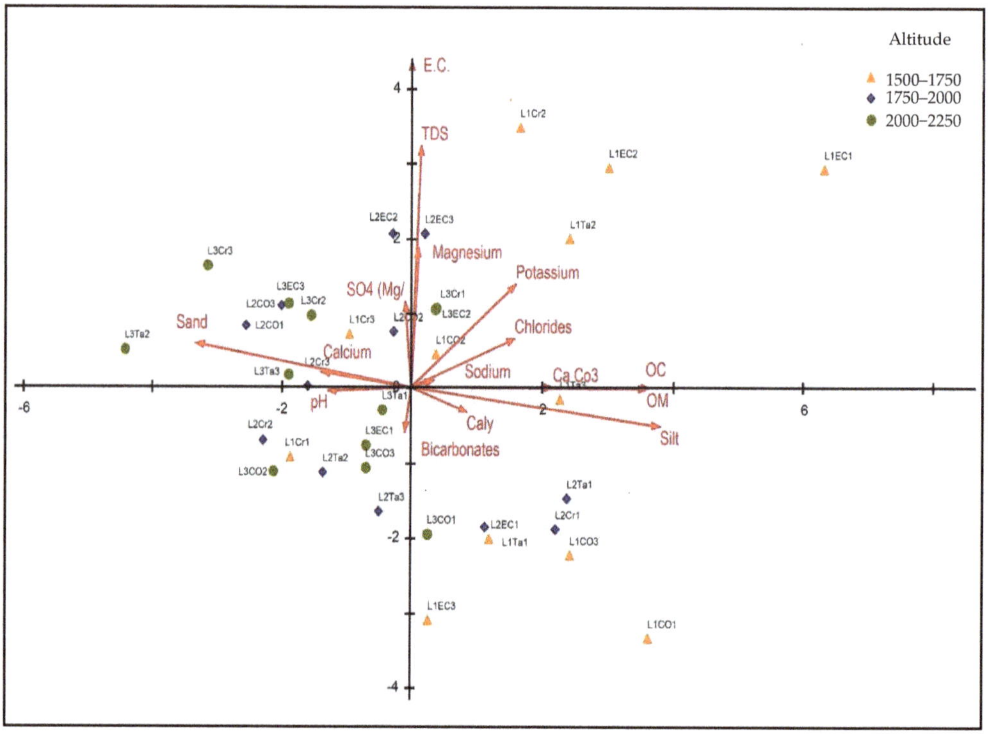

Figure 3. PCA joint plot showing target species at different altitudes in relation to soil parameters. Abbreviations: L1 = altitude 1; L2 = altitude 2; L3 = altitude 3; *Co.* = *Cotineaster orbicularis; Cr.* = *Crataegus sinaica;* Ec. = Echinops spinossissmus; Ta. = Tanacetum sinaicum.

From the obtained results shown in Figure 3, it is clear that electrical conductivity and some free salts, such as K^{+}, Na^{+}, OC, OM, and Mg^{+2}, are negatively correlated with high altitudes (they tend to accumulate at low altitudes). The EC value, OC, and OM content of the soil exhibits a positive relationship with the fertility and nitrogen content in the soil [81,84]. Moreover, the EC value of soil decreased with an increase in the altitude due to the decrease in water content and free salts [85–87]. This was accompanied by a decrease in the ratio of silt and clay contents (lowest values of 24.06% and 4.03%) and an increase in the ratio of sand as it reached its highest value (71.75%) at the highest altitude (2000–2250 m a.s.l.), as shown in Table 2. Collectively, these conditions cause the plants grown at high altitudes to suffer from stress due to deficiencies in water and nutrients, which may affect their growth and distribution [81,84].

Table 2. (**a**) Mean ± SE of soil samples characteristics (pH, E.C, TDS, O.C, O.M, clay, sand, and silt contents) for the studied species at different altitudes. Abbreviations: L1 = altitude 1; L2 = altitude 2; L3 = altitude 3; *Co.* = *Cotineaster orbicularis*; *Cr.* = *Crataegus sinaica*; *Ec.* = *Echinops spinossissmus*; *Ta.* = *Tanacetum sinaicum*. (**b**) Mean ± SE of the chemical analysis of soil samples (anions and cations content) for the studied species at different altitudes. Abbreviations: L1 = altitude 1; L2 = altitude 2; L3 = altitude 3; *Co.* = *Cotineaster orbicularis*; *Cr.* = *Crataegus sinaica*; *Ec.* = *Echinops spinossissmus*; *Ta.* = *Tanacetum sinaicum*.

(a)

Species	pH	E.C. (ds/m)	TDS	O.C. (%)	O.M. (%)	Clay%	Sand%	Silt%
L1 *CO.*	7.43 ± 0.07 A	0.94 ± 0.18 A	408 ± 113 A	1.23 ± 0.16 A	2.12 ± 0.28 A	8.48 ± 0.28 AB	39.82 ± 4.89 ABC	51.71 ± 5.18 AB
L1 *Cr.*	7.53 ± 0.03 A	1.03 ± 0.18 A	657 ± 115 A	0.83 ± 0.13 A	1.43 ± 0.22 A	10.65 ± 1.58 AB	53.64 ± 4.4 ABC	35.71 ± 3.23 ABC
L1 *EC.*	7.57 ± 0.09 A	0.89 ± 0.15 A	569.6 ± 97.9 A	1.01 ± 0.19 A	1.75 ± 0.32 A	16.31 ± 4.04 AB	27.02 ± 5.01 C	56.66 ± 3.41 A
L1 *Ta.*	7.5 ± 0 A	0.9 ± 0.11 A	576.9 ± 71.4 A	1.21 ± 0.12 A	2.09 ± 0.2 A	20.26 ± 4.98 A	37.66 ± 4.4 BC	42.08 ± 0.75 ABC
L2 *CO.*	7.53 ± 0.03 A	0.88 ± 0.05 A	501.5 ± 50.6 A	0.76 ± 0.03 A	1.31 ± 0.05 A	7.35 ± 1.33 AB	64.17 ± 5.41 AB	28.48 ± 6.33 BC
L2 *Cr.*	7.57 ± 0.03 A	0.72 ± 0.04 A	496 ± 53.1 A	1.05 ± 0.19 A	1.81 ± 0.32 A	8.69 ± 2.98 AB	61.2 ± 11.7 AB	30.15 ± 8.76 BC
L2 *EC.*	7.47 ± 0.03 A	0.82 ± 0.14 A	588.4 ± 90.3 A	0.98 ± 0.05 A	1.69 ± 0.09 A	6.93 ± 0.74 AB	58.69 ± 6.9 AB	34.38 ± 6.16 ABC
L2 *Ta.*	7.5 ± 0 A	0.71 ± 0.04 A	484.7 ± 23.5 A	0.93 ± 0.15 A	1.61 ± 0.25 A	5.91 ± 0.21 AB	62.22 ± 1.65 AB	31.87 ± 1.44 ABC
L3 *CO.*	7.6 ± 0 A	0.67 ± 0.04 A	437.5 ± 24.9 A	0.87 ± 0.04 A	1.49 ± 0.08 A	6.37 ± 3.66 AB	51.25 ± 6.73 ABC	25.9. ± 3.28 ABC
L3 *Cr.*	7.47 ± 0.09 A	0.68 ± 0.01 A	540.6 ± 37 A	0.81 ± 0.03 A	1.4 ± 0.06 A	6.58 ± 3.9 AB	65.22 ± 7.9 AB	25.2 ± 4.47 C
L3 *EC.*	7.53 ± 0.03 A	0.6.9 ± 0.1 A	533.8 ± 67 A	0.85 ± 0.08 A	1.47 ± 0.14 A	4.03 ± 0.76 B	65.78 ± 2.93 AB	26.19 ± 2.2 BC
L3 *Ta.*	7.53 ± 0.07 A	0.71 ± 0.07 A	548.1 ± 43.5 A	0.8 ± 0.12 A	1.38 ± 0.2 A	4.19 ± 1.34 B	71.75 ± 6.67 A	24.06 ± 5.4 C

(b)

Species in Different locations	CL (meq/L)	$CaCO_3$ (%)	SO_4 (Mg/L)	HCO_3 (%)	Ca (meq/L)	Mg (meq/L)	Na (meq/L)	K (meq/L)
L1 *CO.*	2.66 ± 0.1 B	9.33 ± 1.17 A	7.2 ± 1.46 BC	0.61 ± 0 A	4.42 ± 0.54 A	2.16 ± 0.52 AB	0.62 ± 0.09 A	0.98 ± 0.3 A
L1 *Cr.*	2.19 ± 0.19 B	5.5 ± 1.04 A	8.44 ± 0.33 ABC	0.61 ± 0 A	5.04 ± 0.03 A	1.69 ± 0.06 AB	0.47 ± 0.06 A	3.12 ± 1.79 A
L1 *EC.*	4.66 ± 0.91 A	9.5 ± 5.57 A	11.57 ± 2.06 AB	0.51 ± 0.1 A	2.83 ± 0.51 A	3.3 ± 0.9 AB	0.6 ± 0.24 A	2.18 ± 1.3 A
L1 *Ta.*	2.28 ± 0.17 B	5 ± 1.32 A	9.56 ± 0.12 ABC	0.51 ± 0.1 A	4.42 ± 0.54 A	1.76 ± 0.04 AB	0.71 ± 0.1 A	2.18 ± 0.54 A
L2 *CO.*	2.66 ± 0.1 B	4.5 ± 1.5 A	11.74 ± 0.32 AB	0.51 ± 0.1 A	4.95 ± 0.94 A	2.78 ± 0.49 AB	0.33 ± 0.03 A	0.64 ± 0.08 A
L2 *Cr.*	2.19 ± 0.1 B	6.5 ± 1.26 A	8.77 ± 1.49 ABC	0.61 ± 0 A	4.48 ± 0.57 A	1.59 ± 0.05 B	0.33 ± 0.03 A	0.77 ± 0.13 A
L2 *EC.*	3.14 ± 0.17 AB	6.83 ± 1.09 A	7.32 ± 1.37 BC	0.61 ± 0 A	4.52 ± 0.59 A	2.74 ± 0.54 AB	0.42 ± 0.06 A	1.54 ± 0.29 A
L2 *Ta.*	2.47 ± 0.34 B	6.67 ± 1.42 A	5.03 ± 0.39 C	0.61 ± 0 A	3.9 ± 0.56 A	1.65 ± 0.02 B	0.53 ± 0.14 A	0.99 ± 0.04 A
L3 *CO.*	2.47 ± 0.1 B	4.67 ± 0.88 A	10.15 ± 1.48 ABC	0.61 ± 0 A	4.42 ± 0.54 A	1.71 ± 0.04 AB	0.48 ± 0.12 A	0.34 ± 0.04 A
L3 *Cr.*	2.47 ± 0.25 B	2.33 ± 0.44 A	13.45 ± 0.43 A	0.61 ± 0 A	4.48 ± 0.57 A	3.26 ± 0.08 A	0.68 ± 0.2 A	0.53 ± 0.31 A
L3 *EC.*	2.85 ± 0.44 AB	5.83 ± 1.09 A	9.45 ± 1.4 ABC	0.61 ± 0 A	4.43 ± 1.09 A	1.69 ± 0.06 AB	1.36 ± 0.92 A	1.41 ± 0.45 A
L3 *Ta.*	1.33 ± 0.53 B	9.33 ± 1.17 A	12.31 ± 0.66 AB	0.61 ± 0 A	4.39 ± 1.05 A	1.65 ± 0.02 B	0.41 ± 0.08 A	1.24 ± 0.37 A

Data are means of three replications ± SE, and at $p < 0.05$; cells with different letters are differ significantly.

3.2. Changes in Photosynthetic Pigments

Chlorophyll plays an important role in light harvesting, and its biosynthesis in chloroplasts is adversely affected when exposed to low-temperature stress [88]. As the altitude increased from 1500 to 2250 m a.s.l., a significant gradual reduction in the content of chlorophylls a and b was observed in the four species (*Co. orbicularis, Cr. sinaica, Ec. spinossissmus,* and *Ta. sinaicum*) in this study (Figure 4). The reduction in chlorophyll pigments may be due to the decreased biosynthesis and/or increased photo-oxidation of chlorophyll. This reduction in chlorophyll contents was concomitant with a decrease in the concentration of Mg^{+2} in the soil at the highest altitude locations (Table 2). Magnesium is an important constituent in the chlorophyll structure and a cofactor of various enzymes in the photosynthetic process [83]. Moreover, the reduction in chlorophyll at high altitudes could indirectly represent the oxidative stress level on plants, which is triggered by the production of ROS. This may inhibit chlorophyll synthesis, as it inhibits the enzymes responsible for chlorophyll biosynthesis or destroys the mesophyll chloroplasts, reducing the number of chloroplasts in the leaves and enhancing their degradation, which leads to a reduction in the photosynthetic rate [89,90]. On the other hand, this reduction was accompanied by a significant increase in the carotenoid content in all species investigated (Figure 4). A more pronounced value was recorded in *Cr. sinaica* at the highest altitude (2000–2250 m a.s.l.). At high altitudes, there are aggressive stressors, such as high light intensity and UV-B radiation [91], which cause the release of electrons from excited chlorophylls and their transfer photosystem I of the photosynthetic process to O_2 to form superoxide radicals which initiate a chain of free radical liberation [92,93]. Carotenoids act as potent antioxidants, as they serve as scavengers of singlet oxygen species, quench the triplet state of chlorophyll molecules [94], and protect the photosynthetic apparatus from damage caused in response to stress conditions [95,96].

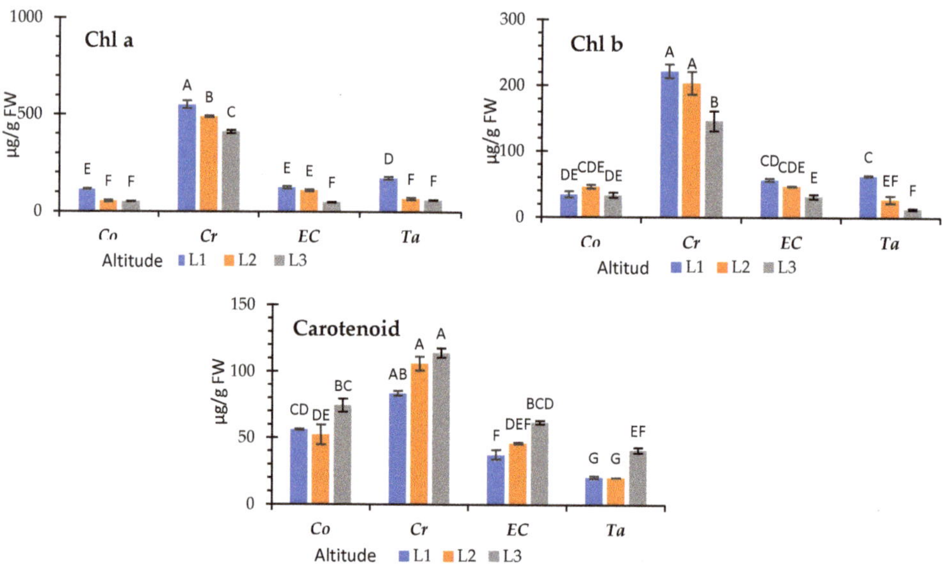

Figure 4. Photosynthetic pigment content (chlorophyll a, chlorophyll b, and carotenoids) in the leaves of four plant species from different altitudes. Each value is the mean of three replicates ± SE. At $p \leq 0.05$, bars with different letters are significantly different. Abbreviations: L1 = altitude 1; L2 = altitude 2; L3 = altitude 3; *Co.* = *Cotineaster orbicularis; Cr.* = *Crataegus sinaica; Ec.* = *Echinops spinossissmus; Ta.* = *Tanacetum sinaicum.*

3.3. *Variation in Total Soluble Sugar, Total Soluble Proteins, and Proline (Osmolytes)*

In natural environments, plants are challenged by adverse abiotic stressors, resulting in various physiological changes that help plants to cope with these stressors. The accumulation of different types of compatible solutes founds in plants is one of the main strategies for coping with various types of environmental stresses [97]. In fact, these solutes help in the maintenance of the integrity of the cellular membrane, protein stabilization, cellular osmotic balance, and ROS scavengers [98].

The data presented in Figure 5 show that with the increase in altitude, there was a significant increase in the content of total soluble sugar in the four investigated species, reaching the highest level of 27.59 µg glucose equivalent/g DW in *Cr. sinaica*, 26.48 µg glucose equivalent/g DW in *Ec. Spinossissmus*, and 25.75 µg glucose equivalent/g DW in *Ta. Sinaicum*, and the lowest level was recorded in *Co. orbicularis*, with 25.10 µg glucose equivalent/g DW. This increase in total soluble sugars may be due to the acceleration of photosynthetic performance. The photosynthetic process typically becomes accustomed to the environment and maintains homeostasis through various adaptations. [99]. The formation of some SMs helps plants cope with stressors, since they act as scavengers of free radicals and protect plant cells from oxidative damage [100]. Furthermore, the SMs pathway is supported by the derivation of photosynthetic products with a variety of intermediates, increasing the photosynthetic rate through a positive-feedback mechanism, so that the soluble sugars, which are the primary products of photosynthesis, increase. In accordance with these results, Hashim et al. [17] stated that endemic species growing naturally at high altitudes tend to increase their total soluble sugar content.

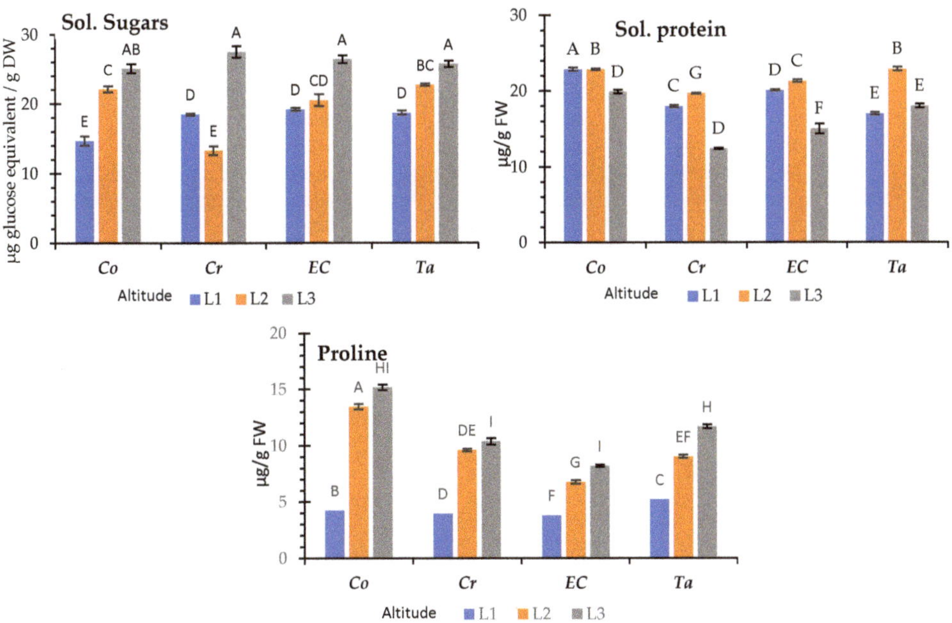

Figure 5. Total soluble sugar, total soluble proteins, and proline contents in the leaves of four plant species from different altitudes. Each value is the mean of three replicates ± SE. Bars with different letters are significantly different at $p \leq 0.05$. Abbreviations: L1 = altitude 1; L2 = altitude 2; L3 = altitude 3; *Co.* = *Cotineaster orbicularis; Cr.* = *Crataegus sinaica; Ec.* = *Echinops spinossissmus; Ta.* = *Tanacetum sinaicum.*

In the current study, the increase in altitude was accompanied by a noteworthy accumulation in proline content in the four species (Figure 5). Proline accumulation was found to be higher in *Co. orbicularis*, which reached 13.46 µg/g Fwt. In addition, the most important roles of proline included increasing the photochemical activities of PS II in thylakoid membranes, adjusting the acidity of cytosol, stabilizing the $NAD^+/NADH$ ratio, and diminishing the lipid peroxidation products, which increased stress tolerance in the investigated species [101]. Proline is a multifunctional amino acid acting as a signaling molecule. It regulates the osmotic pressure inside the cell, restricts the denaturation of proteins, maintains membrane consistency, stabilizes enzymes, protects the cell against stress, limits ROS-induced damage, and maintains nutrient balance through water transport [102,103]. In response to abiotic stress, the accumulation of proline in plants was observed, which serves as a metabolic signal and acts as an osmo-protectant [104].

Protein synthesis or degradation is one of the mechanisms affected by environmental stresses in plants. Additionally, soluble protein content is considered one of the main signs of the physiological and biochemical status of plants grown under stress conditions [105]. In this study, a remarkable increase in total soluble protein contents was observed in the four species studied (Figure 5) at altitude 2 (1750–2000). However, at altitude 3, the total soluble protein content generally decreased with increased stress conditions. The decline in protein content in response to abiotic stress is suggested to be due to diminishing protein synthesis and a switch to an accumulation of amino acids, especially proline, which is in accordance with our results.

3.4. Enzymatic Antioxidant System

As a consequence of the oxidative stress prevailing at the high altitudes, plants should evolve a strong antioxidant defense system to control the production of free radicals, regulate cellular homeostasis, and alleviate oxidants [106]. Therefore, many plants possess both enzymatic and non-enzymatic antioxidants that can attenuate the increased production of ROS [106]. Enzymatic antioxidants studied in the present work included SOD, POX, CAT, and ascorbate APX. Superoxide dismutase is regarded as the first line of defense in the contradiction of ROS by converting superoxide radicals to oxygen and hydrogen peroxide [107]. The present results reveal that plants that are grown at high altitudes generally show lower activity of SOD, thus indicating that a diminished capacity to detoxify the superoxide radical (Figure 6). However, evidence regarding CAT activity under different abiotic stresses shows different results. CAT is one of the antioxidant enzymes that catalyze H_2O_2 to oxygen and water molecules [108].

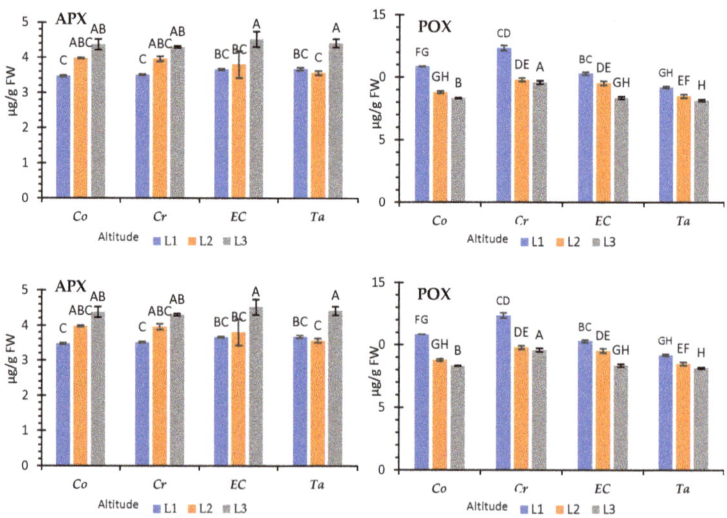

Figure 6. The activities of some antioxidant enzymes, including catalase (CAT), super oxide dismutase (SOD), ascorbate peroxidase (APX), and peroxidase (POX) in the leaves of the target species from different altitudes. Each value is the mean of three replicates ± SE. At $p \leq 0.05$, bars with different letters are significantly different. Abbreviations: L1 = altitude 1; L2 = altitude 2; L3 = altitude 3; *Co.* = *Cotineaster orbicularis*; *Cr.* = *Crataegus sinaica*; *Ec.* = *Echinops spinossissmus*; *Ta.* = *Tanacetum sinaicum*.

In our study, the CAT enzyme's activity was mostly upregulated in the four investigated species grown at differential altitudes, from low to high (Figure 6). Moreover, a higher CAT activity under oxidative stress incites the capability of these plants to scavenge and inhibit ROS overaccumulation. The most pronounced CAT activity was observed in *Co. orbicularis*, which reached 0.843 enzyme activity/g Fwt/min.

In this study, a noticeable increase in the activity of the APX enzyme (Figure 6) was correlated with a decrease in ascorbic acid content (Figure 7). The increase in APX activity was accompanied by a general reduction in POX activity. APX utilizes ascorbate as a specific electron donor to scavenge H_2O_2 in water, while POX scavenges H_2O_2 in the extra-cellular space [92]. The changes in SOD, CAT, APX, and POX activities and ROS accumulation in the investigated four species were not equal due to oxidative stress, indicating that these plants use various strategies in response to stresses. Consequently, the influence of abiotic stresses on antioxidant enzyme activity depends on the type of plant and the degree of stress experienced.

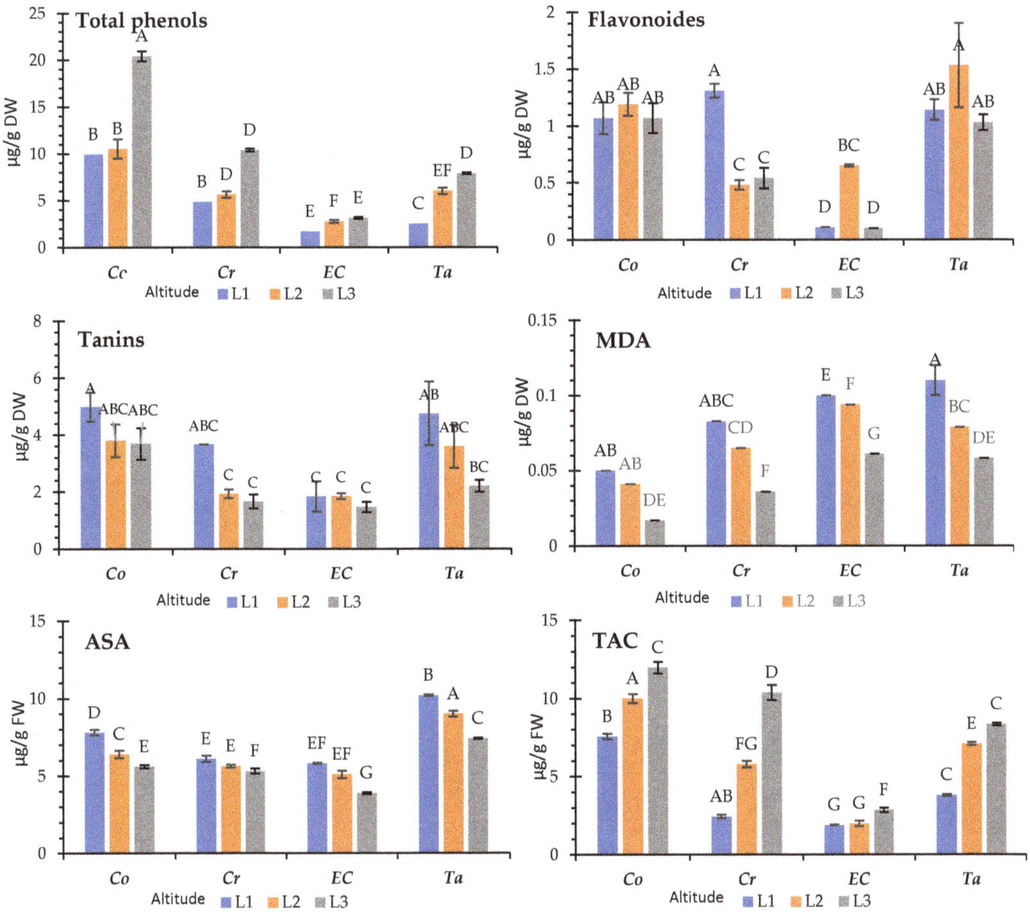

Figure 7. Total phenols, flavonoids, tannins, total antioxidant capacity (TAC), ascorbic acid (ASA), and malondialdehyde (MDA) contents in the leaves of the target species from different altitudes. Each value is the mean of three replicates ± SE. At $p \leq 0.05$, bars with different letters are significantly different. Abbreviations: L1 = altitude 1; L2 = altitude 2; L3 = altitude 3; *Co.* = *Cotineaster orbicularis; Cr.* = *Crataegus sinaica; Ec.* = *Echinops spinossissmus; Ta.* = *Tanacetum sinaicum.*

3.5. Non-Enzymatic Antioxidant Systems

Regarding the abiotic stresses at high altitudes, there is a notable qualitative and quantitative increase in many SMs, including phenolic compounds, flavonoids, tannins, carotenoids, steroids, alkaloids, and terpenoids [8]. In the current study, the total phenols in the four target species increased significantly with increasing altitude (Figure 7), while flavonoids did not show a clear trend, and tannins decreased with increasing altitude. The increase in the total phenols with the increase in altitude (Figure 7) was concomitant with a general reduction in the activity level of the POX enzyme (Figure 6). Phenolic compounds are considered a highly important class of antioxidants, and they are mostly associated with antioxidant activity in plants, as they inhibit lipid peroxidation by scavenging free radicals. Thus, phenolic constituents promote abiotic oxidative stress adaption [109,110]. Moreover, tannins can bind to membranes, forming a tannin–phospholipid complex that may help maintain membrane morphology and permeability [111].

Malondialdehyde (MDA) is the main cytotoxic product of lipid peroxidation, and its abundance signifies the degree of lipid oxidation by oxidants such as free radicals [112]. The present results showed that the level of MDA decreased parallel to a rise in altitude, which indicated that low temperatures reduced the levels of MDA in the four investigated species (Figure 7). The results showed that higher total phenols content was recorded in the two plants of the family Rosaceae, which reached 20.04 mg/g Fwt in *Co. orbicularis* in accordance with more pronounced levels of antioxidant capacity (11.97 mg/g Fwt), based on FRAP assays at the highest altitude in the investigated area (Figure 7). This was followed by *Cr. Sinaica,* which recorded 10.38 mg/g Fwt, while in this study, plants belonging to the family Asteraceae recorded lower values of phenolic compounds and antioxidant capacity, with the lowest value being observed in *Ec. spinossissmus* (Figure 7). At this point, we can say that the plants of the family Rosaceae were more adapted to grow at high altitudes than plants belonging to the family Asteraceae, especially *Co. orbicularis,* which recorded the highest values of CAT and SOD enzyme activity levels; the highest content of total phenols, flavonoids, tannins, and total antioxidant capacity; and the lowest value of MDA, and this was reflected in the higher frequency of *Co. orbicularis* species than other four plants at the highest altitude.

3.6. Principal Component Analysis (PCA)

Applying principal component analysis (PCA) to the present data showed that there are three groups of physiological parameters in relation to the distribution of the four investigated species at three different altitudes. First, photosynthetic pigments (Chl. a, Chl. b, and carotenoids) were positively correlated with *Cr. sainaica* at all altitudes, where the highest altitude (L3) showed a lower correlation. The second group, represented by total soluble sugars, CAT, and APX, was correlated with *Co. orbicularis* and *Ta. sainaicum* at the highest altitude (L3), and with *Ec. spinossissmus* at all altitudes. The third group of physiological parameters was ASA, flavonoids, proline, tannins, and MDA, which correlated positively with *Co. orbicularis* and *Ta. sainaicum* at the second (L2) and third (L3) altitudes (Figure 8).

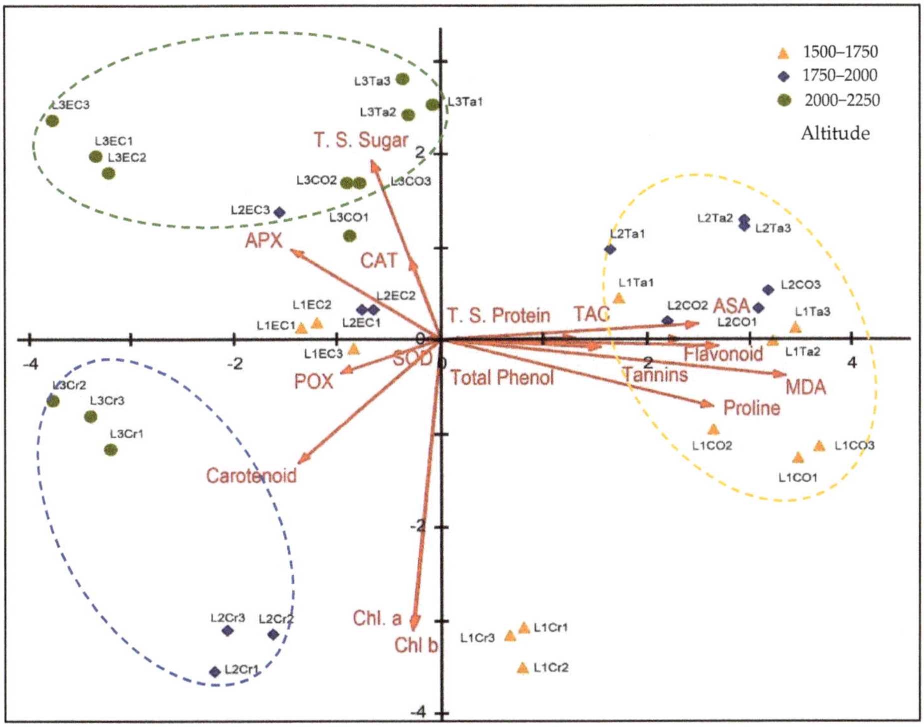

Figure 8. PCA joint plot showing target species at different altitudes in relation to physiological parameters. Abbreviations: L1 = altitude 1; L2 = altitude 2; L3 = altitude 3; *Co.* = *Cotineaster orbicularis*; *Cr.* = *Crataegus sinaica*; *Ec.* = *Echinops spinossissmus*; *Ta.* = *Tanacetum sinaicum*.

4. Conclusions

Plants may respond to various abiotic stresses by escalating SMs or antioxidant capacity. The present investigation concluded that many significant variations were noticed in the antioxidant activity and chemical composition of the four targeted species collected at three different altitudes. Our results demonstrated that *Co. orbicularis* relies on accumulating SMs and increasing the activity of its antioxidant enzymes to adjust to high altitudes, while as a survival strategy for high altitudes, the other investigated species typically rely on antioxidant enzyme activity or the accumulation of some SMs, either in combination or alternatively. This shows that *Co. orbicularis* is more adaptable than the other species to the oxidative stresses caused by the altitudinal effect. It is highly recommended to further investigate *Co. orbicularis's* high adaptation capabilities to different environmental factors. Moreover, our findings are highly relevant for the future cultivation of high-grade medical herbs, given that many medicinal plants are used due to their bioactive ingredients. However, most medicinal plants have more than one class of bioactive secondary metabolites. Thus, investigating how these other groups of molecules change at various altitudes will also be of interest.

Author Contributions: Conceptualization, A.M.H., A.A.J. and M.E.E.; methodology, A.M.H., A.M.A. and A.A.J.; formal analysis, A.M.H. and M.E.E.; investigation, A.M.H. and A.M.A.; resources, I.A.I., A.A.J., Y.A.A. and A.M.H.; data curation, A.M.H. and I.A.I.; writing—original draft preparation, A.M.H., M.E.E., A.A.J. and A.M.A.; writing—review and editing, A.M.H., A.M.A., A.A.J. and I.A.I. All authors have read and agreed to the published version of the manuscript.

Funding: This research received no external funding.

Institutional Review Board Statement: Not applicable.

Informed Consent Statement: Not applicable.

Acknowledgments: The authors would like to thank the Deanship of Scientific Research, Umm Al-Qura University, for supporting this work (grant code: 22UQU4331277DSR05).

Conflicts of Interest: The authors declare no conflict of interest.

References

1. Devi, N.; Hagedorn, F.; Moiseev, P.; Bugmann, H.; Shiyatov, S.; Mazepa, V.; Rigling, A. Expanding Forests and Changing Growth Forms of Siberian Larch at the Polar Urals Treeline during the 20th Century: TREELINE CHANGES IN THE POLAR URALS. *Glob. Change Biol.* **2008**, *14*, 1581–1591. [CrossRef]
2. Okunlola, G.O.; Olatunji, O.A.; Akinwale, R.O.; Tariq, A.; Adelusi, A.A. Physiological Response of the Three Most Cultivated Pepper Species (*Capsicum* spp.) in Africa to Drought Stress Imposed at Three Stages of Growth and Development. *Sci. Hortic.* **2017**, *224*, 198–205. [CrossRef]
3. Qian, C.; Yin, H.; Shi, Y.; Zhao, J.; Yin, C.; Luo, W.; Dong, Z.; Chen, G.; Yan, X.; Wang, X.-R.; et al. Population Dynamics of Agriophyllum Squarrosum, a Pioneer Annual Plant Endemic to Mobile Sand Dunes, in Response to Global Climate Change. *Sci. Rep.* **2016**, *6*, 26613. [CrossRef]
4. Jakada, B.H.; Aslam, M.; Fakher, B.; Greaves, J.G.; Li, Z.; Li, W.; Lai, L.; Ayoade, O.A.; Cheng, Y.; Cao, S.; et al. Identification of SWI2/SNF2-Related 1 Chromatin Remodeling Complex (SWR1-C) Subunits in Pineapple and the Role of Pineapple SWR1 COMPLEX 6 (AcSWC6) in Biotic and Abiotic Stress Response. *Biomolecules* **2019**, *9*, 364. [CrossRef]
5. Li, X.; Guo, C.; Ahmad, S.; Wang, Q.; Yu, J.; Liu, C.; Guo, Y. Systematic Analysis of MYB Family Genes in Potato and Their Multiple Roles in Development and Stress Responses. *Biomolecules* **2019**, *9*, 317. [CrossRef]
6. ErYuan, L.; YaFeng, W.; Yan, X.; Bo, L.; XueMei, S. Growth Variation in Abies Georgei Var. Smithii along Altitudinal Gradients in the Sygera Mountains, Southeastern Tibetan Plateau. *Trees Struct. Funct.* **2010**, *24*, 363–373.
7. Najafabadi, M.; Ehsanzadeh, P. Photosynthetic and Antioxidative Upregulation in Drought-Stressed Sesame (Sesamum Indicum L.) Subjected to Foliar-Applied Salicylic Acid. *Photosynthetica* **2017**, *55*, 611–622. [CrossRef]
8. Yang, L.; Wen, K.-S.; Ruan, X.; Zhao, Y.-X.; Wei, F.; Wang, Q. Response of Plant Secondary Metabolites to Environmental Factors. *Molecules* **2018**, *23*, 762. [CrossRef]
9. Boulos, L. *Flora of Egypt Checklist, Revised Annotated Edition*; Al-Hadara Publishing: Cairo, Egypt, 2009; pp. 198–201.
10. El-Wahab, R.A.; Zaghloul, M.S.; Moustafa, A.A. Conservation of Medicinal Plants in St. Catherine Protectorate, South Sinai, Egypt. In Proceedings of the First International Conference on Strategy of Egyptian Herbaria, Giza, Egypt, 9–11 March 2004; pp. 231–251.
11. Rabei, S.; Khedr, A.; Elgamal, I. Floristic Study of Saint Katherine Protectorate, Sinai: With One New Record to Flora of Egypt. *Taeckholmia* **2021**, *41*, 32. [CrossRef]
12. Omar, K.; Elgamal, I. Assess the Extinction Risk of Mountain Endemic Plants in Egypt Under the Current Climatic Condition: A Case Study of Endemic Silene Species. *Eur. J. Biol. Biotechnol.* **2021**, *2*, 34–47. [CrossRef]
13. Fakhry, A.M.; El-Keblawy, A.; Shabana, H.A.; Gamal, I.E.; Shalouf, A. Microhabitats Affect Population Size and Plant Vigor of Three Critically Endangered Endemic Plants in Southern Sinai Mountains, Egypt. *Land* **2019**, *8*, 86. [CrossRef]
14. Shaltout, K.H.; Ahmed, D.A.; Shabana, H.A. Population Structure and Dynamics of the Endemic Species Phlomis Aurea Decne in Different Habitats in Southern Sinai Peninsula, Egypt. Glob. *Ecol. Conserv.* **2015**, *4*, 505–515. [CrossRef]
15. El-Keblawy, A.A.; Khedr, A.-H.A. Population Structure and Ecological Role of Moringa Peregrina (Forssk.) Fiori. at Its Northwestern Range Edge in the Hajar Mountains. *Plant Biosyst. Int. J. Deal. Asp. Plant Biol.* **2015**, *151*, 29–38. [CrossRef]
16. Elkelish, A.A.; Alhaithloul, H.A.S.; Qari, S.H.; Soliman, M.H.; Hasanuzzaman, M. Pretreatment with Trichoderma Harzianum Alleviates Waterlogging-Induced Growth Alterations in Tomato Seedlings by Modulating Physiological, Biochemical, and Molecular Mechanisms. *Environ. Exp. Bot.* **2020**, *171*, 103946. [CrossRef]
17. Hashim, A.M.; Alharbi, B.M.; Abdulmajeed, A.M.; Elkelish, A.; Hozzein, W.N.; Hassan, H.M. Oxidative Stress Responses of Some Endemic Plants to High Altitudes by Intensifying Antioxidants and Secondary Metabolites Content. *Plants* **2020**, *9*, 869. [CrossRef]
18. Wonsick, M.M.; Pinker, R.T. The Radiative Environment of the Tibetan Plateau. *Int. J. Climatol.* **2014**, *34*, 2153–2162. [CrossRef]
19. Harborne, J.B. Classes and Functions of Secondary Products from Plants. In *Chemicals from Plants*; Institute of Food Research: Norwich, UK, 1999; pp. 1–25. ISBN 978-981-02-2773-9.
20. Nawrot-Chorabik, K.; Sułkowska, M.; Gumulak, N. Secondary Metabolites Produced by Trees and Fungi: Achievements So Far and Challenges Remaining. *Forests* **2022**, *13*, 1338. [CrossRef]
21. Julkunen-Tiitto, R.; Rousi, M.; Bryant, J.; Sorsa, S.; Keinänen, M.; Sikanen, H. Chemical Diversity of Several Betulaceae Species: Comparison of Phenolics and Terpenoids in Northern Birch Stems. *Trees* **1996**, *11*, 16. [CrossRef]
22. Gong, J.; Zhang, Z.; Zhang, C.; Zhang, J.; Ran, A. Ecophysiological Responses of Three Tree Species to a High-Altitude Environment in the Southeastern Tibetan Plateau. *Forests* **2018**, *9*, 48. [CrossRef]

23. Gea-Izquierdo, G.; Fonti, P.; Cherubini, P.; Martin-Benito, D.; Chaar, H.; Canellas, I. Xylem Hydraulic Adjustment and Growth Response of Quercus Canariensis Willd. to Climatic Variability. *Tree Physiol.* **2012**, *32*, 401–413. [CrossRef]
24. Wang, Y.; Gao, S.; He, X.; Li, Y.; Li, P.; Zhang, Y.; Chen, W. Growth, Secondary Metabolites and Enzyme Activity Responses of Two Edible Fern Species to Drought Stress and Rehydration in Northeast China. *Agronomy* **2019**, *9*, 137. [CrossRef]
25. Soliman, M.H.; Alayafi, A.A.M.; El Kelish, A.A.; Abu-Elsaoud, A.M. Acetylsalicylic Acid Enhance Tolerance of Phaseolus Vulgaris L. to Chilling Stress, Improving Photosynthesis, Antioxidants and Expression of Cold Stress Responsive Genes. *Bot. Stud.* **2018**, *59*, 6. [CrossRef] [PubMed]
26. Saleem, M.H.; Ali, S.; Rehman, M.; Rana, M.S.; Rizwan, M.; Kamran, M.; Imran, M.; Riaz, M.; Soliman, M.H.; Elkelish, A.; et al. Influence of Phosphorus on Copper Phytoextraction via Modulating Cellular Organelles in Two Jute (*Corchorus Capsularis* L.) Varieties Grown in a Copper Mining Soil of Hubei Province, China. *Chemosphere* **2020**, *248*, 126032. [CrossRef]
27. Elkelish, A.; Qari, S.H.; Mazrou, Y.S.A.; Abdelaal, K.A.A.; Hafez, Y.M.; Abu-Elsaoud, A.M.; Batiha, G.E.-S.; El-Esawi, M.A.; El Nahhas, N. Exogenous Ascorbic Acid Induced Chilling Tolerance in Tomato Plants Through Modulating Metabolism, Osmolytes, Antioxidants, and Transcriptional Regulation of Catalase and Heat Shock Proteins. *Plants* **2020**, *9*, 431. [CrossRef]
28. Racchi, M. Antioxidant Defenses in Plants with Attention to *Prunus* and *Citrus* spp. *Antioxidants* **2013**, *2*, 340–369. [CrossRef]
29. Keleş, Y.; Everest, A. Relation to Altitude Adaptation and Antioxidant Defence System in Five Shrubs and Trees Species from Middle Taurus Mountains. *Int. J. Nat. Eng. Sci.* **2008**, *2*, 45–49.
30. Catola, S.; Marino, G.; Emiliani, G.; Huseynova, T.; Musayev, M.; Akparov, Z.; Maserti, B.E. Physiological and Metabolomic Analysis of *Punica Granatum* (L.) under Drought Stress. *Planta* **2016**, *243*, 441–449. [CrossRef] [PubMed]
31. Selmar, D.; Kleinwächter, M. Influencing the Product Quality by Deliberately Applying Drought Stress during the Cultivation of Medicinal Plants. *Ind. Crops Prod.* **2013**, *42*, 558–566. [CrossRef]
32. Cheng, L.; Han, M.; Yang, L.; Li, Y.; Sun, Z.; Zhang, T. Changes in the Physiological Characteristics and Baicalin Biosynthesis Metabolism of Scutellaria Baicalensis Georgi under Drought Stress. *Ind. Crops Prod.* **2018**, *122*, 473–482. [CrossRef]
33. Das, K.; Roychoudhury, A. Reactive Oxygen Species (ROS) and Response of Antioxidants as ROS-Scavengers during Environmental Stress in Plants. *Front. Environ. Sci.* **2014**, *2*, 53. [CrossRef]
34. Bozin, B.; Mimica-Dukic, N.; Samojlik, I.; Goran, A.; Igic, R. Phenolics as Antioxidants in Garlic (*Allium Sativum* L., Alliaceae). *Food Chem.* **2008**, *111*, 925–929. [CrossRef]
35. Moussa, Z.; Judeh, Z.M.A.; Ahmed, S.A.; Moussa, Z.; Judeh, Z.M.A.; Ahmed, S.A. *Nonenzymatic Exogenous and Endogenous Antioxidants*; IntechOpen: London, UK, 2019; ISBN 978-1-78985-144-1.
36. Rahimi, Y.; Taleei, A.; Ranjbar, M. Long-Term Water Deficit Modulates Antioxidant Capacity of Peppermint (*Mentha Piperita* L.). *Sci. Hortic.* **2018**, *237*, 36–43. [CrossRef]
37. Khan, M.N.; Zhang, J.; Luo, T.; Liu, J.; Ni, F.; Rizwan, M.; Fahad, S.; Hu, L. Morpho-Physiological and Biochemical Responses of Tolerant and Sensitive Rapeseed Cultivars to Drought Stress during Early Seedling Growth Stage. *Acta Physiol. Plant.* **2019**, *41*, 1–13. [CrossRef]
38. Oral, O.; Kutlu, T.; Aksoy, E.; Fıçıcıoğlu, C.; Uslu, H.; Tuğrul, S. The Effects of Oxidative Stress on Outcomes of Assisted Reproductive Techniques. *J. Assist. Reprod. Genet.* **2006**, *23*, 81–85. [CrossRef]
39. Courrau, J. Evaluating Effectivenes: A Framework for Assessing Management Effectiveness of Protected Areas. Available online: https://www.iucn.org/resources/publication/evaluating-effectiveness-framework-assessing-management-effectiveness (accessed on 4 November 2022).
40. Moustafa, A.E.-R.A.; Zaghloul, M.S. Environment and Vegetation in the Montane Saint Catherine Area, South Sinai, Egypt. *J. Arid Environ.* **1996**, *34*, 331–349. [CrossRef]
41. Moustafa, A.G. Long Term Monitoring of Rosa Arabica Populations as a Threatened Species in South Sinai, Egypt. Available online: https://www.semanticscholar.org/paper/Long-Term-Monitoring-of-Rosa-arabica-Populations-as-Moustafa/a9209f600f5eef6928a0df021c7866cea10018f1 (accessed on 2 April 2020).
42. Moustafa, A.R.A.; Klopatek, J.M. Vegetation and Landforms of the Saint Catherine Area, Southern Sinai, Egypt. *J. Arid Environ.* **1995**, *30*, 385–395. [CrossRef]
43. Danin, A. Desert Rocks—A Habitat Which Supports Many Species That Were New to Science in the Last 40 Years. *Turk. J. Bot.* **2008**, *32*, 459–464.
44. Grainger, J. 'People Are Living in the Park'. Linking Biodiversity Conservation to Community Development in the Middle East Region: A Case Study from the Saint Katherine Protectorate, Southern Sinai. *J. Arid Environ.* **2003**, *54*, 29–38. [CrossRef]
45. Täckholm, V. *Students' Flora of Egypt*, 2nd ed.; Cairo University: Beirut, Lebanon, 1974.
46. Zohary, M. *Flora Palaestina*; Israel Academy of Science and Humanities: Jerusalem, Israel, 1966.
47. POWO Plants of the World Online. *Kew Science*. Available online: https://powo.science.kew.org/ (accessed on 1 November 2022).
48. Boulos, L. *Medicinal Plants of North Africa*; Reference Publications, Inc.: Algonac, MI, USA, 1983.
49. El-Hadidi, M.N.; Batanouny, K.H.; Fahmy, A.G. *The Egyptian Plant Red Data Book*; University of Cairo: Giza, Egypt, 1991.
50. El-Mousallamy, A.M.; Hussein, S.A.; Merfort, I.; Nawwar, M.A. Unusual Phenolic Glycosides from Cotoneaster Orbicularis. *Phytochemistry* **2000**, *53*, 699–704. [CrossRef]
51. Moustafa, A.; Zaghloul, M.; Mansour, S.; Alotaibi, M. Conservation Strategy for Protecting Crataegus x Sinaica against Climate Change and Anthropologic Activities in South Sinai Mountains, Egypt. *Catrina Int. J. Environ. Sci.* **2019**, *18*, 1–6. [CrossRef]

52. Shaltout, K.H.; Eid, E.M.; Al-Sodany, Y.M.; Heneidy, S.Z.; Shaltout, S.K.; El-Masry, S.A. Effect of Protection of Mountainous Vegetation against Over-Grazing and Over-Cutting in South Sinai, Egypt. *Diversity* **2021**, *13*, 113. [CrossRef]
53. Refaat, A.T.; Shahat, A.A.; Ehsan, N.A.; Yassin, N.; Hammouda, F.; Tabl, E.A.; Ismail, S.I. Phytochemical and Biological Activities of Crataegus Sinaica Growing in Egypt. *Asian Pac. J. Trop. Med.* **2010**, *3*, 257–261. [CrossRef]
54. Murch, S.J.; Wierenga, E.J.; El-Demerdash, M.A.; Saxena, P.K. In Vitro Propagation of the Egyptian Medicinal Plant, Echinops Spinosissimus Turra. *Plant Cell Tissue Organ Cult.* **2003**, *74*, 81–86. [CrossRef]
55. Dawidar, A.M.; Metwally, M.A.; Abou-Elzahab, M.; Abdel-Mogib, M. Sesquiterpene Lactones from Echinops Spinosissimus. *Pharmazie* **1990**, *45*, 70–71.
56. Hegazy, M.-E.F.; Hamed, A.R.; Mohamed, T.A.; Debbab, A.; Nakamura, S.; Matsuda, H.; Paré, P.W. Anti-Inflammatory Sesquiterpenes from the Medicinal Herb Tanacetum Sinaicum. *RSC Adv.* **2015**, *5*, 44895–44901. [CrossRef]
57. El-Dahmy, S.I.; Abdel-Ghani, A.E.; Attia, R.A.; Abouelenein, D.D. Chemical Composition and Biological Activities of the Essential Oil of Tanacetum Sinaicum Del. *Grown Egypt. J. Pharmacogn. Phytochem.* **2019**, *8*, 326–331.
58. Estefan, G. *Methods of Soil, Plant, and Water Analysis: A Manual for the West Asia and North Africa Region*, 3rd ed.; International Center for Agricultural Research in the Dry Areas (ICARDA): Beirut, Lebanon, 2017.
59. Dewis, J.; Nations, F.; Freitas, F. Physical and Chemical Methods of Soil and Water Analysis; Food and Agriculture Organization of the United Nations. *FAO Soils Bull.* **1970**, *10*, 275.
60. Johnson, C.M.; Ulrich, A. *Analytical Methods for Use in Plant Analysis*; University of California: Los Angeles, CA, USA, 1959.
61. Shapiro, L.; Brannock, W.W. *A Field Method for the Determination of Calcium and Magnesium in Limestone and Dolomite*; Open-File Report; U.S. Geological Survey: Reston, VA, USA, 1957; Volume 57–99, p. 9. [CrossRef]
62. Blume, H.-P. Page, A. L., R. H. Miller and D. R. Keeney (Ed., 1982): Methods of soil analysis; 2. Chemical and microbiological properties, 2. Aufl. 1184 S., American Soc. of Agronomy (Publ.), Madison, Wisconsin, USA, gebunden 36 Dollar. *Bodenkd* **1985**, *148*, 363–364. [CrossRef]
63. Black, C.A. *American Society of Agronomy. American Society for Testing and Materials Methods of Soil Analysis Part 1, Part 1*; American Society of Agronomy: Madison, WI, USA, 1965.
64. Piper, C.S. Soil and Plant Analysis. *Soil Sci.* **1945**, *59*, 263. [CrossRef]
65. Metzner, H.; Rau, H.; Senger, H. Untersuchungen zur Synchronisierbarkeit einzelner Pigmentmangel-Mutanten von Chlorella. *Planta* **1965**, *65*, 186–194. [CrossRef]
66. Prud'homme, M.-P.; Gonzalez, B.; Billard, J.-P.; Boucaud, J. Carbohydrate Content, Fructan and Sucrose Enzyme Activities in Roots, Stubble and Leaves of Ryegrass (*Lolium Perenne* L.) as Affected by Source/Sink Modification after Cutting. *J. Plant Physiol.* **1992**, *140*, 282–291. [CrossRef]
67. Loewus, F.A. Improvement in Anthrone Method for Determination of Carbohydrates. *Anal. Chem.* **1952**, *24*, 219. [CrossRef]
68. Malik, C.P.; Singh, M.B. *Plant Enzymology and Histo-Enzymology: A Text Manual*; Kalyani Publishers: Delhi, India, 1980.
69. Harborne, A.J. *Phytochemical Methods A Guide to Modern Techniques of Plant Analysis*, 3rd ed.; Springer: Dordrecht, The Netherlands, 1998; ISBN 978-0-412-57260-9.
70. Ejikeme, C.M.; Ezeonu, C.S.; Eboatu, A.N. Determination of Physical and Phytochemical Constituents of Some Tropical Timbers Indigenous to Niger Delta Area of Nigeria. *Eur. Sci. J. ESJ* **2014**, *10*, 247–270.
71. Bradford, M.M. A Rapid and Sensitive Method for the Quantitation of Microgram Quantities of Protein Utilizing the Principle of Protein-Dye Binding. *Anal. Biochem.* **1976**, *72*, 248–254. [CrossRef] [PubMed]
72. Bates, L.S.; Waldren, R.P.; Teare, I.D. Rapid Determination of Free Proline for Water-Stress Studies. *Plant Soil* **1973**, *39*, 205–207. [CrossRef]
73. Heath, R.L.; Packer, L. Photoperoxidation in Isolated Chloroplasts: I. Kinetics and Stoichiometry of Fatty Acid Peroxidation. *Arch. Biochem. Biophys.* **1968**, *125*, 189–198. [CrossRef] [PubMed]
74. Mukherjee, S.P.; Choudhuri, M.A. Implications of Water Stress-Induced Changes in the Levels of Endogenous Ascorbic Acid and Hydrogen Peroxide in Vigna Seedlings. *Physiol. Plant.* **1983**, *58*, 166–170. [CrossRef]
75. Dhindsa, R.S.; Plumb-Dhindsa, P.; Thorpe, T.A. Leaf Senescence: Correlated with Increased Levels of Membrane Permeability and Lipid Peroxidation, and Decreased Levels of Superoxide Dismutase and Catalase. *J. Exp. Bot.* **1981**, *32*, 93–101. [CrossRef]
76. Hermans, C.; Conn, S.J.; Chen, J.; Xiao, Q.; Verbruggen, N. An Update on Magnesium Homeostasis Mechanisms in Plants. *Metallomics* **2013**, *5*, 1170–1183. [CrossRef]
77. Kar, M.; Mishra, D. Catalase, Peroxidase, and Polyphenoloxidase Activities during Rice Leaf Senescence. *Plant Physiol.* **1976**, *57*, 315–319. [CrossRef]
78. Koricheva, J.; Roy, S.; Vranjic, J.A.; Haukioja, E.; Hughes, P.R.; Hänninen, O. Antioxidant Responses to Simulated Acid Rain and Heavy Metal Deposition in Birch Seedlings. Environ. Pollut. *Barking Essex* **1997**, *95*, 249–258. [CrossRef]
79. Kampfenkel, K.; Montagu, M.V.; Inze, D. Effects of Iron Excess on Nicotiana Plumbaginifolia Plants (Implications to Oxidative Stress). *Plant Physiol.* **1995**, *107*, 725–735. [CrossRef] [PubMed]
80. Prieto, P.; Pineda, M.; Aguilar, M. Spectrophotometric Quantitation of Antioxidant Capacity through the Formation of a Phosphomolybdenum Complex: Specific Application to the Determination of Vitamin E. *Anal. Biochem.* **1999**, *269*, 337–341. [CrossRef] [PubMed]
81. Abd El-Wahab, R.; Zayed, A.E.-M.; Moustafa, A.E.-R.; Klopatek, J.; Helmy, M. Landforms, Vegetation, and Soil Quality in South Sinai, Egypt. *Catrina Int. J. Environ. Sci.* **2018**, *1*, 127–138.

82. Sharma, P.; Rana, J.C.; Devi, U.; Randhawa, S.S.; Kumar, R. Floristic Diversity and Distribution Pattern of Plant Communities along Altitudinal Gradient in Sangla Valley, Northwest Himalaya. *Sci. World J.* **2014**, *2014*, e264878. [CrossRef]
83. Guo, W.; Nazim, H.; Liang, Z.; Yang, D. Magnesium Deficiency in Plants: An Urgent Problem. *Crop J.* **2016**, *4*, 83–91. [CrossRef]
84. Salama, F.; Abd El-Ghani, M.; Gadalla, M.; Ramadan, T.; Galal, H.; Gaafar, A. Vegetation Patterns and Floristic Composition along Elevation Gradient on Jabal Musa, South Sinai, Egypt. *Catrina Int. J. Environ. Sci.* **2018**, *17*, 41–57. [CrossRef]
85. Mirzakhaninafchi, H.; Mani, I.; Hasan, M.; Nafchi, A.M.; Parray, R.A.; Kumar, D. Development of Prediction Models for Soil Nitrogen Management Based on Electrical Conductivity and Moisture Content. *Sensors* **2022**, *22*, 6728. [CrossRef]
86. Visconti, F.; de Paz, J.M. Electrical Conductivity Measurements in Agriculture: The Assessment of Soil Salinity. In *New Trends and Developments in Metrology*; Cocco, L., Ed.; IntechOpen: London, UK, 2016; ISBN 978-953-51-2477-1.
87. Rhoades, J.D.; Chanduvi, F.; Lesch, S. *Soil Salinity Assessment: Methods and Interpretation of Electrical Conductivity Measurements*; Food & Agriculture Org: Rome, Italy, 1999.
88. Zhao, Y.; Han, Q.; Ding, C.; Huang, Y.; Liao, J.; Chen, T.; Feng, S.; Zhou, L.; Zhang, Z.; Chen, Y.; et al. Effect of Low Temperature on Chlorophyll Biosynthesis and Chloroplast Biogenesis of Rice Seedlings during Greening. *Int. J. Mol. Sci.* **2020**, *21*, 1390. [CrossRef]
89. Cui, G.; Li, B.; He, W.; Yin, X.; Liu, S.; Lian, L.; Zhang, Y.; Liang, W.; Zhang, P. Physiological Analysis of the Effect of Altitudinal Gradients on Leymus Secalinus on the Qinghai-Tibetan Plateau. *PLoS ONE* **2018**, *13*, e0202881. [CrossRef]
90. dos Santos, T.B.; Ribas, A.F.; de Souza, S.G.H.; Budzinski, I.G.F.; Domingues, D.S. Physiological Responses to Drought, Salinity, and Heat Stress in Plants: A Review. *Stresses* **2022**, *2*, 113–135. [CrossRef]
91. Abd El-Ghani, M.M.; Huerta-Martínez, F.M.; Hongyan, L.; Qureshi, R. Erratum to: Plant Responses to Hyperarid Desert Environments. In *Plant Responses to Hyperarid Desert Environments*; Abd El-Ghani, M.M., Huerta-Martínez, F.M., Hongyan, L., Qureshi, R., Eds.; Springer International Publishing: Cham, Germany, 2017; p. e1. ISBN 978-3-319-59135-3.
92. Hasanuzzaman, M.; Bhuyan, M.H.M.B.; Zulfiqar, F.; Raza, A.; Mohsin, S.M.; Mahmud, J.A.; Fujita, M.; Fotopoulos, V. Reactive Oxygen Species and Antioxidant Defense in Plants under Abiotic Stress: Revisiting the Crucial Role of a Universal Defense Regulator. *Antioxidants* **2020**, *9*, 681. [CrossRef] [PubMed]
93. Caverzan, A.; Casassola, A.; Patussi Brammer, S. Reactive Oxygen Species and Antioxidant Enzymes Involved in Plant Tolerance to Stress. In *Abiotic and Biotic Stress in Plants-Recent Advances and Future Perspectives*; Shanker, A.K., Shanker, C., Eds.; IntechOpen: London, UK, 2016; ISBN 978-953-51-2250-0.
94. Havaux, M. Carotenoid Oxidation Products as Stress Signals in Plants. *Plant J. Cell Mol. Biol.* **2014**, *79*, 597–606. [CrossRef] [PubMed]
95. Uarrota, V.G.; Stefen, D.L.V.; Leolato, L.S.; Gindri, D.M.; Nerling, D. Revisiting Carotenoids and Their Role in Plant Stress Responses: From Biosynthesis to Plant Signaling Mechanisms during Stress. In *Antioxidants and Antioxidant Enzymes in Higher Plants*; Springer International Publishing: Cham, Germany, 2018; pp. 207–232. ISBN 978-3-319-75087-3.
96. Sharma, A.; Kumar, V.; Yuan, H.; Kanwar, M.K.; Bhardwaj, R.; Thukral, A.K.; Zheng, B. Jasmonic Acid Seed Treatment Stimulates Insecticide Detoxification in Brassica Juncea L. *Front. Plant Sci.* **2018**, *9*, 1609. [CrossRef]
97. Chen, H.; Jiang, J.-G. Osmotic Adjustment and Plant Adaptation to Environmental Changes Related to Drought and Salinity. *Environ. Rev.* **2010**, *18*, 309–319. [CrossRef]
98. Kurutas, E.B. The Importance of Antioxidants Which Play the Role in Cellular Response against Oxidative/Nitrosative Stress: Current State. *Nutr. J.* **2015**, *15*, 71. [CrossRef]
99. Lambers, H.; Oliveira, R.S. *Plant Physiological Ecology*; Springer International Publishing: Cham, Germany, 2019; ISBN 978-3-030-29638-4.
100. Guidi, L.; Tattini, M.; Landi, M. How Does Chloroplast Protect Chlorophyll Against Excessive Light? In *Chlorophyll*; Jacob-Lopes, E., Zepka, L.Q., Queiroz, M.I., Eds.; IntechOpen: London, UK, 2017; ISBN 978-953-51-3107-6.
101. Kavi Kishor, P.B.; Suravajhala, P.; Rathnagiri, P.; Sreenivasulu, N. Intriguing Role of Proline in Redox Potential Conferring High Temperature Stress Tolerance. *Front. Plant Sci.* **2022**, *13*, 867531. [CrossRef]
102. Hosseinifard, M.; Stefaniak, S.; Ghorbani Javid, M.; Soltani, E.; Wojtyla, Ł.; Garnczarska, M. Contribution of Exogenous Proline to Abiotic Stresses Tolerance in Plants: A Review. *Int. J. Mol. Sci.* **2022**, *23*, 5186. [CrossRef]
103. Szabados, L.; Savouré, A. Proline: A Multifunctional Amino Acid. *Trends Plant Sci.* **2010**, *15*, 89–97. [CrossRef]
104. Fu, Y.; Ma, H.; Chen, S.; Gu, T.; Gong, J. Control of Proline Accumulation under Drought via a Novel Pathway Comprising the Histone Methylase CAU1 and the Transcription Factor ANAC055. *J. Exp. Bot.* **2018**, *69*, 579–588. [CrossRef]
105. Hakim, M.A.; Juraimi, A.S.; Hanafi, M.M.; Ismail, M.R.; Selamat, A.; Rafii, M.Y.; Latif, M.A. Biochemical and Anatomical Changes and Yield Reduction in Rice (Oryza Sativa L.) under Varied Salinity Regimes. *BioMed Res. Int.* **2014**, *2014*, e208584. [CrossRef] [PubMed]
106. Diaz, J.; Bernal, A.; Pomar, F.; Merino, F. Induction of Shikimate Dehydrogenase and Peroxidase in Pepper (*Capsicum Annuum* L.) Seedlings in Response to Copper Stress and Its Relation to Lignification. *Plant Sci.* **2001**, *161*, 179–188. [CrossRef]
107. Ighodaro, O.M.; Akinloye, O.A. First Line Defence Antioxidants-Superoxide Dismutase (SOD), Catalase (CAT) and Glutathione Peroxidase (GPX): Their Fundamental Role in the Entire Antioxidant Defence Grid. *Alex. J. Med.* **2018**, *54*, 287–293. [CrossRef]
108. Mhamdi, A.; Noctor, G.; Baker, A. Plant Catalases: Peroxisomal Redox Guardians. *Arch. Biochem. Biophys.* **2012**, *525*, 181–194. [CrossRef] [PubMed]
109. Dai, L.-P.; Xiong, Z.-T.; Huang, Y.; Li, M.-J. Cadmium-Induced Changes in Pigments, Total Phenolics, and Phenylalanine Ammonia-Lyase Activity in Fronds of Azolla Imbricata. *Environ. Toxicol.* **2006**, *21*, 505–512. [CrossRef]

110. Krol, A.; Amarowicz, R.; Weidner, S. The Effects of Cold Stress on the Phenolic Compounds and Antioxidant Capacity of Grapevine (*Vitis Vinifera* L.) Leaves. *J. Plant Physiol.* **2015**, *189*, 97–104. [CrossRef]
111. Koh, Y.S.; Wong, S.K.; Ismail, N.H.; Zengin, G.; Duangjai, A.; Saokaew, S.; Phisalprapa, P.; Tan, K.W.; Goh, B.H.; Tang, S.Y. Mitigation of Environmental Stress-Impacts in Plants: Role of Sole and Combinatory Exogenous Application of Glutathione. Front. *Plant Sci.* **2021**, *12*, 791205. [CrossRef]
112. Ozkur, O.; Ozdemir, F.; Bor, M.; Turkan, I. Physiochemical and Antioxidant Responses of the Perennial Xerophyte Capparis Ovata Desf. to Drought. Environ. *Exp. Bot.* **2009**, *66*, 487–492. [CrossRef]

agronomy

MDPI

Review

Climatology, Bioclimatology and Vegetation Cover: Tools to Mitigate Climate Change in Olive Groves

Ana Cano-Ortiz [1], Jose Carlos Piñar Fuentes [1], Felipe Leiva Gea [1], Jehad Mahmoud Hussein Ighbareyeh [1,2], Ricardo Jorje Quinto Canas [3,4], Catarina Isabel Rodrigues Meireles [5], Mauro Raposo [5], Carlos Jose Pinto Gomes [5], Giovanni Spampinato [6], Sara del Río González [7], Carmelo Maria Musarella [1,6] and Eusebio Cano [1,*]

1 Department of Animal and Plant Biology and Ecology Section of Botany, University of Jaen, Campus Universitario Las Lagunillas s/n, 23071 Jaén, Spain
2 Department of Plant Production and Protection, Faculty of Agriculture, Al-Quds Open University, Abu Khatallah Street, Hebron P720, Palestine
3 Faculty of Sciences and Technology, University of Algarve, Campus de Gambelas, 8005-139 Faro, Portugal
4 Centre of Marine Sciences (CCMAR), University of Algarve, Campus de Gambelas, 8005-139 Faro, Portugal
5 Department of Landscape, Environment and Planning, Institute for Mediterranean Agrarian and Environmental Sciences (ICAAM), School of Science and Technology, University of Évora (Portugal), Rua Romão Ramalho, no 59, 7000-671 Évora, Portugal
6 Department of Agraria, Mediterranean University of Reggio Calabria, Feo di Vito, snc-89122 Reggio Calabria, Italy
7 Department of Biodiversity and Environmental Management (Botany), Faculty of Biological and Environmental Sciences, Mountain Livestock Institute (CSIC-ULE), University of León, Campus de Vegazana s/n, 24071 León, Spain
* Correspondence: ecano@ujaen.es

Citation: Cano-Ortiz, A.; Fuentes, J.C.P.; Gea, F.L.; Ighbareyeh, J.M.H.; Quinto Canas, R.J.; Meireles, C.I.R.; Raposo, M.; Gomes, C.J.P.; Spampinato, G.; del Río González, S.; et al. Climatology, Bioclimatology and Vegetation Cover: Tools to Mitigate Climate Change in Olive Groves. *Agronomy* **2022**, *12*, 2707. https://doi.org/10.3390/agronomy12112707

Academic Editors: Alessio Papini, Mushtaq Ahmad, Fazal Ullah and Wajid Zaman

Received: 23 September 2022
Accepted: 28 October 2022
Published: 1 November 2022

Abstract: This work establishes the relationship between bioclimatology and agronomy. Bioclimatic indices are obtained for several areas under olive cultivation and correlated with olive yields. Due to the effect of climate change on cultivation and the high economic losses it produces, we propose a sustainable development model for the territorial classification of crops based on bioclimatic knowledge. Bioclimatic diagrams are prepared to provide information on water stress in crops so that irrigation can be carried out at the most effective time, a measure that has been shown to lead to water and energy savings for growers. In addition to this development model, we propose the application of non-aggressive cultivation techniques such as the use of living plant cover to ensure the protection of the soil and avoid losses due to climate irregularities. Studies conducted up to the present on applied bioclimatology have yielded promising results in the fields of farming and forestry. The maps and bioclimatic indices of Professor Rivas-Martínez, Ic, Io and It/Itc, are essential for bioclimatic classification. The agricultural development model with a bioclimatic basis ensures economic savings for growers and minimizes the environmental impact of cultivation. In the case of olive cultivation we detected that in 2005 all the cultivated areas that were not in their thermoclimatic optimum were damaged by frost. The widespread cultivation of olive groves in the Mediterranean basin, and mainly in the south of Spain, is reason enough to establish a relationship between its production and its bioclimatic environment. The ombroclimatic study in certain localities under olive cultivation shows that areas with Io <2.5 are unproductive (Jodar, Tabernas), and that their low Io value needs to be supplemented with irrigation water. This means extracting water from aquifers for agricultural use, when the current climate irregularities do not allow the excessive use of subsoil water. For the time being the only way of mitigating this situation is with sustainable development, which requires a bioclimatic understanding of the territory; and the use of appropriate cultivation techniques, including herbaceous plant covers. In this last case a knowledge of the plant associations in the phytosociological class *Stellarietea mediae* constitutes the basis for establishing either natural or sown vegetation cover.

Keywords: bioclimatology; agronomy; olive cultivation; climate change; vegetation cover

1. Introduction

This article describes the results of the work carried out by our team in recent years. This research focuses on biogeographic and bioclimatic aspects, climate and soil bioindicators, vegetation and vegetation dynamics, and landscape study. Based on our findings we have devised agricultural and forestry management models that have led us to a worldwide sustainability model with proposals for management to mitigate the ongoing profound and irregular global climate change. This has been achieved by conducting studies in such widely differing locations as Palestine [1–5], Italy, Spain, Portugal [6–13], the Caribbean, Mexico and Brazil [14–19]. To pursue sustainable development, it is necessary to conduct prior research into the methodology to ensure the intended aim is attained. These studies must be specific to each territory. The following factors must be analyzed in order to propose a management model: physical and chemical environmental factors, biogeography, bioclimatology, vegetation (biological indicators) and vegetation series [20].

Bioclimatology is an ecological science that has assumed greater importance in recent years and whose aim is to highlight the relationship between living beings (biology) and climate (physics) [21–23]. It is distinguished from climatology in that the information, indices and units it uses are related to and delimited by species and phytocoenoses/plant communities. The development of bioclimatology as a key discipline in the service of vegetation, agriculture and forestry sciences is one of the most important scientific innovations in recent times; the progress in this science has made it possible to identify several plant communities, and particularly to identify more accurately the main vegetation bands observed on an ascending altitudinal scale, in addition to establishing agricultural and forestry models for sustainable development [9,24,25]. Advances in geobotanical knowledge have shown promising results for the management of the natural environment [26].

Precipitation and temperature are the most important factors conditioning the existence of certain plant ecosystems [27]. Thus, each biogeographic region or group of regions has a particular altitudinal zonation of plant ecosystems as the mean annual temperature progressively decreases with altitude (thermoclimate) [28].

As climate (temperature and precipitation) is correlated with the altitudinal zonation of plant ecosystems, it can be seen that certain rhythms or changes occur everywhere on Earth based on temperature and precipitation [29]. Both the physical continent or bioclimatic belts and the biological plant content or vegetation series can be recognized as a function of these changes using the concept of universal vegetation series jurisdiction, and can therefore be applied to any place on the planet.

Bioclimatology as a science has undergone major advances in the last 30 years thanks to Professor Rivas-Martínez, who has created a bioclimatic classification of the Earth since 1996 [25,27–30].

Climatology and bioclimatology are essential for understanding and explaining the natural environment, which is currently undergoing the effects of climate change. It is therefore crucial to know what resources are available to humankind to mitigate this change, as it is leading to serious natural disasters involving economic damage and loss of human life. The third assessment report revealed that temperatures had increased by approximately 0.6 °C (between 0.4 °C and 0.8 °C), affecting physical and biological systems in different parts of the Earth [31]. The fourth assessment report contains clear evidence that climate change and global warming are now a reality that affects us all [32]. Global concentrations of carbon dioxide, methane and nitrous oxide in the atmosphere in recent years have risen substantially as a result of human activities since 1750. The linear projection for the next 100 years is estimated at 0.74 °C (between 0.56 °C and 0.92 °C), compared to 0.6 °C in the third report. According to the sixth report from the Intergovernmental Panel on Climate Change, global warming is now an unequivocal reality [33]. This increase in temperature occurs all over the planet but more markedly in more northerly latitudes [34–37].

Bioclimatology, vegetation, vegetation cover and agriculture are all terms that must be clearly understood, as their effective management can mitigate climate change. This poses the need for a worldwide economic model based on sustainable development and a rational use of natural resources.

Agriculture is conditioned by cyclones and anticyclones, which in turn shape the various macrobioclimates and bioclimates throughout the world (thermoclimates and ombroclimates). These all affect both the quantity and quality of crop yields, so a knowledge of local bioclimatic indicators is critical for improving agriculture. The use of bioclimatic indices—continentality, ombrothermic and thermicity indices—should be obligatory before establishing a certain crop, as there is a close correlation between these indices and agricultural yields.

Bioclimatology reveals the close relationship between living beings and physical factors (climate), which is evidently strongly linked to and connected with agriculture and vegetation ecosystems, as the concepts of bioclimate and bioclimatic belt (thermotype and ombrotype) are essential for planning crops. It is important to bear in mind that the aim is to achieve maximum yields in terms of quantity and quality with the lowest environmental and economic cost, which is ultimately what both growers and society demand. This cannot be done without taking into account bioclimatology as a critical science for agricultural planning. The use of bioindicators and bioclimatic indices must therefore be incorporated into agricultural management, and this requires obtaining a bioclimatic interpretation of the territory as an essential framework in which to establish the crop.

Once the bioclimate has been determined for each ecosystem and crop, a landscape analysis must be conducted in concordance with the bioclimate of the territory. This will contribute to achieving successful reforestation that will act as a CO_2 sink and potentially mitigate climate change, always assuming that there is a parallel reduction in CO_2 emissions and a change in energy policies. This landscape analysis leads us to propose the use of the concept of vegetation series/set of plant communities, which—as we have said—has universal jurisdiction, as a basis for establishing a sustainable development model [9,38–42].

The term "vegetation dynamics" is applied in vegetation series when the dynamic is natural and the process is driven by the ecosystem, but not when there is excessive human intervention that is contrary to the natural dynamic. When this dynamic is artificially accelerated (human action), the vegetation series undergoes drastic modifications leading to a situation where the series cannot recover and therefore disappears, and causing other new formations to appear which transform the landscape. Each biogeographic territory with its distinctive bioclimate and soil therefore has a specific vegetation series with its own dynamic, making it necessary to establish a specific management model [43,44].

When establishing crops in the process of territorial management, which, as we have mentioned, should be done on a bioclimatic basis considering the vegetation series, it is essential to determine the range of bioclimatic values in which the crop is at its optimum. Another factor to be taken into account is the cultivation technique, with a view to increasing quantity and quality and minimizing environmental, economic and human impact.

In essence, agriculture can be externally sustained or sustainable. Sustained agriculture is practiced with an external energy input, and can be described as agricultural production with no regard for conservation and a constant supply of products (nutrients) in order to boost yields. The aim is to achieve the highest possible yields whatever the environmental cost, and of course at a high price for the grower in monetary terms; in contrast, sustainable agriculture consists of using the resource without depleting it and without the need to incorporate nutrients into the soil [9,45,46].

Until 1970, traditional agriculture was the norm in many countries; however, the all-encompassing technological revolution that has been in force until the present day has since become widely accepted both economically and socially. The result is the uncontrolled use of pesticides, herbicides and phytosanitary products which cause a serious environmental impact that can occasionally give rise to insects and plants that are resistant to treatment without having achieved any significant improvement in living standards in the farming sector. The cost of using these chemical products is the same or in some cases greater than the economic benefit gained from the treatment, without even considering the added environmental cost [47–51].

Until a few years ago cultivation was predominantly traditional with models acquired over many centuries. Cultivation has always been somewhat plural, involving a variety of different crops, rather than the monoculture which has now become standard in some cases, a practice that delivers the land into the hands of speculators.

Non-standard management models evolved to create specific models in each territory, in which aspects such as pruning, tilling, type of planting and the method of obtaining the agricultural product were specific to each town/district [52]. This practice has now been lost in the unending quest for greater yields, leading to a sustained agriculture characterized by a loss of biodiversity and of the stability of the agroecosystem.

By sustainability we mean the capacity to persist over time without detriment to natural (soil, diversity, fauna, etc.) and cultural capital; this includes management models that enable production and renovation over time. Among the different definitions of sustainable agriculture, we choose the description of the American Society of Agronomy, namely, "A sustainable agriculture is one that, over the long term, enhances environmental quality and the resource base on which agriculture depends; provides for basic human food and fiber needs; is economically viable; and enhances the quality of life for farmers and society as a whole". In reality, farmers should aim to abandon the criterion of maximum yield per hectare and improve the productivity of their operation through better management of productive factors [53,54].

Although some authors consider traditional and sustainable agriculture to be one and the same, this view is erroneous. Traditional agriculture is based on a type of knowledge that has been acquired over millennia and has been rigorously tried and tested in an empirical way, and cannot be substituted by the practices of current biological agriculture, with its highly simplistic model. It can be maintained that traditional management forms constitute a coherent ecosystem, whereas today's sustainable agriculture does not fully follow the traditional model.

The practice of externally sustained agriculture causes profound transformations which in many cases lead to a loss of biodiversity on a massive scale. The decline in floristic diversity produces a loss of fauna, affects biological control mechanisms, and causes an explosion of insect plagues. The reasons for this decrease in diversity include inappropriate tilling techniques, loss of plant cover, lack of organic fertilizers, use of rollers (soil compaction) and indiscriminate use of herbicides, all of which ultimately results in increased erosion and soil loss [54]. It is therefore recommended to use agricultural practices that are non-aggressive with the environment such as traditional and sustainable cultivation, at least for marginal areas; for non-marginal areas with high production it is advisable to apply a cultivation that combines both techniques, i.e., planned cultivation with an ecological/botanical basis [55,56].

Non-aggressive cultivation is now possible thanks to advances in both bioclimatic and soil-related botanical research [57] which reveals the nutritional status of the soil and allows the appropriate doses of fertilizer to be used. This avoids the overuse of these substances, whose excess contaminates aquifers and implies financial losses for farmers [58].

2. Materials and Methods

The bioclimatology of southern Spain is studied here (Mediterranean Basin), focusing on areas that are predominantly dedicated to olive groves. This is done using the bioclimatic studies of several authors [25,27,30,51,59–61], and the study on vegetation cover by Cano-Ortiz [57]. With this research and with previous research conducted by [62], we obtain the bioclimatic indices and maps. Considering the productive and non-productive areas of olive groves in Spain, we propose a sustainable development model that ensures productivity in spite of climate change, based on the knowledge of the physical, bioclimatic and biogeographic factors and vegetation series [23].

For this, we use climatic data from 45 meteorological stations, with special attention to the Jodar, Aracena, Tabernas and Torredonjimeno stations. Bioclimatic diagrams are developed for these stations, which serve to establish the different hydric periods: excess water, regulation, dry. In the regulation period, the plants use the water accumulated in the soil, and do not enter into hydric stress until the dry period. We apply current knowledge on bioclimatology, and we elaborate the different bioclimatic indices: ombrothermic index (Io), continentality index (Ic), thermicity and compensated thermicity index (It/Itc) and Iosc2 and Iosc3 summer indices. Using the values of these indices in the different meteorological stations, and data on olive grove production in these same places, the optimal production areas for these crops are obtained [63,64], and we analyze herbaceous plant covers as CO_2 sinks.

3. Results and Discussion

3.1. Bioclimatic Analysis of Specific Territories

We analyze the bioclimatic studies in specific cases in various locations with olive cultivation (Table 1) as an essential prior step for territorial classification. The analysis involves formulating various bioclimatic indices and combining them with biological indices to create the bioclimatic maps and diagrams that are needed to generate the management model. The method consists of preparing tables with bioclimatic indices and the respective bioclimatic maps of thermo- and ombrotypes, which we then apply to determine the optimum crops in each of these locations. The aim of this method is to minimize the economic and environmental costs of production, and, when applied to certain crops such as olive groves and vineyards in the Mediterranean Basin, to suggest promising approaches for sustainable development [64–66] (Figures 1–3).

Figure 1. Consequences of an agricultural planning that does not take into account bioclimatology. Olive grove damaged by frost in 2005 due to being located in an enclosed valley affected by temperature inversion.

Table 1. Values of bioclimatic indices for the southern Iberian Peninsula (Andalusia). P. Mean annual precipitation. Io. Ombrothermic index. Ic. Continentality index (mean thermal interval). It/Itc. Thermicity index/compensated thermicity index. Alt. Altitude. Iosc2, Iosc3. Summer ombrothermic indices.

Location	P	Io	Ic	It/Itc	Alt	Iosc2	Iosc3	Iosc3/I
1. Vadillo-Castil (J)	1182.3	8.42	17.6	199	970	0.48	1.10	229
2. Pontones (J)	1148.7	6.98	18.0	214	740	0.53	0.92	1.73
3. La Iruela (J)	850.1	4.78	19.2	343	933	0.27	0.71	2.62
4. B. Moraleda (J)	612.1	3.33	19.7	370	887	0.36	0.66	1.83
5. B. Segura-Perales (J)	612.7	3.52	20.7	253	760	0.43	0.64	1.48
6. Siles (J)	785.7	4.33	21.6	260	826	0.39	0.75	1.92
7. V. del Arzobispo (J)	698.2	3.54	19.4	297	685	0.23	0.54	2.34
8. Villacarrillo (J)	610.8	3.26	20.4	284	794	0.26	0.52	2.00
9. P. del Dañador (J)	612.3	3.75	18.8	260	700	0.40	0.73	1.85
10. B. E.- Centenillo (J)	679.1	3.87	19.4	271	824	0.18	0.45	2.50
11. P. Rumblar (J)	657.4	3.28	18.6	353	300	0.18	0.45	2.50
12. Torredonjimano (J)	648.9	3.25	19.7	329	591	0.33	0.55	1.66
13. P. Tranco de Beas (J)	849.6	4.65	18.2	298	600	0.41	0.81	1.97
14. Beas de Segura (J)	658.0	3.20	19.7	322	577	0.30	0.58	1.93
15. Cazorla. ICONA (J)	792.4	4.68	19.2	257	885	0.32	0.74	2.31
16. Huelma-Solera (J)	526.9	3.20	20.8	293	1084	0.40	0.74	1.85
17. P. de la Bolera (J)	653.7	3.83	17.0	284	980	0.50	0.90	1.80
18. Jimena (J)	600.3	3.20	20.0	303	590	0.32	0.69	2.15
19. A. Real-Charilla (J)	681.0	4.05	17.3	270	920	0.42	0.68	1.61
20. P.del Jandula (J)	505.2	2.50	19.5	339	360	0.17	0.41	2.43
21. Arjona (J)	609.9	2.97	19.5	406	410	0.25	0.41	1.64
22. Bailen (J)	581.7	2.70	20.1	369	369	0.14	0.32	2.34
23. Andújar (J)	463.9	2.13	19.1	371	212	0.17	0.28	1.66
24. Jaén. Instituto (J)	578.3	2.85	18.8	345	510	0.20	0.40	2.00
25. Linares (J)	642.2	3.12	18.3	339	419	0.21	0.52	2.48
26. L. Torubias (J)	490.4	2.34	20.3	355	290	0.14	0.31	2.26
27. Mancha Real (J)	551.6	3.02	18.1	299	753	0.37	0.69	1.87
28. P. Guadalmena (J)	517.9	2.74	20.0	308	602	0.37	0.55	1.48
29. L. Fuente Higuera (J)	471.0	2.36	18.6	342	300	0.15	0.30	2.00
30. Ubeda (J)	579.6	3.03	18.4	313	748	0.22	0.44	2.00
31. La P. de Segura (J)	674.7	3.55	19.9	305	584	0.35	0.64	1.82
32. Cabra de S. Cristo (J)	449.9	2.62	17.7	275	938	0.35	0.65	1.86
33. Ubeda P. Guadiana (J)	404.6	2.24	19.6	285	420	0.23	0.46	2.00
34. P. Guadalmellato (CO)	698.4	3.42	18.0	349	200	0.40	0.64	1.60
35. C. La Jarosa (CO)	831.1	4.46	16.5	315	340	0.25	0.31	1.24
36. Pantano P. Nuevo (CO)	760.6	4.03	17.9	321	410	0.16	0.59	3.68
37. Villaralto (CO)	501.4	2.66	18.8	310	583	0.42	0.60	1.42
38. Pozoblanco (CO)	514.4	2.66	19.9	311	649	0.34	0.66	1.94
39. Aldea de Cuenca (CO)	559.0	2.98	18.1	307	571	0.23	0.67	2.91
40. H. del Duque. Aer. (CO)	476.9	2.68	18.4	284	540	0.42	0.68	1.61
41. Pedroche (CO)	506.8	2.60	19.2	322	621	0.38	0.63	1.65
42. La Rambla (CO)	527.4	2.81	17.9	308	200	0.30	0.41	1.37
43. Castro del Rio (CO)	470.7	2.45	20.1	305	210	0.17	0.36	2.12
44. Montoro (CO)	572.4	2.72	18.1	360	195	0.19	0.37	1.94
45. Pozoblanco. Cerro (CO)	594.7	2.98	19.0	339	500	0.32	0.52	1.62

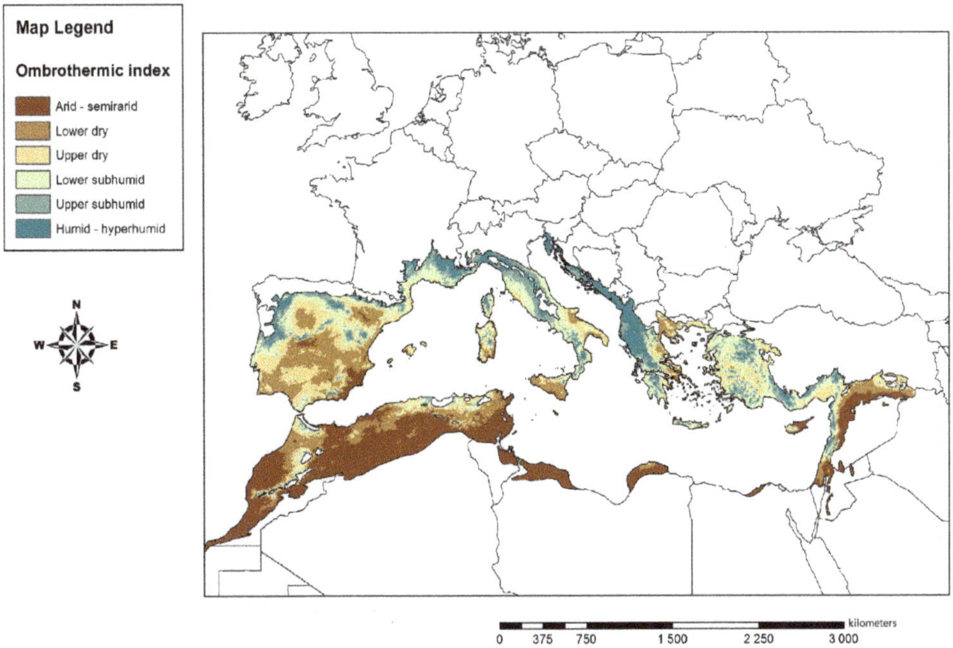

Figure 2. Map of ombrotypes in the Mediterranean Basin. The optimum ombrotype for olive cultivation is the dry-subhumid. Geographic Coordinates Mediterranean Basin: 38° N 17° E.

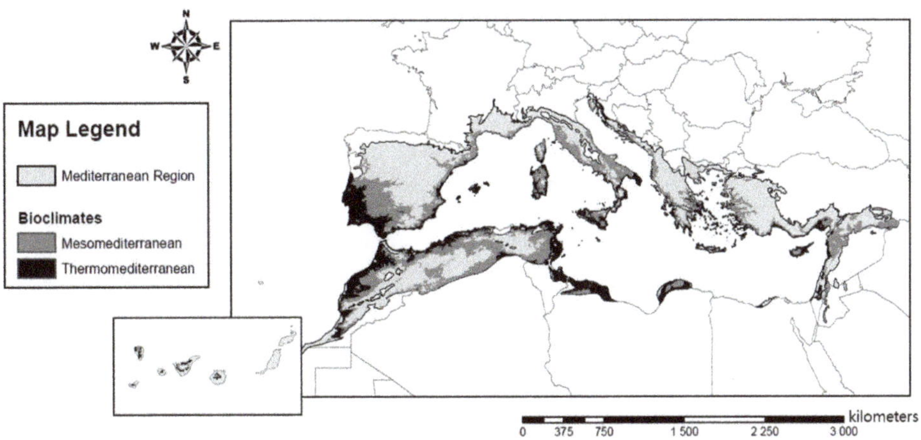

Figure 3. Map of thermotypes in the Mediterranean Basin. The optimum thermotype for olive cultivation is the thermo- and meso-Mediterranean. Geographic Coordinates Mediterranean Basin: 38° N 17° E.

3.2. Bioclimatic Analysis of the Southern Iberian Peninsula

The bioclimatic analysis is conducted using the meteorological stations in the southern Iberian Peninsula, obtaining for each one the values of T (temperature), P (precipitation), Io (ombrothermic index), Ic (continentality index), It/Itc (thermicity index and compensated thermicity), Iosc2 (summer ombrothermic index of July and August), Iosc3 (summer ombrothermic index of June, July and August) and PVA (Period of Vegetative Activity).

The analysis of the climatic and bioclimatic parameters reveals the following: (1) Most of the territory sampled has a PVA of 12 months. There is therefore no stop due to cold. (According to Montero [67], a stop due to cold is understood as a condition where the mean monthly temperature falls below 7.5 °C.) This coincides exactly with the territories more to the south and southwest of the province of Jaén (Southern Spain), an area dominated by olive cultivation with a mainly Thermo-Mediterranean thermotype and a dry ombrotype. In contrast, places with a PVA of 8–9 are dominant in the northwest of the province of Jaén and Granada, with the particularity that both territories have an upper meso-Mediterranean thermotype. However, the high value of the ombroclimatic index—over 4—is explained by the screening effects of the Segura, Las Villas and Cazorla mountain ranges (Table 1) (Figure 1) against cyclones.

Several factors have conditioned the cultivation of different varieties of olive in each area, and yet these varieties are not always at their ecological optimum from the bioclimatic point of view. Although the influence of bioclimatology on olive yields has been noted in previous studies [62], the bioclimatic characterization of the different varieties has subsequently been undertaken by our research team. Another objective of this work is to relate the distribution of the olive varieties with the different bioclimatic indices and to characterize each variety by its bioclimatic requirements in order to provide new criteria for the agricultural planning of crops [68,69]. It is therefore necessary to conduct biogeographical and soil studies of potential growing sites and use this information to propose a model for agricultural management [23].

Environmental factors→Bioclimatology and biogeography→Vegetation series→Crop type

A correct territorial classification can be achieved by applying this model to rain fed crops. In previous works we established a close correlation between kg/hectare yield for olive groves and the values of bioclimatic indices. This allowed us to establish that the locations of maximum yield for the Picual variety belong to the upper thermo-Mediterranean thermotype and the upper dry ombrotype, and to the lower meso-Mediterranean thermotype. We subsequently extended these studies to other olive varieties, namely "Hojiblanca", "Lechín", "Morisca", "Manzanilla", "Gordal" and "Verdiales" for Spain [62,69].

3.3. Bioclimate—Plant Cover: Mitigating the Damage Caused by Climate Change

In the last 50 years a new type of agriculture has become prevalent, and is now widely accepted both economically and socially. This has resulted in the uncontrolled use of pesticides, herbicides and phytosanitary products, causing substantial environmental damage, with the resulting harm for the population [54,70].

A crop is unproductive when the cost of production is equal to or higher than the income received by the grower. This expense includes the entire array of activities necessary for production plus the environmental cost, which in most cases is impossible to quantify. Crops are considered unproductive when their cost is higher than the income they generate, and crops that are apparently productive with modern technology should also be seen as unproductive if the environmental cost is excessively high. An excessive cost should be considered as any situation in which the crop leads to the irreversible loss of a certain resource (although cases of contamination can be permitted in which the system is able to self-regenerate), thereby refuting the expression "the polluter pays", as resources may be lost for which it is impossible to set a price because they have such a high ecological and natural worth. Cultivation is therefore not valid if it leads to irreversible losses of soil and biodiversity or the contamination of water and land. It is essential to consider these last two issues when establishing a crop. All situations corresponding to any of these cases must undergo technical agrarian reform, and be subject to agricultural planning to ensure that the crop is maintained while respecting the principles outlined above, as well as being productive and even benefiting from potential increases in yields. These yields can also be of high quality, as quantity and quality are not incompatible. This implies taking into consideration the bioclimatic aspects of the territory [45,71].

Agricultural planning must be done based on respect for the environment and the optimization of the crop in question, as all crops have an ecological and bioclimatic optimum. In this situation any potential impact on the environment would be decreased. All this would involve the creation of specific agricultural and non-standard models which would lead us in some cases to maintain certain crops or to replace them with other alternative crops that are more productive and less contaminating.

The severity of the damage to a crop due to climate change depends on whether it is in its bioclimatic optimum. The soil water reserve must be optimized by determining the relationship between the water availability for the plant in the soil, potential evapotranspiration (ETP) and residual evapotranspiration (e) by means of a bioclimatic diagram.

The application of bioclimatic diagrams to such diverse localities as Aracena, Jodar, Montiel, Tabernas and Torredonjimeno reveals significant differences in terms of the date on which the crop begins to undergo water stress. This may be mitigated if the relation between the summer ombrothermic indices Iosc3/Iosc2 >1 [72], meaning that there is compensation thanks to the June rains. The crop is productive when Io >2.5 (Table 2).

Table 2. Values of indices for the comparative analysis between meteorological stations: It/Itc = thermicity index and compensated thermicity; Io = ombrothermic index; Ic = continentality index; Iosc2, Iosc3 = summer ombrothermic index.

	It/Itc	Io	Ic	Ios2	Ios3	Ios3/Ios2
Aracena	281/281	5.88	17.8	0.29	0.81	2.79
Arjona	321/336	2.84	19.50	0.11	0.30	2.72
Jodar	328/343	2.31	20.0	0.20	0.38	1.90
O. Montiel	195/209	3.30	20.9	0.48	0.86	1.79
Tabernas	388/388	1.14	16.4	0.06	0.10	1.66
Torredonjimeno	322/331	3.25	19.7	0.33	0.55	1.66

The use of cultivation techniques that prolong the time the water reserves remain in the soil can also mitigate the effects of drought. This can only be done if we know the date on which the crop begins to undergo stress, as the period of vegetative activity is identified when the availability D > than the potential evapotranspiration ETP. The regulation period occurs when D < ETP, and the dry period if the residual evapotranspiration is e > D (Figure 4).

The bioclimatic diagram for the station of Jodar with a meso-Mediterranean bioclimate and a dry-semi-arid ombrotype shows how the water availability D for an agricultural plantation or a forest formation begins to undergo water stress in May, when the residual evapotranspiration becomes greater than the water availability D, and there is a long period of drought and a short regulation period. This date can be determined by applying the bioclimatic diagram of Montero Burgos and González Rebollar [67], somewhat modified by using only water-related parameters: P = precipitation, ETP = potential evapotranspiration, D = water availability, R = soil water reserve, e = residual evapotranspiration.

In the case of Aracena, with an upper meso-Mediterranean upper subhumid bioclimate, the water availability D becomes less than the evapotranspiration in the first days of August, although the availability D is less than the ETP from the end of May, implying a long regulation period and a dry period lasting barely one month; this is because there is strong compensation thanks to rains in June, Iosc3/Iosc2 = 2.79 (Figure 5).

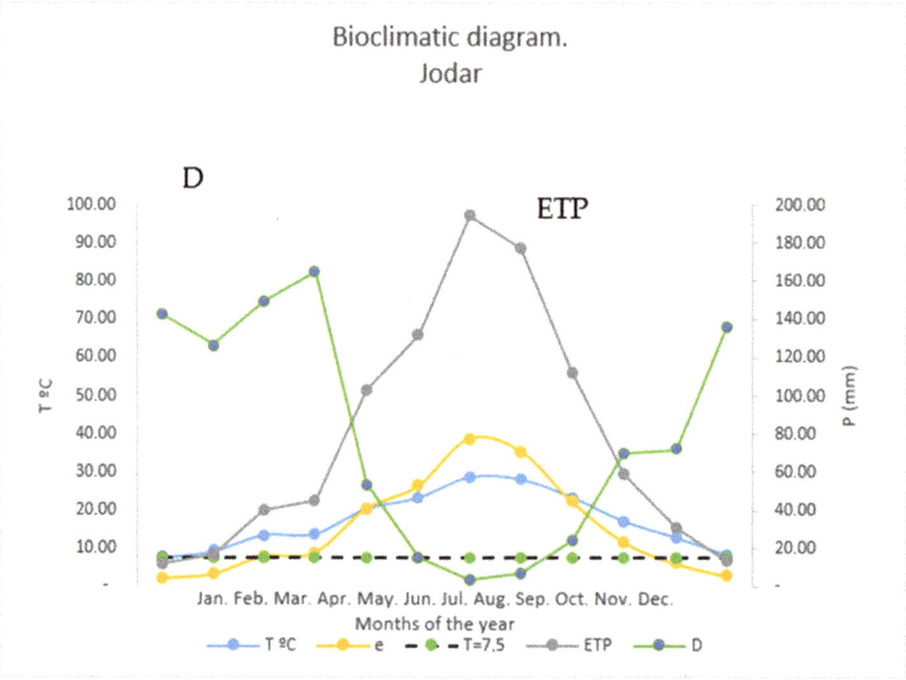

Figure 4. Bioclimatic diagram for Jodar: P = precipitation. ETP = potential evapotranspiration. D = water availability. R = soil water reserve. e = residual evapotranspiration.

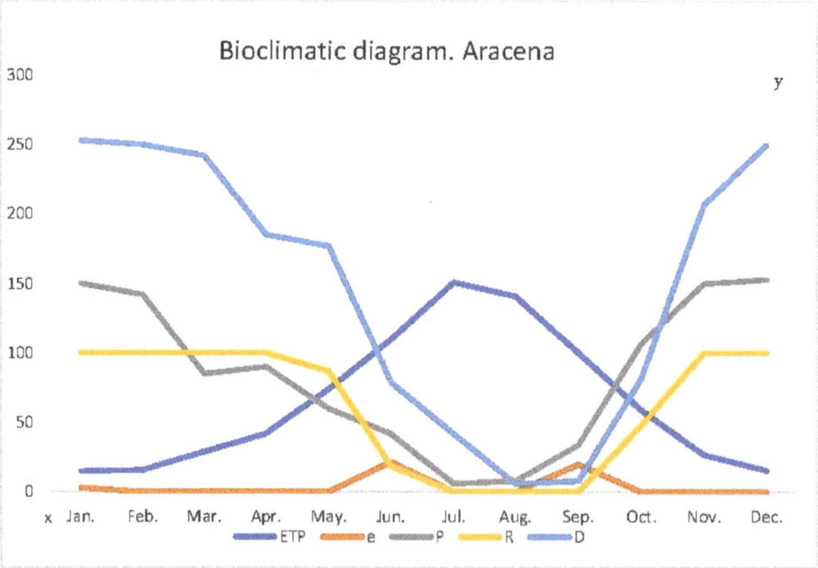

Figure 5. Bioclimatic diagram for Aracena: P = precipitation, ETP = potential evapotranspiration, D = water availability, R = soil water reserve, e = residual evapotranspiration (mm. water); Y: mm. Water; X: Month.

In the case of Tabernas, with an upper thermo-Mediterranean lower semi-arid climate, the water availability D < ETP in February. The reserve values R and water availability D for the crop are very low—even zero—as a result of scarce precipitation, and there may be a regulation period in the case of an olive-type crop (Figure 6) (Table 3).

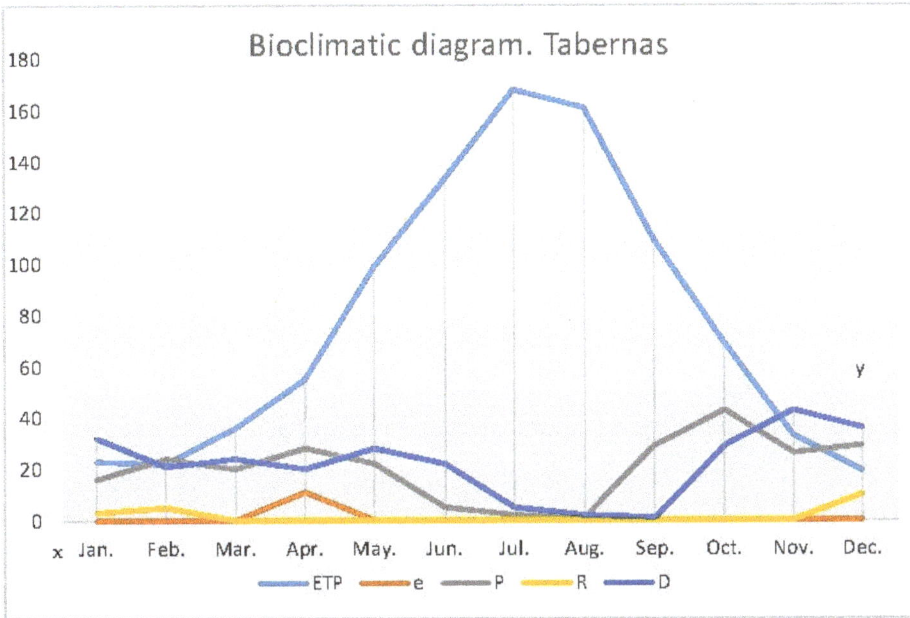

Figure 6. Bioclimatic diagram for Tabernas: P = precipitation, ETP = potential evapotranspiration, D = water availability, R = soil water reserve, e = residual evapotranspiration (mm. water); Y: mm. Water; X: Month year.

Table 3. Parameters for Tabernas: potential evapotranspiration (ETP), residual evapotranspiration (e), precipitation (P), soil water reserve (R), water availability for plants (D).

Tabernas	ETP	e	P	R	D
January	23	4.6	16	3	32
February	22	4.4	24	5	21
Mach	36	7.2	20	0	24
April	55	11	28	0	20
May	99	19.8	22	0	28
June	133	26.6	5	0	22
July	168	33.6	2	0	5
August	161	32.2	1	0	2
September	109	21.8	29	0	1
October	69	13.8	43	0	29
November	33	6.6	26	0	43
December	19	3.8	29	10	36

In the case of Torredonjimeno with a lower meso-Mediterranean upper dry bioclimate, D < ETP from mid-May and never falls below the residual evapotranspiration, meaning that there is no real dry period (Figure 7).

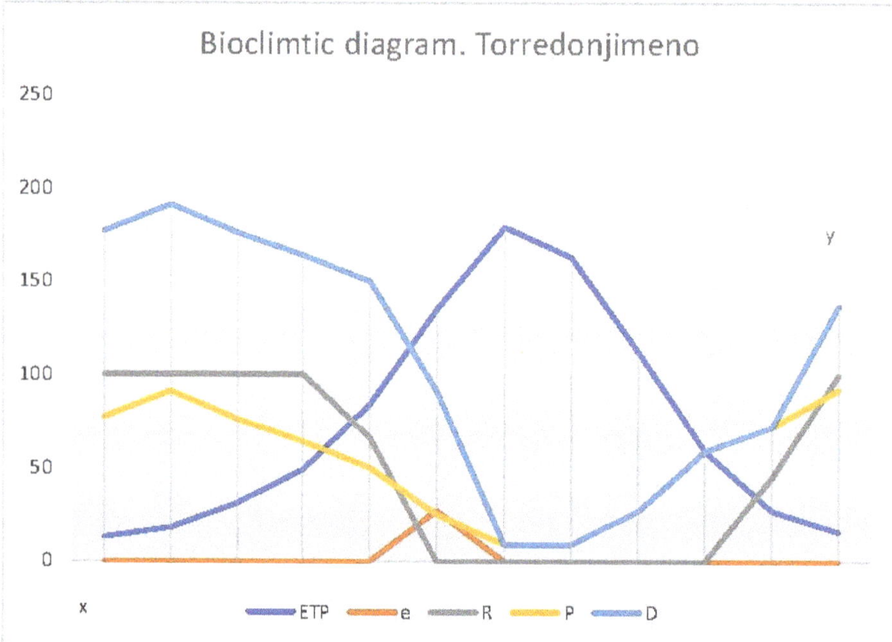

Figure 7. Bioclimatic diagram for Torredonjimeno: P = precipitation, ETP = potential evapotranspiration, D = water availability, R = soil water reserve, e = residual evapotranspiration (mm. water).

In this analysis it can be seen that the regulation period and the dry period differ from one locality to another, with four months of drought in Aracena and eight to ten months in Tabernas, whereas Torredonjimeno has no drought period at all.

The strategy involves increasing the regulation period, the phase between D < ETP and e > D; this period can be extended with organic soil amendments and the use of vegetation cover.

This work is based on the knowledge of the climate and bioclimate and the flora and soil bioindicators, and starts with a thorough study of these bioindicators and their comparison with numerous soil analyses carried out in the same location. This allows us to implement effective treatments that offer an indispensable resource for managing vegetation cover in order to maximize its benefits and generate its use by identifying these bioindicators and understanding what they reveal.

For all these reasons we have studied the occurring plant associations from the phytosociological point of view, due to its efficacy and value [26,73,74], but also prioritizing soil factors and observing their correlations with the floristic-phytosociological component [75]. We take into account the prevailing climate factors in the study areas, which are highly influential in determining the presence of these associations.

Plant communities in cultivated areas may be of two types: first, heathland or thicket which acts as a biodiversity center and an element for controlling erosion, as it reduces soil loss when it forms part of verges or boundaries; and herbaceous grasslands, whose function is also to decrease erosion, although they may be of very different types depending on the soil and climate characteristics. However, it can be said that all the communities act as vegetation cover and intervene in the control of erosion [75]. The use of herbaceous plant cover in preference to thicket is advisable, as herbaceous plants do not compete with crops for water and nutrients, or only minimally so (Figure 8).

Figure 8. Sustainable agriculture, olive cultivation with leguminous and Poaceae species.

Living vegetation cover can therefore be described as beneficial, although the use of plant cover is not new as it was used by the Romans in their vineyards and has been studied and applied since the early 20th century.

Herbaceous and woody plant cover has been shown to act as a CO_2 sink; consequently, in addition to preventing erosion, it mitigates climate change [41,42,76]. According to the Spanish Federation of Municipalities and Provinces [77] and NEIKER-Tecnalia [78], agricultural croplands and forests incorporate CO_2 into the soil and act as carbon sinks. This points to the importance of land use in mitigating climate change; it is estimated that agricultural croplands incorporate up to 1.98 T/ha per year [79]. The presence of herbaceous species among the rows of fruit trees, grapevines and olive trees (vegetation cover) is a generalized practice in integrated production and would produce an additional increase in carbon uptake that would be equal to or greater than that induced by permanent pasture (around 0.4 t/ha per year), according to the report issued by the National Institute for Agrarian Reform (INRA). This is recommended provided the water resource is not very deficient, as it minimizes the use of herbicides, protects the soil against erosion and compaction and has positive effects on biodiversity. It is estimated that the establishment of certain herbaceous crops such as the cardoon (*Cynara cardunculus* L.) accounts for a net carbon sequestration of 3.2–3.7 t/ha per year, assuming that this biomass offsets up to 65–75% of the CO_2 emissions produced by the use of fossil fuel [39]. Mota et al. studied the absorption of CO_2 by barley, oats and wheat in the region of Murcia using a density of 100 plants per m^2 for the specific case of *Hordeum*, and obtained values of incorporation of carbon of 325 g C/m^2 per year [80], a density equal to low cover. Our sampling plots covered that range between 25% and 100%, with a mean height of the dominant species of 0.35 m for the *Hordeum* plots. We obtained 1000 individuals per m^2 in the sampled plots, which represents the incorporation of 3250 gr/m^2 of carbon per year, with a cover of 100% and a mean height of 0.35 m. The calculation of the biovolume (Bv) [81] (Figure 9) for the different percentages of cover is as follows:

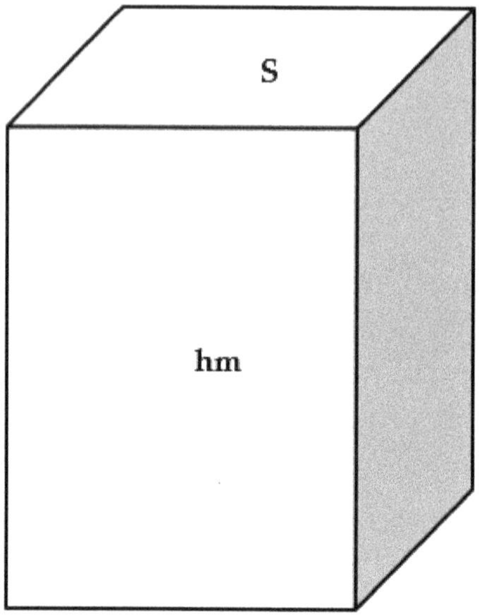

Figure 9. Bv = S × hm × Cm: Bv = biovolume, S = surface, Cm = medium herbaceous cover, hm = average height of the dominant species.

To calculate the Biovolume, a pyramid or cube is used, whose base must be the sampling surface in m^2 (S); Cm is the soil covered by vegetation, and hm is the average height of the dominant species.

$$Bv100 = 0.35; Bv75 = 0.26; Bv50 = 017; Bv25 = 0.08$$

The correspondence of each biovolume with the amount of carbon incorporated is the following:

$$Bv100 = 0.35 = 3250 \text{ gr } C/m^2; Bv75 = 0.26 = 2414.2 \text{ gr } C/m^2; Bv50 = 017 = 1578.5 \text{ gr } C/m^2; Bv25 = 0.08 = 812.5 \text{ gr } C/m^2$$

This represents 32.5 tons/hectare per year for 100% cover.

Cultivation and erosion are two very closely related terms, as the erosion of the plot is determined to a greater or lesser degree by the cultivation technique applied. Tilling is a traditional but aggressive technique that can lead to losses of 20–40 tons per hectare per year, compared to 2.5 tons per hectare per year in plots with interspersed strips of vegetation. This erosion is aggravated by the current climate irregularities. Tilling has customarily been practiced in order to increase the availability of water for cultivation and control weeds, and requires ploughing the plot several times. This strategy is now totally counterproductive as it does not increase water availability but quite the reverse; in a tilled olive grove the capacity of water to infiltrate in the soil is reduced, leading to the occurrence of a "tillage pan" caused by repeated transit (compacting) by heavy machinery. Tilling produces disaggregation, loss of structure, and loss of gasses and organic matter. It has been demonstrated that a crop tilled for ten years has a 30% loss of organic matter, which requires having to add high doses of fertilizers to achieve acceptable yields (Figure 10). However, proper use of organic manure can help improve soil physico-chemical properties and microbial biomass in an arid climate [82,83].

Figure 10. Ecological fertilisation with fertilisers in the form of crop residues and manure.

The chemical control of weeds has resulted in a situation where the floristic composition in the phytocoenoses of herbaceous plants has been modified and where the elimination of certain species has led to the spread of other much more invasive species [80,84,85]. This is often due to a lack of knowledge about the use of herbicides [86–88], as chemical compounds are applied that are not appropriate for a particular species. We should be careful not to assume that all the flora present in olive groves is harmful and our enemy. Evidently this cannot be framed as a battle between what we call "weeds" and humans; it is therefore advisable to ensure a greater understanding of these weeds and their botanical-ecological behavior in order to establish control mechanisms.

4. Conclusions

The main aim of this work is to highlight the importance of bioclimatology in the management of territory. Both the bioclimatic and agronomic studies of cultivation techniques have proven to be indispensable for a sustainable development that mitigates climate change. The study shows that there is an incorporation of carbon into the soil that may approximate 32.5 tons/hectare per year for 100% cover.

Now is the time for all countries to act in a coordinated way to generate economic models that maintain the quantity and quality of yields without continuing to degrade the natural environment. Our studies in a variety of territories demonstrate that it is possible to obtain models of agrarian and forestry production that reverse climate change. These models allow production to be optimized with the minimum environmental cost. Moreover, the model allows for crop changes in the face of new climate scenarios.

In certain areas on Earth such as Andalusia, the effects of climate change on agriculture are becoming catastrophic, as temperatures and rainfall irregularities impact crops and produce lower yields; this situation is exacerbated in the case of intensive cultivation with inappropriate agricultural techniques. The placement of crops based on bioclimatic criteria and the use of autochthonous vegetation covers are two important techniques for mitigating climate change. It is therefore essential to organize agriculture according to the bioclimatic profile of the territory. It is also important to use vegetation covers that allay acute rises in temperature and reduce evapotranspiration, both of which can be achieved through sustainable development approaches.

Our results show that there are positive and negative temperature trends in Andalusia, and it is precisely the upward trends that dominate in the Guadalquivir valley, a predominantly agricultural area with a prevalence of olive groves. This implies rising temperatures, a situation which, coupled with high losses in vegetation cover, exacerbates water loss and ultimately leads to lower yields.

Author Contributions: Conceptualization, A.C.-O., C.J.P.G., F.L.G. and E.C.; Data curation, A.C.-O., C.M.M., R.J.Q.C., J.C.P.F., S.d.R.G., J.M.H.I., G.S., C.J.P.G. and E.C.; Formal analysis, E.C.; Investigation, A.C.-O. and E.C.; Methodology, A.C.-O., C.M.M. and E.C.; Project administration, E.C.; Resources, E.C.; Supervision, A.C.-O., C.M.M. and E.C.; Validation, A.C.-O., C.M.M., J.C.P.F., S.d.R.G., C.J.P.G., M.R. and E.C.; Software, J.C.P.F. and S.d.R.G.; Visualization, A.C.-O., C.M.M., R.J.Q.C., C.I.R.M. and E.C.; Writing—original draft, A.C.-O. and E.C.; Writing—review and editing, A.C.-O., C.M.M., R.J.Q.C., J.C.P.F., S.d.R.G., C.J.P.G. and E.C. All authors have read and agreed to the published version of the manuscript.

Funding: This research received no external funding.

Institutional Review Board Statement: Not applicable.

Informed Consent Statement: Not applicable.

Conflicts of Interest: The authors declare no conflict of interest.

References

1. Piñar Fuentes, J.C.; Leiva, F.; Cano-Ortiz, A.; Musarella, C.M.; Quinto Canas, R.; Pinto Gomes, C.; Cano, E. Impact of the Management of Grass Cover During with Herbicides in the Biodiversity, Cover and Humidity of the Soil in the Olive Groves of the South of the Iberian Peninsula During the Period 2006–2016. *Agronomy* **2021**, *11*, 412. [CrossRef]
2. Ighbareyeh, J.M.H.; Cano-Ortiz, A.; Cano, E. Biological and bioclimatic basis to optimize plant production: Increased economic areas of Palestine. *Agric. Sci. Res. J.* **2014**, *4*, 10–20.
3. Ighbareyeh, J.M.H.; Cano Ortiz, A.; Cano, E. Analysis of the physical factors of Palestinian bioclimate. *Am. J. Chim. Change* **2014**, *3*, 223–231. [CrossRef]
4. Ighbareyeh, J.M.H.; Cano Ortiz, A.; Cano, E.; Ighbareyeh, M.M.H.; Suliemieh, A.A.A. Assessing Crop Yield sustainability under the Climatic and Bioclimatic Change in the Area of Palestine. *Am. J. Clim. Change* **2015**, *4*, 48–56. [CrossRef]
5. Ighbareyeh, J.M.H.; Ighbareyeh, M.M.H.; Suliemieh, A.A.A.; Abdel-Qader, D.; Cano, E.; Cano Ortiz, A. Peach (*Prunus persica* L. Batsch) production and environmental conditions in Jenin, Palestine. *Net J. Agric. Sci.* **2019**, *7*, 78–84. [CrossRef]
6. Ighbareyeh, J.M.H.; Suliemieh, A.A.-R.A.; Sheqwarah, M.; Cano-Ortiz, A.; Carmona, E.C. Flora and Phytosociological of Plant in Al-Dawaimah of Palestine. *Res. J. Ecol. Environ. Sci.* **2022**, *2*, 58–91. [CrossRef]
7. Quinto Canas, R.; Cano-Ortiz, A.; Musarella, C.M.; del Río, S.; Raposo, M.; Fuentes, J.C.P.; Gomes, C.P. *Quercus rotundifolia* Lam. Woodlands of the Southwestern Iberian Peninsula. *Land* **2021**, *10*, 268. [CrossRef]
8. Canas, R.Q.; Cano-Ortiz, A.; Spampinato, G.; del Río, S.; Raposo, M.; Fuentes, J.C.P.; Gomes, C.P. Contribution to the Knowledge of Rocky Plant Communities of the Southwest Iberian Peninsula. *Plants* **2021**, *10*, 1590. [CrossRef]
9. Cano-Ortiz, A.; Musarella, C.M.; Piñar Fuentes, J.C.; Quinto Canas, R.; Pinto Gomes, C.J.; Spampinato, G.; Ighbareyeh, J.M.H.; del Río, S.; Cano, E. Forest and Arborescent Scrub Habitats of Special Interest for SCIs in Central Spain. *Land* **2021**, *10*, 183. [CrossRef]
10. Cano, E.; Piñar, F.J.C.; Cano-Ortiz, A.; Leiva, G.F.; Ighbareyeh, J.M.H.; Quinto Canas, R.J.; Pinto Gomes, C.J.; Spampinato, G.; Del Río González, S.; Musarella, C.M. Chapter 17-Bioclimatology and botanical resources for sustainable development. In *Natural Resources Conservation and Advances for Sustainability*; Jhariya, M.K., Meena, R.S., Banerjee, A., Meena, S.N., Eds.; Elsevier: Amsterdam, The Netherlands, 2022; pp. 377–388. ISBN 978-0-12-822976-7. [CrossRef]
11. Piñar Fuentes, J.C.; Cano-Ortiz, A.; Musarella, C.M.; Quinto Canas, R.; Pinto Gomes, C.J.; Spampinato, G.; del Río, S.; Cano, E. Bioclimatology, Structure, and Conservation Perspectives of *Quercus pyrenaica*, *Acer opalus* subsp. *Granatensis*, and *Corylus avellana* Deciduous Forests on Mediterranean Bioclimate in the South-Central Part of the Iberian Peninsula. *Sustainability* **2019**, *11*, 6500. [CrossRef]
12. Quinto Canas, R.; Mendes, P.; Cano-Ortiz, A.; Musarella, C.; Pinto-Gomes, C. Forest fringe communities of the southwestern Iberian Peninsula. *Rev. Chapingo Serte Cienc. For. Ambiente* **2018**, *24*, 415–434. [CrossRef]
13. Quinto-Canas, R.; Mendes, P.; Meireles, C.; Musarella, C.; Pinto-Gomes, C. The *Agrostion castellanae* Rivas Goday 1957 Corr. Rivas Goday & Rivas-Martínez 1963 Alliance in the Southwestern Iberian Peninsula. *Plant Sociol.* **2018**, *55*, 21–29. [CrossRef]
14. Piñar Fuentes, J.C.; Raposo, M.; Pinto Gomes, C.J.; del Río González, S.; Spampinato, G.; Cano, E. New Contributions to the *Ericion umbellatae* Alliance in the Central Iberian Peninsula. *Sustainability* **2021**, *13*, 5639. [CrossRef]
15. Cano-Ortiz, A.; Musarella, C.; Fuentes, C.; Pinto-Gomes, C.; Río, S.; Cano, E. Diversity and conservation status of mangrove communities in two areas of Mesocaribea biogeographic region. *Curr. Sci.* **2018**, *115*, 534–540. [CrossRef]
16. Cano-Ortiz, A.; Musarella, C.M.; Piñar, J.C.; Spampinato, G.; Veloz, A.; Cano, E. Vegetation of the dry bioclimatic areas in the Dominican Republic. *Plant Biosyst.* **2015**, *149*, 451–472. [CrossRef]
17. Cano Ortiz, A.; Musarella, C.M.; Piñar Fuentes, J.C.; Pinto Gomes, C.J.; Cano, E. Distribution patterns of endemic flora to define hotspots on Hispaniola. *Syst. Biodivers.* **2016**, *14*, 261–275. [CrossRef]

18. Cano Ortiz, A.; Quinto, C.R.; Pinar, F.J.C.; del Rio, S.; Pinto, G.C.J.; Cano, E. Endemic Hemicryptophyte Grasslands of the High Mountains of the Caribbean. *Res. J. Ecol. Environ. Sci.* **2022**, *2*, 1–20. [CrossRef]
19. Kegler, A.; Diesel, S.; Wasum, R.A.; Herrero, L.; Del Río, S.; Penas, Á. Contribution to the phytosociological survey of the primary forests in the NE of Rio Grande do Sul (Brazil). *Plant Biosyst.* **2010**, *144*, 53–84. [CrossRef]
20. Cano Ortiz, A.; Musarella, C.M.; Pinto Gomes, C.J.; Quinto Canas, R.; Piñar Fuentes, J.C.; Cano, E. Phytosociological Study, Diversity and Conservation Status of the Cloud Forest in the Dominican Republic. *Plants* **2020**, *9*, 741. [CrossRef]
21. Cano Ortiz, A.; Musarella, C.M.; Piñar Fuentes, J.C.; Pinto Gomes, C.J.; Quinto Canas, R.; Del Río, S.; Cano, E. Indicative value of the dominant plant species for a rapid evaluation of the nutritional of soils. *Agronomy* **2020**, *11*, 1. [CrossRef]
22. Lv, J.; Xie, Y.; Luo, H. Erosion Process and Temporal Variations in the Soil Surface Roughness of Spoil Heaps under Mul-ti-Day Rainfall Simulation. *Remote Sens.* **2020**, *12*, 2192. [CrossRef]
23. Burrascano, S.; Sabatini, F.M.; Blasi, C. Testing indicators of sustainable forest management on understorey composition and diversity in southern Italy through variation partitioning. *Plant Ecol.* **2011**, *212*, 829–841. [CrossRef]
24. Cano, E.; Ruiz, L.; Cano-Ortiz, A.; Nieto, J. *Bases Para el Establecimiento de Modelos de Gestión Agrícola y Forestal in Memorian Al prof*; Isidoro, R.M., Ed.; Servicio Publicaciones Universidad de Jaén: Jaén, Spain, 2002; pp. 131–142.
25. Cano, E.; Cano-Ortiz, A.; Musarella, C.M.; Piñar Fuentes, J.C.; Ighbareyeh, J.M.H.; Leiva, G.F.; Del Río, S. Mitigating climate change through bioclimatic applications and cultivation techniques in agriculture (Andalusia, Spain). In *Sustainable Agriculture, Forest and Environmental Management*; Jhariya, M.K., Banerjee, A., Meena, R.S., Yadav, D.K., Eds.; Springer Nature Singapore Pte Ltd: Singapore, 2019; pp. 31–69. [CrossRef]
26. Rivas-Martínez, S. Clasificación Bioclimática de la Tierra. *Folia Botánica Matritensis* **1996**, *16*, 1–32.
27. Blasi, C.; Biondi, E.; Izco, J. 100 years of plant sociology: A celebration. *Plant Biosyst.* **2011**, *145* (Suppl. 1), 1–3. [CrossRef]
28. Rivas-Martínez, S.; Loidi Arregui, J. Bioclimatoloy of the Iberian Peninsula. *Itinera Geobot.* **1999**, *13*, 41–47.
29. Pesaresi, S.; Biondi, E.; Casavecchia, S. Bioclimates of Italy. *J. Maps* **2017**, *13*, 955–960. [CrossRef]
30. Pesaresi, S.; Galdenzi, D.; Biondi, E.; Casavecchia, S. Bioclimate of Italy: Application of the worldwide bioclimatic classification system. *J. Maps* **2014**, *10*, 538–553. [CrossRef]
31. Rivas-Martínez, S.; Penas, Á.; Díaz González, T.E.; Ladero Álvarez, M.; Asensi Marfil, A.; Díez Garretas, B.; Molero Mesa, J.; Valle Tendero, F.; Cano, E.; Costa Talens, M.; et al. Mapa de series, geoseries y geopermaseries de vegetación de España. Parte II. *Itinera Geobot.* **2011**, *18*, 425–800.
32. IPCC. Climate Change. Synthesis Report. In *A Contribution of Working Groups I, II, and III to the Third Assessment Report of the Intergovernmental Panel on Climate Change*; Watson, R.T., the Core Writing Team, Eds.; Cambridge University Press: Cambridge, UK; New York, NY, USA, 2001; p. 398.
33. IPCC. Climate change. The physical science basis. In *Contribution of Working Group I to the Fourth Assessment Report of the Intergovernmental Panel on Climate Change*; Solomon, S., Qin, D., Manning, M., Chen, Z., Marquis, M., Averyt, K.B., Tignor, M., Miller, H.L., Eds.; Cambridge University Press: Cambridge, UK; New York, NY, USA, 2007; pp. 1–135.
34. IPCC. *Climate Change 2021: The Physical Science Basis. Contribution of Working Group I to the Sixth Assessment Report of the Intergovernmental Panel on Climate Change*; Masson-Delmotte, V., Zhai, P., Pirani, A., Connors, S.L., Péan, C., Berger, S., Caud, N., Chen, Y., Goldfarb, L., Gomis, M.I., et al., Eds.; Cambridge University Press: Cambridge, UK; New York, NY, USA, 2021; *In press*. [CrossRef]
35. Del Río, S.; Cano-Ortiz, A.; Herrero, L.; Penas, A. Recent trends in mean maximum and minimum air temperaturas over Spain (1961–2006). *Theor. Appl. Climatol.* **2012**, *149*, 605–626. [CrossRef]
36. Del Río, S.; Anjum, L.M.; Cano-Ortiz, A.; Herrero, L.; Hassan, A.; Penas, A. Recent mean temperature trends in Pakistan and links with teleconnection patternes. *Int. J. Climatol.* **2013**, *33*, 277–290. [CrossRef]
37. Del Río, S.; Canas, R.; Cano, E.; Cano-Ortiz, A.; Musarella, C.M.; Pinto Gomes, C.; Penas, A. Modelling the impacts of climte change on hábitat suitabilityy and vulnerability in deciduous forest in Spain. *Ecol. Indic.* **2021**, *131*, 108202. [CrossRef]
38. Lovelli, S.; Perniola, M.; Scalcione, E.; Troccoli, A.; Ziska, L.H. Future climate change in the Mediterranean area: Implications for water use and weed management. *Ital. J. Agron.* **2012**, *7*, 44–49. [CrossRef]
39. Macías, F. Sumideros de carbono para el forzamiento climático Antropoceno. Una visión de alternativas de actuación desde la ciencia del suelo. *Edafología* **2004**, *11*, 7–25.
40. Mota, C.; Alcaraz-López, C.; Iglesias, M.; Martínez-Ballesta, M.C.; Carvajal, M. *Investigación Sobre la Absorción de CO2 por los Cultivos Más Representativos de la Región de Murcia, Ed*; CSIC: Madrid, Spain, 2011; pp. 1–41.
41. Rosas, C.A. Sumideros de carbono. ¿Solución a la mitigación de los efectos del cambio climático? *Ecosistemas* **2018**, *11*, 1–6.
42. Spampinato, G.; Massimo, D.E.; Musarella, C.M.; De Paola, P.; Malerba, A.; Musolino, M. Carbon Sequestration by Cork Oak Forests and Raw Material to Built up Post Carbon City. In *New Metropolitan Perspectives. ISHT 2018. Smart Innovation, Systems and Technologies*; Calabrò, F., Della, S.L., Bevilacqua, C., Eds.; Springer: Cham, Switzerland, 2019; Volume 101. [CrossRef]
43. Spampinato, G.; Malerba, A.; Calabrò, F.; Bernardo, C.; Musarella, C.M. Cork Oak Forest Spatial Valuation Toward Post Carbon City by CO_2 Sequestration. In *New Metropolitan Perspectives. NMP 2020. Smart Innovation, Systems and Technologies*; Bevilacqua, C., Calabrò, F., Della, S.L., Eds.; Springer: Cham, Switzerland, 2021; Volume 178. [CrossRef]
44. Marshall, J.; Brown, V.; Boatman, N.; Lutman, P.; Squire, G. The impact of herbicides on weed abundance and biodiversity PN0940. A Report for the UK Pesticides Safety Directorate. *IACR Long Ashton. Res. Stn.* **2001**, *1*, 1–147.
45. Marshall, E.J.P. Biodiversity, herbicides and non-target plants. *Brighton. Crop. Prot. Conf. Weeds.* **2001**, *2*, 855–862.

46. Cano, E. Una Agricultura Respetuosa con el Medio Ambiente y Productiva es Posible: La Bioclimatología como Herramienta Viabilizadora. *Foro* **2021**, *5*, 17–23. Available online: https://www.revistaforo.com/2021/0502-03 (accessed on 15 May 2021).
47. Kumar, A.; Sharma, P.; Joshe, S. Assessing the Impacts of Climate Change on Land Productivity in Indian Crop Agriculture: An Evidence from Panel Data Analysis. *J. Agric. Sci. Technol.* **2016**, *18*, 1–13.
48. Tang, Y.C.; Mao, S.F.; Ma, X.Q.; Qiu, M.M.; Ma, K.; Zhu, M.X.; Wang, Z.J. The Influence of Three Different Types of Herbicides on Biodiversity. *Adv. Mater. Res.* **2013**, *838*, 2417–2426.
49. Vyas, M.D.; Jain, A.K. Effect of pre-and post-emergence herbicides on weed control and productivity of soybean (*Glycine max*). *Indian J. Agron.* **2003**, *48*, 309–311.
50. Wang, S.; Fu, B.J.; Gao, G.Y.; Yao, X.L.; Zhou, J. Soil moisture and evapotranspiration of different land cover types in the Loess Plateau, China. *Hydrol. Earth Syst. Sci.* **2012**, *16*, 2883–2892. [CrossRef]
51. Zhang, Y.K.; Schilling, K.E. Effects of land cover on water table, soil moisture, evapotranspiration, and groundwater re-charge: A field observation and analysis. *J. Hydrol.* **2006**, *319*, 328–338. [CrossRef]
52. Chamizo, S.; Meijide, A.; Serrano-Ortiz, P.; Sánchez-Cañete, E.P.; López-Ballesteros, A.; Kowalski, A.S. The influence of weeds on evapotranspiration and water use efficiency in an irrigated Mediterranean olive orchard. *EGU Gen. Assem. Conf. Abstr. Geophys. Res. Abstr.* **2018**, *20*, EGU2018–EGU14431.
53. Rivas-Martínez, S. Mapa de series, geoseries y geopermaseries de vegetación de España. Parte I. *Itinera Geobot.* **2007**, *17*, 5–436.
54. Cano, E.; Cano Ortiz, A. *Bioclimatología Aplicada a la Agronomía in Nuevas Tendencias en Olivicultura*; Servicio Publublicaciones Universidad Jaén: Jaén, Spain, 2016; pp. 7–69. ISBN 978-84-8439-000-0.
55. Cano Ortiz, A. *Bioindicadores y Cubiertas Vegetales en el Olivar in Nuevas Tendencias en Olivicultura*; Servicio Publicaciones Universidad Jaén: Jaén, Spain, 2016; pp. 70–117. ISBN 978-84-8439-000-0.
56. Perrino, E.V.; Musarella, C.M.; Magazzini, P. Management of grazing Italian river buffalo to preserve habitats defined by Directive 92/43/EEC in a protected wetland area on the Mediterranean coast: Palude Frattarolo, Apulia, Italy. *Euro-Mediterr. J. Environ. Integr.* **2021**, *6*, 32. [CrossRef]
57. Perrino, E.V.; Calabrese, G. Vascular flora of vineyards in the DOC area "Gioia del Colle" (Apulia, Southern Italy): Preliminary data. *Nat. Croat.* **2018**, *27*, 41–55. [CrossRef]
58. Cano-Ortiz, A. Bioindicadores ecológicos y manejo de cubiertas vegetales como herramienta para la implantación de una agricultura sostenible. Ph.D. Thesis, Universidad de Jaén, Jaén, Spain, 2007.
59. El-Hajj, A.; Halwani, B.; Halwani, J. Hydro-Geochemical Study of the Coastal Aquifer in Tripoli (Lebanon). *Res. J. Ecol. Environ. Sci.* **2022**, *2*, 103–117. [CrossRef]
60. Rivas-Martínez, S.; Díaz, T.E.; Fernández-González, F.; Izco, J.; Loidi, J.; Lousa, M.; Penas, A. Vascular Plant Communities of Spain and Portugal (addenda to the syntaxonomical checklist of 2001, part I). *Itinera Geobot.* **2002**, *15*, 5–432.
61. Rivas-Martínez, S.; Díaz, T.E.; Fernández-González, F.; Izco, J.; Loidi, J.; Lousa, M.; Penas, A. Vascular Plant Communities of Spain and Portugal (addenda to the syntaxonomical checklist of 2001, part II). *Itinera Geobot.* **2002**, *15*, 433–922.
62. Rivas-Martínez, S.; Rivas-Saenz, S. Orldwide Bioclimatic Classification System. Phytosociological Research Center, Spain. (1996–2020). Available online: http://www.globalbioclimatics.org (accessed on 15 May 2022).
63. Cano, E.; García Fuentes, A.; Torres, J.A.; Salazar, C.; Melendo, M.; Pinto Gomes, C.; Valle, F. Phytosociologie appliquée a la planification agricole. *Colloq. Phytosociol.* **1997**, *27*, 1007–1022.
64. Cano Ortiz, A.; Ighbareyeh, J.; Cano, E. Bioclimatic Applications and Soil Indicators for Olive Cultivation (South of the Iberian Peninsula). *Glob. Adv. Res. J. Agric. Sci.* **2014**, *3*, 433–438.
65. Cano, E.; Velóz Ramirez, A.; Cano-Ortiz, A.; Esteban, F.J. Distribution of Central American *Malastomataceae*: Biogeographical Analysis of the Caribbean Islands. *Acta Bot. Gallica.* **2009**, *156*, 527–557. [CrossRef]
66. Cano, E.; Cano-Ortiz, A.; Del Río, S.; Alatorre, J.; Velóz, A. Bioclimatic map of the. Dominican Republica. *Plant Sociol.* **2012**, *49*, 81–90. [CrossRef]
67. Cano-Ortiz, A.; Piñar Fuentes, J.C.; Quinto Canas, R.; Pinto Gomes, C.J.; Cano, E. Analysis of the Relationship Between Bioclimatology and Sustainable Developement. In *New Metropolitan Persperctives*; Calabrò, B., Spina, D., Eds.; Springer Nature: Berlin/Heidelberg, Germany, 2020; Volume 178, pp. 1291–1301, NMP 2020, SIST.
68. Burgos, J.L.; González, R.J.L. *Diagramas Bioclimáticos*; Ministerio de Agricultura Pesca y Alimentación: Madrid, Spain, 1983; pp. 1–382.
69. Cano, E.; Ruiz, L.; Melendo, M.; Nieto, J.; Cano Ortiz, A. Bases bioclimáticas para la planificación del olivar en el centro-sur de la Península Ibérica (España, Portugal). In *Actas IFOAN*; Sociedad Española de Agricultura Ecológica. SEAE: Valencia, Spain, 2004; pp. 305–311. ISBN 84-609-7732-3.
70. Cano, E.; Cano Ortiz, A. Bioclimatología y Bioindicadores del olivar: Bases fundamentales para un desarrollo sostenible. In *Asociación Grupo de Estudios Avanzados-Grupo Textura*; Andalucía, E.O., Ed.; Universidad de Sevilla: Sevila, Spain, 2011; pp. 83–97. ISBN 978-84-695-7405-8. Available online: https://investigacion.us.es/sisius/sis_showpub.php?idpers=4244 (accessed on 28 October 2022).
71. Reyes Palomino, S.E.; Cano CCoa, D.M. Efectos de la agricultura intensiva y el cambio climático sobre la biodiversidad. *Rev. De Investig. Altoandinas* **2022**, *24*, 53–64. [CrossRef]

72. Cisneros, O.; Modrego, P.; Rueda, J.; Ágreda, T.; Águeda, B.; Alonso, R.; Gordo, J.; Sombrero, A.; De Benito, A.; Tenorio, J.L.; et al. Evaluación de diferentes alternativas de forestación y agricultura como sumideros de carbono en Castilla y León. In *5º Congreso Forestal;* S.E.C.F.-Junta de Castilla y León, Ed.; Sociedad Española de Ciencias Forestales: Avila, Spain, 2009; pp. 2–10.
73. Cano, E.; García Fuentes, A.; Torres, J.A.; Pinto Gomes, C.J.; Cano Ortiz, A.; Montilla, R.J.; Muñoz, J.J.; Ruiz, L.; Rodríguez, A. Estudio de los quejigares de Sierra Morena oriental (Jaén). *Lagascalia* **2004**, *24*, 51–61.
74. Blasi, C.; Di Pietro, R.; Fortini, P. A phytosociological analysis of abandoned terraced olive grove shrublands in the Tyrrhenian district of Central Italy. *Plant Biosyst.* **2000**, *134*, 305–331. [CrossRef]
75. Rühl, J.; Pasta, S.; Schnittler, M. A chronosequence study of vegetation dynamics on abandoned vine and caper terraces of Pantelleria Island (Sicily). *Arch. Nat. Conserv. Landsc. Res.* **2006**, *45*, 71–94.
76. Calabrese, G.; Perrino, E.V.; Ladisa, G.; Aly, A.; Tesfmichael Solomon, M.; Mazdaric, S.; Benedetti, A.; Ceglie, F.G. Short-term effects of different soil management practices on biodiversity and soil quality of Mediterranean ancient olive orchards. *Org. Agric.* **2015**, *5*, 209–223. [CrossRef]
77. Patti, M.; Musarella, C.M.; Laface, V.L.A.; Cano-Ortiz, A.; Quinto-Canas, R.; Spampinato, G. The Use of Plants for Building Purposes in the Popular Tradition. In *New Metropolitan Perspectives. NMP 2022. Lecture Notes in Networks and Systems;* Calabrò, F., Della Spina, L., Piñeira Mantiñán, M.J., Eds.; Springer: Cham, Switzerland, 2022; Volume 482. [CrossRef]
78. Federación Española de Municipios y Provincias. *Los Sumideros de Carbono a Nivel Local. Edita;* Federación Española de Municipios y provincias: Vigo, Spain, 2010; p. 162.
79. NEIKER-Tecnalia. *Sumideros de Carbono de la Comunidad Autónoma del País Vasco: Capacidad de Secuestro y Medidas Para su Promoción;* Servicio Central de Publicaciones del Gobierno Vasco: Vitoria-Gasteiz, Spain, 2014; p. 212.
80. Nagy, D.U.; Rauschert, E.S.; Henn, T.; Cianfaglione, K.; Stranczinger, S.; Pal, R.W. The more we do, the less we gain? Balancing effort and efficacy in managing the *Solidago gigantea* invasion. *Weed Res.* **2020**, *60*, 232–240. [CrossRef]
81. Agencia Europea del Medio Ambiente. El Suelo y el Cambio Climático. 2015. Recuperado de. Available online: http://www.eea.europa.eu/es/senales/senales-2015/articu-los/el-suelo-y-el-cambio-climatico (accessed on 15 August 2022).
82. García Fuentes, A.; Cano, E. Fitosociología aplicada al conocimiento de los herbazales: Nuevo método para el cálculo del valor forrajero. *Ecol. Mediterránea* **1993**, *19*, 19–28. [CrossRef]
83. Bhanwaria, R.; Singh, B.; Musarella, C.M. Effect of Organic Manure and Moisture Regimes on Soil Physiochemical Properties, Microbial Biomass C_{mic}:N_{mic}:P_{mic} Turnover and Yield of Mustard Grains in Arid Climate. *Plants* **2022**, *11*, 722. [CrossRef]
84. Panuccio, M.R.; Mallamaci, C.; Attinà, E.; Muscolo, A. Using Digestate as Fertilizer for a Sustainable Tomato Cultivation. *Sustainability* **2021**, *13*, 1574. [CrossRef]
85. Švehláková, H.; Turčová, B.; Rajdus, T.; Plohák, P.; Nováková, J. Effective combination of management methods suppresses invasive Jerusalem artichoke. *IOP Conf. Ser. Earth Environ. Sci.* **2021**, *900*, 012045. [CrossRef]
86. Zhang, Y.; Yang, X.; Zhu, Y.; Li, L.; Zhang, Y.; Li, J.; Qiang, S. Biological control of *Solidago canadensis* using a bioherbicide isolate of *Sclerotium rolfsii* SC64 increased the biodiversity in invaded habitats. *Biol. Control.* **2019**, *139*, 104093. [CrossRef]
87. Powles, S.B.; Yu, Q. Evolution in Action: Plants Resistant to Herbicides. *Annu. Rev. Plant Biol.* **2010**, *61*, 317–347. [CrossRef]
88. Hermosín, M.C.; Rodríguez-Linaza, L.C.; Cornejo, J.; Ordóñez-Fernández, R. Efecto de uso de agroquímicos en el olivar sobre la calidad de las aguas. *Sostenibilidad Prod. Oliv. Andal.* **2009**, 127–160.

agronomy

MDPI

Article

Floristic Diversity of Jabal Al-Ward, Southwest Tabuk Region, Kingdom of Saudi Arabia

Ahmed Elkordy [1,*], Iman H. Nour [2], Faten Y. Ellmouni [3], Najla A. Al Shaye [4], Dhafer A. Al-Bakre [5] and Ahmed EL-Banhawy [6]

[1] Botany and Microbiology Department, Faculty of Science, Sohag University, Sohag 82524, Egypt
[2] Botany and Microbiology Department, Faculty of Science, Alexandria University, Alexandria 21511, Egypt
[3] Botany Department, Faculty of Science, Fayoum University, Fayoum 63514, Egypt
[4] Department of Biology, College of Science, Princess Nourah bint Abdulrahman University, Riyadh 11564, Saudi Arabia
[5] Biology Department, College of Science, University of Tabuk, Tabuk 47512, Saudi Arabia
[6] Botany and Microbiology Department, Faculty of Science, Suez Canal University, Ismailia 41522, Egypt
[*] Correspondence: aelkordy@science.sohag.edu.eg

Abstract: Jabal Al-Ward is one of the Hijazi mountains situated between Al-Ulā and Al-Wajh, southwest Tabuk Province, Saudi Arabia's northwesterly border region. It is considered the highest mountain in this area and is enriched in wildlife. For the first time, the present research aimed to investigate the floristic composition, phytogeographical distribution, and plant diversity in Jabal Al-Ward. One hundred ninety-eight species representing 47 plant families have been identified. The Asteraceae, Poaceae, and Fabaceae represented more than a third of the region's floristic composition. The perennial species (53.5%) were dominant over the annuals (46.46%). This is a prominent feature in Jabal Al-Ward, where the perennial species may be more tolerant of climatic changeability than the annuals. Seven life form categories were found; therophytes (46.46%) showed to be the most common life form. In addition, there were four main phytogeographical groups: Mono-regional, Bi-regional, Pluri-regional, and Worldwide. The Mono-regional and Bi-regional categories had the highest participation, with 38.5% and 37.4%, respectively. Thirty-six species (18.2%) were found to be native to the Saharo-Arabian region. The Saharo-Arabian region was combined with eight more regions, including Saharo-Arabian/Sudano-Zambesian (12.6%), Irano-Turanian/Saharo-Arabian (9.1%), Mediterranean/Saharo-Arabian (5.6%), Irano-Turanian/Mediterranean/Saharo-Arabian (4.5%), Irano-Turanian/Saharo-Arabian/Sudano-Zambesian (2%), Euro-Siberian/Irano-Turanian/Mediterranean/Saharo-Arabian and Saharo-Arabian/Sudanian (1% each), Mediterranean/Saharo-Arabian/Sudano-Zambesian, and Irano-Turanian/Mediterranean/Saharo-Arabian/Saharo-Zambesian (0.5% each). The current study demonstrated the highest species richness compared to earlier research on various locations in Tabuk Province. In future work, the species and endemic richness along elevation gradients should be studied in Jabal Al-Ward. As well as the IUCN status of each taxon, the DNA barcoding of endangered species will be of great significance if applied in the surveyed area.

Keywords: Arabian Peninsula; chorology; flora of Saudi Arabia; flora of West Asia; Hijazi Mountains; Jabal Al-Ward; Tabuk; Saudi Arabia

Citation: Elkordy, A.; Nour, I.H.; Ellmouni, F.Y.; Al Shaye, N.A.; Al-Bakre, D.A.; EL-Banhawy, A. Floristic Diversity of Jabal Al-Ward, Southwest Tabuk Region, Kingdom of Saudi Arabia. *Agronomy* **2022**, *12*, 2626. https://doi.org/10.3390/agronomy12112626

Academic Editors: Alessio Papini, Mushtaq Ahmad, Fazal Ullah and Wajid Zaman

Received: 7 October 2022
Accepted: 21 October 2022
Published: 25 October 2022

1. Introduction

The Kingdom of Saudi Arabia's total area is around 2,250,000 km^2. This vast area holds diverse heterogenic landscapes. It is considered an arid or semi-arid region, where the xerophytic plants represent the highest proportion of Saudi Arabian flora [1]. The country's topography reflects the diversified flora explored by Collenette [2] and Chaudhary [3–5]. The flora of Saudi Arabia is one of the wealthiest biodiversity areas in the Arabian Peninsula, with substantial crops and medicinal plants [6]. Saudi Arabia contained 2223 species and 816 genera assigned to 129 families, with about 147 endemic species [5]. The flora's taxa are of combined origin; North African, East African, Mediterranean, and Irano-Turanian plants [7,8].

The Tabuk region occupies approximately 146,072 km^2 and is situated in the north-western part of Saudi Arabia, with diverse ecosystems and habitats [9,10]. One of the most important agricultural areas in Saudi Arabia is the Tabuk region, which produces about a quarter of the country's olive and dates. A floristic survey was conducted in eight discrete regions of Tabuk by Rajasab [11] to reveal 198 plant species belonging to 52 families. Eighty-two species representing 30 families were recognized by Moawed and Ansari [12] in the coastal areas of the Red Sea (from Ras Hemaid to Umluj road across Al Wajh and Dhiba regions) in Tabuk province. Al-Mutairi et al. [13] recorded ninety-six species exemplifying thirty-eight families at four localities in Tabuk (Sharma, Alqan, Al-Lwaz Mountains, and Alzetah). A total of 102 species belonging to 34 families were recorded in the Alaqan region by Moawed [14]. At the Harrat ArRahah, a volcanic field in South Tabuk, 135 species represent 34 families by Fakhry and Al-Kenany [15]. In addition, Alghanem et al. [16] identified 30 perennial species (15 families) at Al-Wadi Al-akhder. At the same time, 69 weed species representing 20 families were detected in farms of the Tabuk region [17]. Al Mutairi [9] reported one hundred and sixty-three species assigned to forty-one families at five sites (Aldesah, Alzetah, Alawz, Harra and Sharma). Ansari et al. [18] have estimated the identified species cited in the different studies of Tabuk and mentioned two hundred twenty-seven species in their comprehensive review. These species belonged to 157 genera and 45 angiosperms' families, including 12 endangered species. However, *Astragalus collenettiae* Hedge & Podlech was the only endemic species recorded by Fakhry & Kenany [15] in the Harrat ArRahah.

Achillea fragrantissima (Forssk.) Sch.Bip., *Artemisia judaica* L., *Ochradenus baccatus* Delile and *Rhazya stricta* Decne. Were mentioned to have potent anti-microbial activity against nosocomial pathogens [19]. Al-Mutairi et al. [13] and Moawed [14] reported that the dominant plant families were: Asteraceae, Brassicaceae, Chenopodiaceae, Fabaceae, Lamiaceae, Poaceae, Resedaceae, and Zygophyllaceae. Moreover, the plant life form categories, phytogeographical affinities, and the economic importance of the vegetation in this area have been studied [9,13,14,20].

Al-Mutairi et al. [13] highlighted the importance of exploring floristic compositions for analyzing socioeconomic and vegetation in this unique area. Moreover, Al-Mutairi et al. [9,13] stated that Tabuk is a biodiversity hot spot region and should be subjected to a conservation program to protect its natural resources. Assessing the floristic composition in a particular area provides insights into species, habitats, or ecosystem diversity. Thus, suitable conservation strategies should be implemented to save the threatened taxa and their habitats [21]. In addition, studies on regional floras are crucial to perform cross-biome comparisons. These studies also serve as the basis for further analyses of the plant communities' function, structure, and evolution in order to better understand the delimitation and mapping of biomes [22,23].

Jabal Al-Ward is one of the Hijazi mountains located between Al-Ulā and Al-Wajh, southwest Tabuk province, where this region is about 900 and 1600 m asl. However, Jabal Al-Ward is regarded as the highest mountain in this region, with up to 2096 m asl [24]. Generally, Jabal Al-Ward is enriched in wildlife, including flora, amphibians, and reptiles [25,26]. This region mainly consists of some valleys that cut across the mountains and act as large natural drainages [24].

Despite the previous trials to describe the floristic composition of the Tabuk region [9,11–17,19,20], Jabal Al-Ward has remained unsatisfactorily studied. Therefore, the present study contributed to investigating and evaluating the floristic composition, phytogeographical distribution, and plant diversity of the wild plant taxa growing in the Jabal Al-Ward region, southwest Tabuk, Saudi Arabia, for the first time.

2. Materials and Methods

2.1. Study Area

This study was performed in the Jabal Al-Ward region (Figure 1) during most plant species' active growing seasons.

Figure 1. Map of the Kingdom of Saudi Arabia showing the study area's location. The red rectangle refers to Jabal Al-Ward region.

The excursions were conducted during the growing seasons of 2019 and 2020. The eremic vegetation was surveyed in 25 localities, in the mountain ridges, distributed to cover the study area, and their elevations ranged from 1435 m to 1807 m asl. The geographical coordinates were recorded for each locality (Table 1). Each locality was visited at least three times to ensure that the required field data was entirely recorded. Sampling surveys using quadrats of 50 m × 50 m is recommended for arid habitats [27,28]; however, the quadrate size 10 m × 10 m was suitable for the current study, as the species number was constant in the latter quadrate size.

The quadrate was randomly distributed at each site to record the vegetation of the studied area. Species included in the quadrate's border were counted in the sampling. The plants with their vegetative branches spreading over the quadrate's sides were also considered, even though their roots were anchored outside the borders.

The study area lies within the eremic subtropical dry zone, characterized by hot summers and warm winters [29]. The Tabuk region had 33.5 mm of annual precipitation on average from 1978 to 2013, which is less than the 250 mm worldwide average for arid regions [30]. Data concerning monthly climatology of min-, mean- and max-temperature, and precipitation during the period of 1991–2020 at Tabuk, Saudi Arabia, were obtained from the Climate Change Knowledge Portal [31]. The average annual rainfall from 1991 to 2020 was 5.725 mm, and the minimum was detected in August and September (0.42 and 0.4 mm, respectively), while the maximum was in November (14.95 mm). The average monthly minimum temperature was 6.9 °C in January, whereas the average maximum was 37.81 °C in August (Figure 2).

Table 1. Coordinates of 25 sampling sites in Jabal Al-Ward.

Locations	Coordinates	
	Latitude (N)	Longitude (E)
1	26°26′52.7″	37°15′51.9″
2	26°26′53.7″	37°15′52.7″
3	26°26′54.5″	37°15′53.3″
4	26°26′54.9″	37°15′54.2″
5	26°26′55.1″	37°15′55.1″
6	26°26′55.5″	37°15′56.1″
7	26°26′56.4″	37°15′56.5″
8	26°26′57.3″	37°15′57.2″
9	26°26′57.2″	37°15′58.2″
10	26°26′56.9″	37°15′59.4″
11	26°26′56.4″	37°16′00.8″
12	26°26′56.0″	37°16′02.4″
13	26°26′55.6″	37°16′03.5″
14	26°26′55.7″	37°16′04.7″
15	26°26′56.4″	37°16′05.2″
16	26°26′57.3″	37°16′05.3″
17	26°26′58.2″	37°16′05.6″
18	26°26′58.9″	37°16′06.1″
19	26°26′58.6″	37°16′07.1″
20	26°26′57.9″	37°16′07.9″
21	26°26′57.4″	37°16′08.5″
22	26°26′56.6″	37°16′09.1″
23	26°26′56.0″	37°16′09.8″
24	26°26′55.5″	37°16′10.7″
25	26°26′55.2″	37°16′11.8″

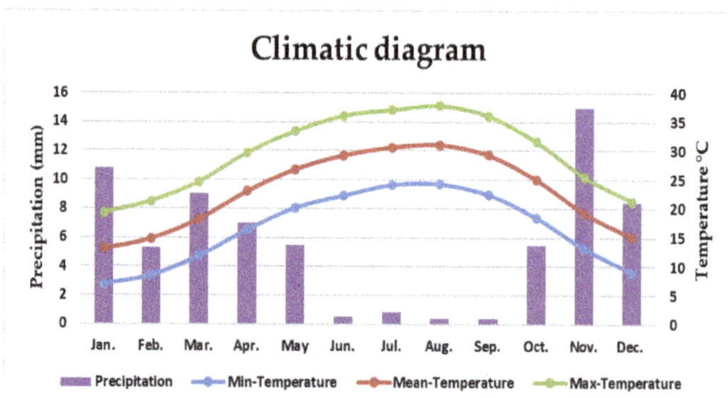

Figure 2. Climatic diagram showing min-, mean- and max-temperature, and precipitation during (1991–2020) in Tabuk, Saudi Arabia. The chart is based on the Climate Change Knowledge Portal data (CCKP 2021).

2.2. Plant Collection and Species Identification

Identification of the collected taxa followed Collenette [2], Cope [32], Mighaid [33], and Chaudhary [3–5]. Nomenclature was revised for accepted names through the Plants of the World Online (POWO) [34] and Taxonomic Name Resolution Service (TNRS) [35]. Voucher specimens were deposited in the herbarium of the Biology Department, College of Sciences, Taibah University Al-Ulā branch. Life form categories were recognized after Govaerts et al. [36], who updated Raunkiaer's [37] classification. Phytogeographical categories were distinguished based on Wickens [38], Zohary [39], and White and Léonard [40]. The generic coefficient (GC) was calculated according to Jaccard [41], which is the ratio of the total number of genera to the total number of species. The statistical analysis was performed using Microsoft Excel to create the charts and histograms.

3. Results

Floristic Composition

This study identified 198 vascular plant species in Jabal Al-Ward, and no gymnosperms were listed. The recorded taxa belonged to 152 genera in 47 plant families. The Dicotyledonous families accounted for 83%, while the Monocotyledonous represented 17% (Figure 3). The Monocotyledonous families are Asparagaceae, Asphodelaceae, Colchicaceae, Cyperaceae, Iridaceae, Liliaceae, and Poaceae. Perennial species constituted 53.54% (106 species); meanwhile, the annuals recorded 46.46% (92 species).

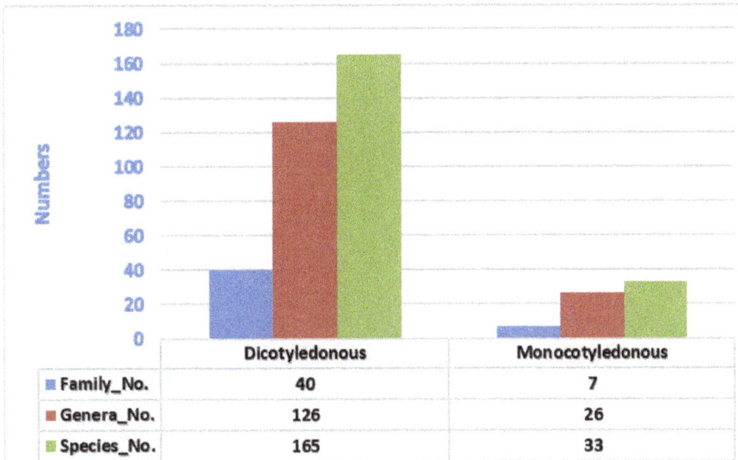

Figure 3. Histogram of the family, genera, and species number based on class Dicotyledonous and Monocotyledonous in Jabal Al-Ward.

The floristic analysis of the Jabal Al-Ward region revealed the presence of the following families: Asteraceae (29 species), Poaceae (25 species), Fabaceae (23 species), and Brassicaceae (14 species); they account for (14.57%, 12.56%, 11.55%, and 7.53%) of the total flora of the study area respectively. Amaranthaceae, Lamiaceae, Solanaceae, and Zygophyllaceae were represented by 9–7 species, accounting for 8% of the entire flora. Apocynaceae, Boraginaceae, and Euphorbiaceae were represented by 6–5 species. Caryophyllaceae, Geraniaceae, Plantaginaceae, and Resedaceae were represented by four species. Two to three species represented ten plant families, while twenty-two plant families were represented by at least one plant species (Table 2, Figure 4). The species richness in Jabal Al-Ward accounted for 8.9%, genera 18.6%, and families 36.43% of the total flora of Saudi Arabia. The generic coefficient (%) is 76.8% in the study area, while Saudi Arabia accounted for 36.7%. *Anthemis scrobicularis* Yavin was the only endemic species recognized in the study area.

Table 2. List of plant taxa recorded in Jabal Al-Ward. Life span: Annual (ANN), Perennial (PER). Life form: Chamaephytes (CH), Cryptophytes (CR), Helophytes (HEL), Hemicryptophytes (HEM), Parasites (PA), Phanerophytes (PH), Therophytes (TH). Chorotypes: American (AM), Cosmopolitan (COSM), Euro-Siberian (ES), Irano-Turanian (IT), Mediterranean (ME), Neotropical (NEO), Paleotropical (PAL), Pantropical (PAN), Saharo-Arabian (SA), Sudano-Zambesian (SZ), and Sudanian (SUD).

Plant Family	Taxa	Life Form	Chorology	Life Span
Acanthaceae	*Blepharis edulis* (Forssk.) Pers.	CH	IT + SA + SZ	PER
	Hygrophila auriculata (Schumach.) Heine	TH	PAN	ANN
Aizoaceae	*Aizoon canariense* L.	TH	SZ	ANN
Amaranthaceae	*Aerva javanica* (Burm.f.) Juss. ex Schult.	CH	PAL	PER
	Alternanthera pungens Kunth	TH	NEO	ANN
	Amaranthus hybridus L.	TH	PAN	ANN
	Bassia muricata (L.) Asch.	TH	IT + SA	ANN
	Bassia eriophora (Schrad.) Asch.	CH	IT + SA	PER
	Chenopodium murale L.	TH	COSM	ANN
	Halocnemum strobilaceum (Pall.) M.Bieb.	CH	IT + ME + SA	PER
	Haloxylon salicornicum (Moq.) Bunge ex Boiss.	CH	SA	PER
	Caroxylon imbricatum (Forssk.) Moq. Syn. *Salsola imbricata* Forssk.	PH	SZ	PER
Apiaceae	*Deverra tortuosa* (Desf.) DC.	CH	SA	PER
	Ferula sinaica Boiss.	HEM	IT	PER
Apocynaceae	*Calotropis procera* (Aiton) Dryand.	PH	SA + SZ	PER
	Apteranthes europaea (Guss.) Murb. Syn. *Caralluma europaea* (Guss.) N.E.Br.	CH	ME + SA	PER
	Leptadenia pyrotechnica (Forssk.) Decne.	PH	SA	PER
	Marsdenia tenacissima (Roxb.) Moon Syn. *Pergularia tomentosa* L.	CH	SA + SZ	PER
	Rhazya stricta Decne.	CH	SA + SZ	PER
	Solenostemma oleifolium (Nectoux) Bullock & E.A.Bruce ex Maire Syn. *Solenostemma argel* (Delile) Hayne	CH	SA	PER
Asparagaceae	*Bellevalia flexuosa* Boiss.	CR	ME	PER
Asphodelaceae	*Asphodelus tenuifolius* Cav.	CR	SA + SZ	PER
Asteraceae	*Achillea fragrantissima* (Forssk.) Sch.Bip.	CH	IT + SA	PER
	Achillea falcata L.	CH	IT	PER
	Anthemis scrobicularis Yavin	TH	IT + ME + SA	ANN
	Anthemis melampodina (Boiss.) Eig subsp. *deserti*	TH	ME + SA	ANN
	Artemisia herba-alba Asso	CH	SA	PER
	Artemisia judaica L.	CH	SA	PER
	Atractylis carduus (Forssk.) C. Chr.	TH	ME	ANN
	Calendula tripterocarpa Rupr.	TH	PAN	ANN
	Centaurea procurrens Sieber ex Spreng.	CH	ME	PER
	Centaurea sinaica DC.	TH	ES + IT	ANN
	Echinops glaberrimus DC.	HEM	SA	PER
	Echinops spinosissimus Turra	CH	IT + SA	PER

Table 2. *Cont.*

Plant Family	Taxa	Life Form	Chorology	Life Span
	Filago desertorum Pomel	TH	IT + SA	ANN
	Helichrysum stoechas (Ten.) Nyman subsp. *barrelieri*	HEM	ME	PER
	Ifloga spicata (Forssk.) Sch.Bip.	TH	SA	ANN
	Ifloga spicata (Forssk.) Sch. Bip. subsp. *albescens Chrtek*	TH	ME + SA	ANN
	Lactuca serriola L.	TH	IT	ANN
	Pseudognaphalium luteoalbum (L.) Hilliard & B.L.Burtt Syn. *Laphangium luteoalbum* (L.) Tzvelev	HEL	IT + ME + SA	PER
	Launaea procumbens (Roxb.) Ramayya & Rajagopal	TH	SA	ANN
	Launaea spinosa (Forsk.) Sch.Bip. ex Kuntze	CH	SA	PER
	Onopordum boissierianum Raab-Straube & Greuter Syn. *Onopordum sibthorpianum* Boiss. & Heldr.	HEM	ME + SA	PER
	Picris cyanocarpa Boiss.	TH	SA	ANN
	Pulicaria incisa (Lam.) DC.	CH	SA + SZ	PER
	Pulicaria undulata (L.) C.A.Mey.	CH	SA + SZ	PER
	Ramaliella tortuosissima (Boiss.) Zaika, Sukhor. & N.Kilian Syn. *Scorzonera tortuosissima* Boiss.	CH	IT + SA	PER
	Senecio glaucus L.	TH	IT + SA	ANN
	Sonchus oleraceus (L.) L.	TH	COSM	ANN
	Tanacetum sinaicum (Fresen.) Delile ex K.Bremer & Humphries	CH	IT	PER
	Urospermum picroides (L.) Scop. ex F.W.Schmidt	TH	IT + ME	ANN
Boraginaceae	*Alkanna orientalis* (L.) Boiss.	CH	IT + ME	PER
	Arnebia hispidissima (Lehm.) A.DC.	TH	SA + SZ	ANN
	Echium horridum Batt.	TH	ME + SA	ANN
	Gastrocotyle hispida (Forssk.) Bunge	TH	IT + SA	ANN
	Trichodesma africanum (L.) Sm.	TH	SA + SZ	ANN
Heliotropiaceae	*Heliotropium bacciferum* Forssk.	CH	SA + SUD	PER
	Heliotropium crispum Desf.	HEM	IT + SA	PER
	Heliotropium europaeum L.	TH	ES + IT + ME	ANN
Brassicaceae	*Coincya tournefortii* (Gouan) Alcaraz, T.E.Díaz, Rivas Mart. & Sánchez-Gómez Syn. *Brassica tournefortii* Gouan	TH	ME + SA	ANN
	Diplotaxis harra (Forssk.) Boiss.	TH	IT + SA	ANN
	Diplotaxis acris (Forssk.) Boiss.	TH	IT + ME	ANN
	Enarthrocarpus strangulatus Boiss.	TH	SA	ANN
	Eremobium aegyptiacum (Spreng.) Asch. ex Boiss.	HEM	SA	PER
	Eremobium aegyptiacum var. *lineare* (Delile) Zohary	HEM	SA	PER
	Farsetia aegyptia Turra	CH	SA + SZ	PER
	Lepidium didymum L.	TH	SA	ANN
	Matthiola arabica Boiss.	TH	SA	ANN
	Morettia canescens Boiss.	CH	ME	PER
	Morettia parviflora Boiss.	TH	SZ	ANN

Table 2. *Cont.*

Plant Family	Taxa	Life Form	Chorology	Life Span
	Rhamphospermum arvense (L.) Andrz. ex Besser Syn. *Sinapis arvensis* L.	TH	ME	ANN
	Sisymbrium irio L.	TH	ES + IT + ME + SA	ANN
	Zilla spinosa (L.) Prantl subsp. *spinosa*	CH	ES + IT + ME + SA	PER
Caryophyllaceae	*Gymnocarpos decandrus* Forssk.	CH	SA	PER
	Gypsophila viscosa Murray	TH	IT	ANN
	Polycarpaea robbairea (Kuntze) Greuter & Burdet	CH	SA + SZ	PER
	Silene villosa Forssk.	TH	SA	ANN
Cistaceae	*Helianthemum lippii* (L.) Dum.Cours.	CH	SA + SZ	PER
Cleomaceae	*Cleome amblyocarpa* Barratte & Murb.	TH	SA + SZ	ANN
	Cleome arabica L.	CH	SUD	PER
Colchicaceae	*Colchicum ritchii* R.Br.	CR	SA	PER
	Colchicum tunicatum Feinbrun	CR	IT	PER
Convolvulaceae	*Convolvulus arvensis* L.	CR	PAL	PER
Crassulaceae	*Umbilicus intermedius* Boiss.	CR	IT + ME	PER
Cucurbitaceae	*Citrullus colocynthis* (L.) Schrad.	HEM	IT + ME + SA + SZ	PER
	Cucumis prophetarum L.	HEM	SA	PER
Cyperaceae	*Cyperus rotundus* L.	CR	COSM	PER
Ephedraceae	*Ephedra aphylla* Forssk.	PH	SA	PER
Euphorbiaceae	*Chrozophora oblongifolia* (Delile) A.Juss. ex Spreng.	CH	SZ	PER
	Euphorbia granulata Forssk.	TH	SZ	ANN
	Euphorbia peplus L.	TH	COSM	ANN
	Euphorbia prostrata Aiton	TH	COSM	ANN
	Euphorbia retusa Forssk.	TH	COSM	ANN
Fabaceae	*Vachellia tortilis* (Forssk.) Galasso & Banfi Syn. *Acacia tortilis* (Forssk.) Hayne	PH	SZ	PER
	Alhagi graecorum Boiss.	CH	IT + ME	PER
	Astragalus eremophilus Boiss.	TH	IT + ME + SA	ANN
	Astragalus palaestinus Eig	TH	ME	ANN
	Astragalus spinosus (Forssk.) Muschl.	CH	IT + SA	PER
	Astragalus tribuloides Delile	TH	IT + SA	ANN
	Crotalaria aegyptiaca Benth.	HE	SZ	PER
	Hippocrepis cyclocarpa Murb.	TH	ME	ANN
	Hippocrepis unisiliquosa L.	TH	IT + SA	ANN
	Lotus corniculatus L.	HEM	COSM	PER
	Lotus glinoides Delile	TH	SZ	ANN
	Lotus lanuginosus Vent.	HEM	SA	PER
	Medicago laciniata (L.) Mill.	TH	SA	ANN
	Medicago polymorpha L.	TH	COSM	ANN
	Melilotus indicus(L.) All.	TH	PAL	ANN
	Onobrychis ptolemaica (Delile) DC.	HEM	IT	PER
	Ononis natrix L.	CH	ME	PER
	Retama raetam (Forssk.) Webb	PH	SA	PER
	Senna italica Mill.	CH	SZ	PER
	Tephrosia purpurea (L.) Pers.	CH	SA + SZ	PER
	Trigonella anguina Delile	TH	SA	ANN

Table 2. *Cont.*

Plant Family	Taxa	Life Form	Chorology	Life Span
	Trigonella glabra Thunb. subsp. *glabra* Syn. *Trigonella hamosa* Del. ex Smith	TH	ME + SA	ANN
	Trigonella stellata Forssk.	TH	IT + SA	ANN
Geraniaceae	*Erodium laciniatum* (Cav.) Willd.	TH	SA	ANN
	Erodium oxyrhinchum M.Bieb.	TH	SA + SZ	ANN
	Erodium oxyrhinchum (Boiss.) Schönb.-Tem.subsp. *bryoniifolium*	TH	IT	ANN
	Monsonia nivea (Decne.) Webb	CR	SA + SZ	PER
Iridaceae	*Moraea sisyrinchium* (L.) Ker Gawl.	CR	IT + ME	PER
Lamiaceae	*Pseudodictamnus undulatus* (Benth.) Salmaki & Siadati Syn. *Ballota undulata* (Sieber ex Fresen.) Benth.	CH	ME	PER
	Lavandula coronopifolia Poir.	CH	SA + SZ	PER
	Lavandula pubescens Decne.	CH	SA + SZ	PER
	Mentha longifolia (L.) Huds.	CH	PAL	PER
	Mentha longifolia (Briq.) Harley subsp. *typhoides*	HEM	PAL	PER
	Otostegia fruticosa (Forssk.) Schweinf. ex Penzig	CH	SA	PER
	Stachys aegyptiaca Pers.	CH	ME + SA	PER
	Teucrium polium L.	CH	IT + ME	PER
Liliaceae	*Gagea commutata* K. Koch	CR	IT	PER
	Gagea reticulata (Pall.) Schult. & Schult.f.	CR	IT	PER
Malvaceae	*Malva neglecta* Wallr.	TH	ES + IT	ANN
	Malva parviflora L.	TH	IT + ME	ANN
Molluginaceae	*Glinus lotoides* L.	TH	PAL	ANN
Moraceae	*Ficus palmata Forssk.*	PH	SUD	PER
Nitrariaceae	*Peganum harmala* L.	HEM	IT + ME + SA	PER
Nyctaginaceae	*Boerhavia repens* L.	CH	PAL	PER
Orobanchaceae	*Cistanche tubulosa* (Schenk) Wight	PA	IT + SA + SZ	PER
Papaveraceae	*Fumaria parviflora* Lam.	TH	COSM	ANN
Plantaginaceae	*Kickxia floribunda* (Boiss.) Täckh. & Boulos	CH	SA	PER
	Nanorrhinum hastatum (R.Br. ex Benth.) Ghebr.	CH	SA + SUD	PER
	Plantago ciliata Desf.	TH	IT + ME + SA	ANN
	Plantago lanceolata L.	TH	IT + ME + SA	ANN
Poaceae	*Avena barbata* Pott ex Link	TH	ME	ANN
	Bromus scoparius L.	TH	IT + ME + SA	ANN
	Bromus danthoniae Trin.	TH	IT	ANN
	Bromus madritensis L.	HEM	IT + ME	PER
	Cenchrus echinatus L.	TH	AM	ANN
	Centropodia fragilis (Guinet & Sauvage) Cope	TH	ME + SA	ANN
	Cutandia memphitica (Spreng.) Benth.	TH	IT + ME	ANN
	Cynodon dactylon (L.) Pers.	CR	PAN	PER
	Enneapogon desvauxii P.Beauv.	HEM	IT + SA	PER
	Eragrostis aegyptiaca (Willd.) Delile	TH	SZ	ANN
	Eremopyrum bonaepartis (Spreng.) Nevski	TH	IT	ANN
	Lamarckia aurea (L.) Moench	TH	IT + ME	ANN
	Lolium perenne L.	HEM	COSM	PER
	Panicum turgidum Forssk.	CR	SA + SZ	PER
	Parapholis incurva (L.) C.E.Hubb.	TH	IT + ME	ANN

Table 2. *Cont.*

Plant Family	Taxa	Life Form	Chorology	Life Span
	Phalaris minor Retz.	TH	IT + ME	ANN
	Phalaris paradoxa L.	TH	IT + ME	ANN
	Poa annua L.	TH	ES + IT + ME	ANN
	Polypogon monspeliensis (L.) Desf.	TH	IT + ME + SA	ANN
	Rostraria pumila (Desf.) Tzvelev	TH	ME + SA	ANN
	Schismus barbatus (L.) Thell.	TH	IT + SA	ANN
	Stipellula capensis (Thunb.) Röser &Hamasha Syn. *Stipa capensis* Thunb.	TH	IT + SA	ANN
	Stipa lagascae Roem. & Schult.	HEM	IT + ME	PER
	Stipa obtusa (Nees & Meyen) Hitchc.	HEM	SA + SZ	PER
	Stipagrostis ciliata (Desf.) De Winter	HEM	IT + SA + SZ	PER
Polygonaceae	*Rumex spinosus* L. Syn. *Emex spinosa* (L.) Campd.	TH	PAN	ANN
	Rumex vesicarius L.	TH	ME + SA + SZ	ANN
Portulacaceae	*Portulaca oleracea* L.	TH	ME + SA	ANN
Primulaceae	*Anagallis arvensis* L.	TH	ES + IT + ME	ANN
Rhamnaceae	*Ziziphus spina-christi* (L.) Desf.	PH	SA + SZ	PER
Resedaceae	*Caylusea hexagyna* (Forssk.) M.L.Green	TH	SA + SZ	ANN
	Ochradenus baccatus Delile	PH	SA	PER
	Reseda muricata C. Presl	CH	SA	PER
	Reseda sphenocleoides Deflers	CH	SA + SZ	PER
Rubiaceae	*Galium spurium* L.	TH	ES + IT + ME	ANN
Scrophulariaceae	*Verbascum sinaiticum* Benth.	HEM	IT + SA	PER
Solanaceae	*Datura innoxia* Mill.	TH	COSM	ANN
	Hyoscyamus desertorum (Asch. ex Boiss.) Täckh.	TH	SA	ANN
	Hyoscyamus pusillus L.	TH	IT	ANN
	Lycium depressum Stocks	PH	IT	PER
	Lycium shawii Roem. & Schult.	PH	SA + SZ	PER
	Solanum incanum L.	CH	SUD	PER
	Solanum americanum Mill.	HEM	ES + IT + ME	PER
	Withania somnifera (L.) Dunal	CH	IT + ME	PER
Tamaricaceae	*Tamarix aphylla* (L.) H.Karst.	PH	IT + SA + SZ	PER
Typhaceae	*Typha domingensis* Pers.	HEL	PAN	PER
Urticaceae	*Forsskaolea tenacissima* L.	CH	SA + SZ	PER
	Parietaria alsinifolia Delile	TH	SA	ANN
Zygophyllaceae	*Fagonia arabica* L.	CH	SA	PER
	Fagonia bruguieri DC.	CH	SA	PER
	Fagonia glutinosa Delile	CH	SA	PER
	Fagonia mollis Delile	CH	SA	PER
	Seetzenia lanata (Willd.) Bullock	TH	SUD	ANN
	Tetraena simplex (L.) Beier & Thulin	TH	SA + SZ	ANN
	Tribulus terrestris L.	TH	ES + IT + ME	ANN

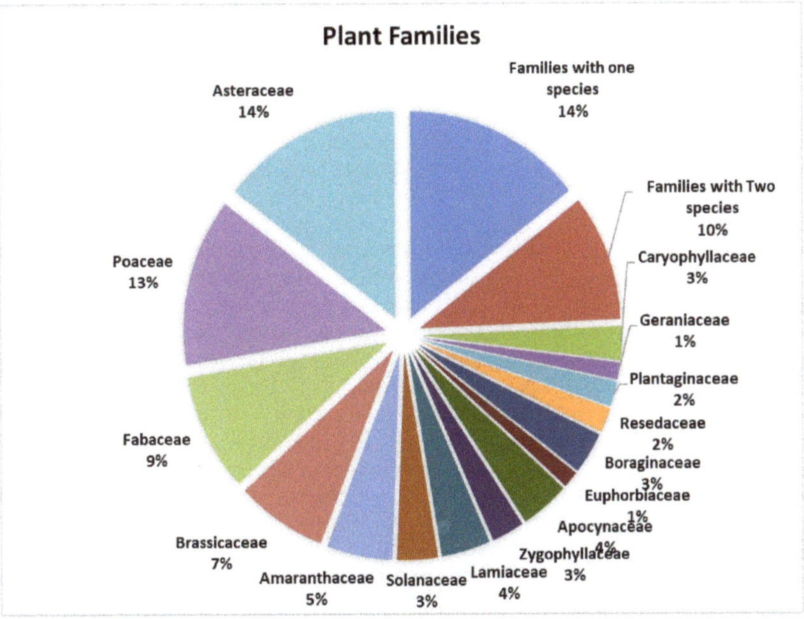

Figure 4. The proportional contribution of plant families in Jabal Al-Ward.

Seven life form categories were observed in the current study. The most frequent group was the therophytes, with 92 species, followed by chamaephytes, which were identified as 55 different species. There were 22 species of hemicryptophytes. Thirteen species were cryptophytes and twelve species were identified as phanerophytes. Three species were helophytes and one parasitic species was recorded in the Jabal Al-Ward area. Percentages of each life form relevant to the total number of species are illustrated in Figure 5.

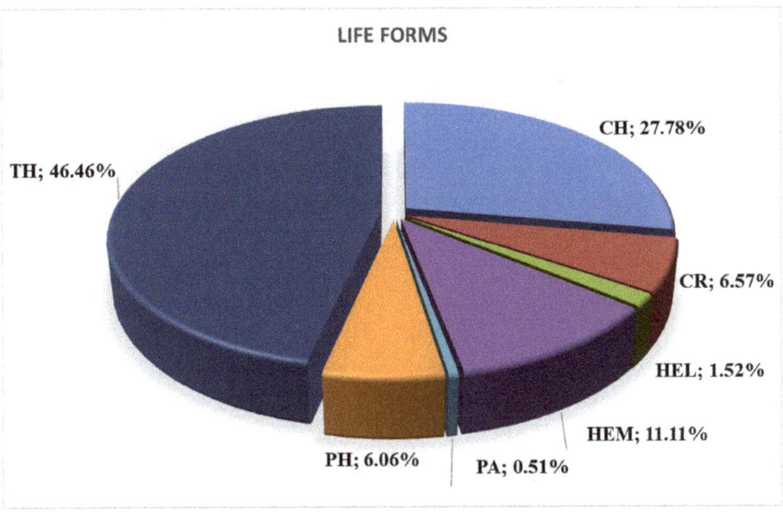

Figure 5. Life form spectrum of the recorded species in Jabal Al-Ward region. TH = Therophytes, CH = Chamaephytes, HEM = Hemicryptophytes, PH = Phanerophytes, GE = Geophytes, CR = Cryptophytes, HEL = Helophytes, PA = Parasites.

The chorological analysis of Jabal Al-Ward vegetation showed that the Mono-regional group was represented by the highest number of taxa 76 (38.5%). It comprised seven chorotypes; Saharo-Arabian (18.2%) with 36 species accounted for the highest representation. The Saharo-Arabian region was combined with eight more regions, including Saharo-Arabian/Sudano-Zambesian (12.6%), Irano-Turanian/Saharo-Arabian (9.1%), Mediterranean/Saharo-Sindian (5.6%), Irano-Turanian/Mediterranean/Saharo-Arabian (4.5%), Irano-Turanian/Saharo-Arabian/Sudano-Zambesian (2%), Euro-Siberian/Irano-Turanian/Mediterranean/Saharo-Arabian, and Saharo-Arabian/Sudanian (1% each), Mediterranean/Saharo-Arabian/Sudano-Zambesian, and Irano-Turanian/Mediterranean/Saharo-Arabian/Sudano-Zambesian (0.5% each).

Followed by Irano-Turanian (7.10%), the latter chorotype was recorded by 14 species (7.10%). Additionally, it was found in eight different arrangements with other floristic regions, such as: Irano-Turanian/Saharo-Arabian (9.1%), Irano-Turanian/Mediterranean (8.10%), Irano-Turanian/Mediterranean/Saharo-Arabian (4.5%), Euro-Siberian/Irano-Turanian/Mediterranean (3%), Irano-Turanian/Saharo-Arabian/Sudano-Zambesian (2%), Euro-Siberian/Irano-Turanian and Euro-Siberian/Irano-Turanian/Mediterranean/Saharo-Arabian (1% each), and Irano-Turanian/Mediterranean/Saharo-Arabian/Sudano-Zambesian (0.5%). Fifteen species were documented to belong to the Mediterranean region and achieved 5.60% as a Mono-regional element. Furthermore, seven mixed regions were indicated, such as Irano-Turanian/Mediterranean (8.1%), Mediterranean/Saharo-Arabian (5.6%), Irano-Turanian/Mediterranean/Saharo-Arabian (4.5%), Euro-Siberian/Irano-Turan ian/Mediterranean (3%), Euro-Siberian/Irano-Turanian/Mediterranean/Saharo-Arabian (1% each) and Mediterranean/Saharo-Arabian/Sudano-Zambesian and, Irano-Turanian/M-editerranean/Saharo-Arabian/Sudano-Zambesian (0.5% each). The Sudano-Zambesian chorotype has ten species representing 5.1% and it is found with four other mixed regions Finally, *Cenchrus echinatus* L. is an alien species representing the American type with 0.5%.

Afterward, the Bi-regional elements have six main chorotypes with 74 taxa (37.40%). The highest element was Saharo-Arabian/Sudano-Zambesian with 25 species representing 12.6%, followed by Irano-Turanian/Saharo-Arabian with 18 species (9.1%), Irano-Turanian/Mediterranean with 16 species (8.10%), then Mediterranean/Saharo-Arabian with 11 species (5.6%). Conversely, only 1% of the total number of species belonged to the Euro-Siberian/Irano-Turanian (*Centaurea sinaica* DC. and *Malva neglecta* Wallr.) and Saharo-Arabian/Sudanian (*Heliotropium bacciferum* Forssk. and *Nanorrhinum hastatum* (R.Br. ex Benth.) Ghebr.).

Subsequently, the Worldwide assemblage contained 25 species and occupied 12.60% of the recorded taxa. The Cosmopolitan types scored 5.60% for 11 species. The lowest scores were recorded for Paleotropical, Pantropical, and Neotropical at 3.50%, 3.0%, and 0.50%, respectively.

Finally, the Pluri-regional attained six chorotypes, accounting for 11.50% with 23 species. The Irano-Turanian/Mediterranean/Saharo-Arabian, Euro-Siberian/Irano-Turanian/Mediterranean, Euro-Siberian/Irano-Turanian/Mediterranean, and Irano-Turanian/Saharo-Arabian/Sudano-Zambesian exhibited 3.5%, 3%, and 2%, respectively. While one species was categorized for each of the following chorotypes: Mediterranean/Saharo-Arabian/Sudano-Zambesian (*Rumex vesicarius* L.) and Irano-Turanian/Mediterranean/Saharo-Arabian/Sudano-Zambesian (*Citrullus colocynthis* (L.) Schrad.) (Figure 6, Table 3).

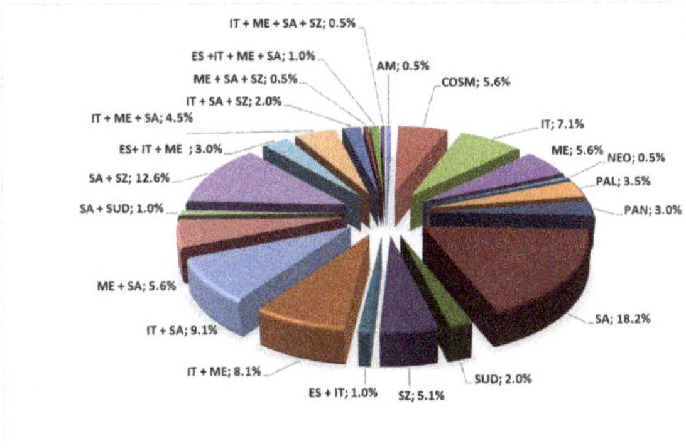

Figure 6. Floristic regions and chorotypes from the recorded taxa in Jabal Al-Ward. Note: American (AM), Cosmopolitan (COSM), Euro-Siberian (ES), Irano-Turanian (IT), Mediterranean (ME), Neotropical (NEO), Paleotropical (PAL), Pantropical (PAN), Saharo-Arabian (SA), Sudano-Zambesian (SZ), and Sudanian (SUD).

Table 3. The number of the recorded taxa belonging to the main floristic categories and their relevant percentages (%) in Jabal Al-Ward. American (AM), Cosmopolitan (COSM), Euro-Siberian (ES), Irano-Turanian (IT), Mediterranean (ME), Neotropical (NEO), Paleotropical (PAL), Pantropical (PAN), Saharo-Arabian (SA), Sudano-Zambesian (SZ), and Sudanian (SUD).

Chorotype	Number of Species	Percentage (%)
	Mono-Regional	
AM	1	0.50%
IT	14	7.10%
ME	11	5.60%
SA	36	18.20%
SUD	4	2.00%
SZ	10	5.10%
Subtotal	76	38.50%
	Bi-regional	
ES + IT	2	1.00%
IT + ME	16	8.10%
IT + SA	18	9.10%
ME + SA	11	5.60%
SA + SZ	25	12.60%
SA + SUD	2	1.00%
Subtotal	74	37.40%
	Pluri-regional	
ES + IT + ME	6	3.00%
IT + ME + SA	9	4.50%
IT + SA + SZ	4	2.00%
ME + SA + SZ	1	0.50%
ES + IT + ME + SA	2	1.00%
IT + ME + SA + SZ	1	0.50%
Subtotal	23	11.50%
	Worldwide	
COSM	11	5.60%
NEO	1	0.50%
PAL	7	3.50%
PAN	6	3.00%
Subtotal	25	12.60%
Total	198	100%

4. Discussion

Tabuk province is characterized by diverse habitats, including coastal plains, valleys, dunes, wadis, salt pans (sabkhas), and mountain ridges [42]. These habitats represent biogeographic units that guarantee a perfect distribution of this region's flora and fauna life regime [10]. Mountains are of great floristic and ecological importance in the Middle East. They provide a shelter for many species, which qualifies them as centers of species richness [43].

The Tabuk region has a high diversity of plants [9,20]. However, the present study demonstrates that the highest number of species (198 species representing 47 families) were recorded in one region, compared with the former studies performed in discrete areas of Tabuk [9,11–17].

The most dominant families are the Asteraceae, Poaceae, and Fabaceae. They represented more than a third of Jabal Al-Ward's floristic composition. The dominance of these families over the others is attributed to their ability to seed dispersal efficiency of their taxa, leading to a high diversity and wide distribution [44]. Al-Mutairi [13] mentioned that the Asteraceae is dominant in a different area of the Tabuk region. Additionally, our findings concur with several floristic studies conducted at different Saudi Arabian localities [45–47]. However, fourteen percent of the total number of families were represented by only one species: Aizoaceae, Asparagaceae, Asphodelaceae, Cistaceae, Convolvulaceae, Crassulaceae, Cyperaceae, Ephedraceae, Iridaceae, Molluginaceae, Moraceae, Nitrariaceae, Nyctaginaceae, Orobanchaceae, Papaveraceae, Portulacaceae, Primulaceae, Rhamnaceae, Rubiaceae, Scrophulariaceae, Tamaricaceae, and Typhaceae. The three genera with the highest species richness are *Astragalus* (Fabaceae), *Euphorbia* (Euphorbiaceae), and *Fagonia* (Zygophyllaceae); four species have been reported for each genus. The generic coefficient is 76.8% in Jabal Al-Ward, which is more than twice that of Saudi Arabia. Based on the high percentage of generic coefficient, it can be concluded that there are more intergeneric competitions in the area [48]. Furthermore, Al-Nafie [49] mentioned that the existence of a certain number of species corresponding to different taxa is a salient characteristic of this region, showing high plant diversity. Members of the recognized families are also well known for their higher economic and medicinal potentiality [50–58]. *Anthemis scrobicularis* was the only endemic species recorded in the surveyed area [59]. Many studies also reported limited endemism in the various regions of the country [15,49,60–62].

In the current study, the perennial species were dominant over the annuals (53.54% to 46.46%). The perennial/annual ratio is affected by abiotic factors such as the aridity of the area and the amount of precipitation [62,63]. This is a notable characteristic of Jabal Al-Ward because perennial plants may be more tolerant of climatic changeability than annual plants [64]. The predominant life forms among the others are the therophytes (46.46%), followed by chamaephytes (27.78%) and phanerophytes (6.06%). These findings are in line with some research on desert ecosystems [13–15,44–47,60,65–67]. Thakur [48], Das, et al. [68] and Fadl et al. [60] reported that the dominance of the therophytic pattern and phanerophytes are associated with lower altitudes, warm-dry arid climate, and habitat disturbances. On the other hand, at higher altitudes, hemicryptophytes and cryptophytes overwhelm cold conditions. The relatively high occurrence of chamaephytes (27.78%) is also associated with the desert habitat, topography, and biotic factors [69]. However, floristic research documented the predominance of other life forms over therophytes in Jabal Fayfa, Mecca, and Taif [61,62,70]. Lower frequency was observed for the helophytes (1.52%) in the current study, as they are indicators of the wetlands [71,72]. *Cistanche tubulosa* (Schenk) Wight was the only parasitic species documented in Jabal Al-Ward, where it grows mainly on the root of the *Calotropis procera* (Aiton) Dryand. Tree, which is commonly distributed in some arid areas [73].

The identified species were classified into four major phytogeographical groups: Mono-regional, Bi-regional, Pluri-regional, and Worldwide. It was remarkable that the Mono-regional and Bi-regional groups achieved the highest representation at 38.5% and 37.4%, respectively. The presence of interzonal habitats contributes to the higher prevalence of Bi-regional elements [39]. The abundance of Pluri-regional and global taxa is another feature of the Saudi Arabian flora [49]. The existence of species growing in many regions of the world, or the widely distributed plants indicates the relationship between the Arabian Peninsula and these regions (Euro-Siberian, Irano-Turanian, Mediterranean, Sudanian, and Sudano-Zambesian). The dispersal of these species in the Arabian Peninsula is influenced by climate change and the long-distance migration of people, animals, and birds [70].

Saudi Arabia's vegetation belongs to Saharo-Arabian region [39]. Thirty-six species belonged to the Saharo-Arabian region, with the best representation accounting for 18.2%. According to Zohary [74], Saharo-Sindian is corresponding to Saharo-Arabian. Abdel Khalik et al. [62] point out that the Saharo-Arabian and Sudano-Zambezian species are good indicators of desert habitats. This explains the present study's relatively high contribution of Saharo-Arabian/Sudano-Zambesian. The Saharo-Arabian taxa in Saudi Arabia were developed gradually and discontinuously from some adjoining regions under the aridity's effect, such as the Sudanian, Mediterranean, and Irano-Turanian regions [39]. Therefore, our findings are consistent with most studies carried out in the various Saudi Arabian provinces, especially Tabuk [12–14,45,65].

The Irano-Turanian chorotype was recorded by fourteen species (7.10%) and it was found in eight different arrangements with other floristic regions. The Irano-Turanian region is one of the world's largest and most diverse floristic regions and serves as a source forthe halophytes (such as *Halocnemum strobilaceum* (Pall.) M.Bieb., *Haloxylon salicornicum* (Moq.) Bunge ex Boiss., *Caroxylon imbricatum* (Forssk.) Moq., *Heliotropium bacciferum* Forssk., *H. crispum* Desf., *H. europaeum* L., *Tamarix aphylla* (L.) H.Karst.), and xerophytes (representing the rest of the studied taxa) for adjacent countries [75,76]. The region is also characterized by relatively low precipitation and a long dry season. The maximum temperature was measured in the summer, similar to the hottest spots in the Sahara, while the minimum winter temperatures were lower than those recorded in the Mediterranean region [77]. These characters make this region a transition between the Saharo-Arabian and the Mediterranean.

Fifteen species were documented to belong to the Mediterranean region and achieved 5.60% as a Mono-regional element. According to Abdel Khalik et al. [62], the survival of Mediterranean species in the study area indicates a more mesic environment.

Recent studies conducted in the Arabian Peninsula have relied on DNA barcode markers to assess genetic variation among taxa in various geographical regions [78,79]. Projects are ongoing worldwide to create DNA barcode libraries for vascular plant flora and make these data accessible to conserve and utilize biodiversity [80]. DNA barcodes are used to enhance our understanding of how species evolve and interact and how we can delay their extinction [81,82].

5. Conclusions

The current study demonstrated the highest species richness compared to former studies performed in separate locations in Tabuk Province. One hundred ninety-eight plant species belonging to forty-seven plant families were identified in Jabal Al-Ward. The Asteraceae, Poaceae, and Fabaceae represented more than a third of Jabal Al-Ward's floristic composition. Annual species (46.46% of the total) were subordinate to perennial species (53.54%). Perennial species may be more tolerant of climatic changeability than annuals. This is a prominent feature in Jabal Al-Ward. Therophytes (46.46%) exemplify the most predominant life forms over the others, as they are associated with lower altitudes, warm-dry arid climate, and habitat disturbances. The recorded plant species were classified into four phytogeographical categories, with the highest representations being for the Mono-regional and Bi-regional groups. The phytogeographical distribution of the species revealed that thirty-six species (18.2%) belonged to the Saharo-Arabian region. The Saharo-Arabian and Sudano-Zambezian species are reliable indicators of desert habitats, which explained the highest contribution of Saharo-Arabian Sudano-Zambesian in the present study. In

Saudi Arabia, the Saharo-Arabian taxa were developed gradually and discontinuously from some adjoining regions such as the Sudanian, Mediterranean, and Irano-Turanian regions.

Author Contributions: Conceptualization, A.E. and A.E.-B.; methodology, A.E., A.E.-B., N.A.A.S. and D.A.A.-B.; Software, F.Y.E. and A.E.-B.; data curation, A.E., F.Y.E., A.E.-B. and I.H.N.; investigation, A.E., F.Y.E., A.E.-B. and I.H.N.; resources, A.E., A.E.-B., N.A.A.S. and D.A.A.-B.; Formal analysis, A.E., F.Y.E., A.E.-B. and I.H.N.; writing—original draft preparation, I.H.N., A.E., F.Y.E. and A.E.-B.; writing—review and editing, I.H.N., F.Y.E., A.E.-B., A.E., N.A.A.S. and D.A.A.-B.; funding acquisition, A.E., I.H.N., F.Y.E., N.A.A.S., D.A.A.-B. and A.E.-B. All authors have read and agreed to the published version of the manuscript.

Funding: This research received no external funding.

Data Availability Statement: The data presented in this study are available in this article.

Conflicts of Interest: The authors declare no conflict of interest.

References

1. Zahran, M. *Vegetation Types of Saudi Arabia*; King Abdel Aziz University Press: Jeddah, Saudi Arabia, 1982.
2. Collenette, I.S. *Wildflowers of Saudi Arabia*; National Commission for Wildlife Conservation and Development: Riyadh, Saudi Arabia, 1999.
3. Chaudhary, S.A. *Flora of the Kingdom of Saudi Arabia*; Agriculture and Water Research Center, Riyadh, Saudi Arabia: Riyadh, Saudi Arabia, 2000; Volume II.
4. Chaudhary, S.A. *Flora of the Kingdom of Saudi Arabia*; Agriculture and Water Research Center, Riyadh, Saudi Arabia: Riyadh, Saudi Arabia, 1999; Volume I.
5. Chaudhary, S.A. *Flora of the Kingdom of Saudi Arabia*; Agriculture and Water Research Center, Riyadh, Saudi Arabia: Riyadh, Saudi Arabia, 2001; Volume III.
6. Rahman, M.A.; Mossa, J.S.; Al-Said, M.S.; Al-Yahya, M.A. Medicinal Plant Diversity in the Flora of Saudi Arabia 1: A Report on Seven Plant Families. *Fitoterapia* **2004**, *75*, 149–161. [CrossRef] [PubMed]
7. Alfarhan, A.H. A Phytogeographical Analysis of the Floristic Elements in Saudi Arabia. *Pak. J. Biol. Sci.* **1999**, *2*, 702–711.
8. Alsherif, E.A.; Ayesh, A.M.; Rawi, S.M. Floristic Composition, Life Form and Chorology of Plant Life at Khulais Region, Western Saudi Arabia. *Pak. J. Bot.* **2013**, *45*, 29–38.
9. Al-Mutairi, K.A. Do Spatially Structured Soil Variables Influence the Plant Diversity in Tabuk Arid Region, Saudi Arabia? *Sustainability* **2022**, *14*, 2611. [CrossRef]
10. Al Saud, M.M. Physical Characteristics. In *Sustainable Land Management for NEOM Region*; Al Saud, M.M., Ed.; Springer International Publishing: Cham, Switzerland, 2020; pp. 1–232.
11. Rajasab, A. *A Pictorial Guide to the Plants of Tabuk Region, Saudi Arabia and Their Traditional Uses*; LAP Lambert Academic Publishing: Saarbrücken, Germany, 2011.
12. Moawed, M.M.; Ansari, A.A. Wild Plants Diversity of Red Sea Coastal Region, Tabuk, Saudi Arabia. *J. Chem. Pharm. Res.* **2015**, *7*, 220–227.
13. Al-Mutairi, K.A.; Al-Shami, S.A.; Khorshid, Z.; Moawed, M.M. Floristic Diversity of Tabuk Province, North Saudi Arabia. *J. Anim. Plant Sci.* **2016**, *26*, 1019–1025.
14. Moawed, M.M. Plant Flora of Alaqan Region, Tabuk Province, Saudi Arabia. *Egypt. J. Exp. Biol.* **2016**, *12*, 107–113.
15. Fakhry, A.; Kenany, E. Floristic Diversity and Endangered Species in Harrat ArRahah Solidified Lava Area, Southern Tabuk, Saudi Arabia. *Egypt. J. Exp. Biol.* **2019**, *15*, 217–226. [CrossRef]
16. Alghanem, S.M.; Al-Atwi, H.Q.; Al-Saiari, M.O.; Karim, A.; Al-Balawi, M.; Al-Zahrani, A.; Al-Sayed, A.M. Floristic Diversity and Perennial Vegetation Analysis of Al-Wadi Al-Akhder, Tabuk Region, Saudi Arabia. *Int. J. Plant Sci. Ecol.* **2020**, *6*, 31–38.
17. Al-Balawi, S.M.; Alwahibi, M.; Al-Harbi, N.A. Diversity and Floristic Study of Weeds Species in Tabuk Region Farms in Saudi Arabia. *Int. J. Bot. Stud.* **2021**, *6*, 1560–1566.
18. Ansari, A.A.; Siddiqui, Z.H.; Alatawi, F.A.; Alharbi, B.M.; Alotaibi, A.S. An Assessment of Biodiversity in Tabuk Region of Saudi Arabia: A Comprehensive Review. *Sustainability* **2022**, *14*, 10564. [CrossRef]
19. Fatima, F.; Bhat, S.H.; Ullah, M.F.; Abu-Duhier, F.; Husain, E. In-Vitro Antimicrobial Activity of Herbal Extracts from Tabuk Region (Kingdom of Saudi Arabia) against Nosomial Pathogens: A Preliminary Study. *Glob. J. Health Sci.* **2018**, *10*, 83. [CrossRef]
20. Al-Mutairi, K.A. Influence of Soil Physical and Chemical Variables on the Species Composition and Richness of Plants in the Arid Region of Tabuk, Saudi Arabia. *Ekologia* **2017**, *36*, 112–120. [CrossRef]
21. Fernandes, M.F.; Cardoso, D.; de Queiroz, L.P. An Updated Plant Checklist of the Brazilian Caatinga Seasonally Dry Forests and Woodlands Reveals High Species Richness and Endemism. *J. Arid Environ.* **2020**, *174*, 104079. [CrossRef]
22. Hughes, C.E.; Pennington, R.T.; Antonelli, A. Neotropical Plant Evolution: Assembling the Big Picture. *Bot. J. Linn. Soc.* **2013**, *171*, 1–18. [CrossRef]

23. Schrire, B.D.; Lavin, M.; Lewis, G.P. Global Distribution Patterns of the Leguminosae: Insights from Recent Phylogenies. *Biol. Skr.* **2005**, *55*, 375–386.
24. Fiema, Z.T.; Nehmé, L.; Kennedy, W.; Al-Talhi, D. The Al-Ulā–Al-Wajh Survey Project: The 2013 Reconnaissance Season. *ATLAL-J. Saudi Arab. Archaeol.* **2020**, *29*, 109–119.
25. Aloufi, A.A.; Amr, Z.S.; Baker, M.A.A.; Hamidan, N. Diversity and Conservation of Terrestrial, Freshwater, and Marine Reptiles and Amphibians in Saudi Arabia. *Amphib. Reptil. Conserv.* **2019**, *13*, 181–202.
26. Aloufi, A.A.; Amr, Z.S.; Abu Baker, M.A. Reptiles and Amphibians of Al Madinah Al Munawwarah Province, Saudi Arabia. *Russ. J. Herpetol.* **2021**, *28*, 123–137. [CrossRef]
27. Hao, L.; Qingdong, S.; Imin, B.; Kasim, N. Methodology for Optimizing Quadrat Size in Sparse Vegetation Surveys: A Desert Case Study from the Tarim Basin. *PLoS ONE* **2020**, *15*, e0235469. [CrossRef]
28. Hu, H.H.; Liu, B.; Liang, Y.S.; Ye, J.F.; Saqib, S.; Meng, Z.; Lu, L.M.; Chen, Z.D. An Updated Chinese Vascular Plant Tree of Life: Phylogenetic Diversity Hotspots Revisited. *J. Syst. Evol.* **2020**, *58*, 663–672. [CrossRef]
29. Al-Harbi, N.A. Ecology of Indigenous Plants in Abandoned Agricultural Areas in Tabuk Region, Saudi Arabia. *Open J. Ecol.* **2022**, *12*, 162–174. [CrossRef]
30. Alsubeai, A.; Burckhard, S.R. Interannual Climate Variability in Tabuk, Saudi Arabia: Impacts on Annual and Seasonal Precipitation. *Atmos. Clim. Sci.* **2021**, *11*, 645–657. [CrossRef]
31. CCKP Climate Change Knowledge Portal. Available online: https://climateknowledgeportal.worldbank.org/country/saudi-arabia/climate-data-historical (accessed on 19 September 2022).
32. Cope, K.A. *Key to the Grasses of the Arabian Peninsula*; Arab Bureau of Education for the Gulf States: Riyadh, Saudi Arabia, 1985.
33. Mighaid, A.M. *Flora of Saudi Arabia*, 4th ed.; King Abdul Aziz University Press: Jeddah, Saudi Arabia, 1996.
34. POWO Plants of the World Online. Facilitated by the Royal Botanic Gardens, Kew. Available online: http://www.Plantsoftheworldonline.Org/%0A2021 (accessed on 7 November 2021).
35. Boyle, B.L.; Matasci, N.; Mozzherin, D.; Rees, T.; Barbosa, G.C.; Kumar Sajja, R.; Enquist, B.J. Taxonomic Name Resolution Service, Version 5.0. Available online: https://tnrs.biendata.org/ (accessed on 25 July 2022).
36. Govaerts, R.; Frodin, D.G.; Radcliffe-Smith, A.; Carter, S. *World Checklist and Bibliography of Euphorbiaceae (with Pandaceae)*; Royal Botanic Gardens: Kew, UK, 2000; ISBN 1900347857.
37. Raunkiaer, C. *The Life Forms of Plants and Statistical Plant Geography*; Clarendon Press: Oxford, UK, 1934.
38. Wickens, G. The Flora of Jebel Marra (Sudan Republic) and Its Geographical Affinities. *Kew Bull. Addit. Ser.* **1978**, *144*, 496. [CrossRef]
39. Zohary, M. *Geobotanical Foundations of the Middle East*; Gustav Fischer: Stuttgart, Germany, 1973.
40. White, F.; Leonard, J. Phytogeographical Links between Africa and Southwest Asia. *Flora Veg. Mund.* **1991**, *9*, 229–246.
41. Jaccard, P. The Distribution of the Flora in the Alpine Zone. *N. Phytol.* **1912**, *11*, 37–50. [CrossRef]
42. Kumar, A. Palynology of the Recent Intertidal Sediments of the Southern Red Sea Coast of Saudi Arabia. *Palynology* **2021**, *45*, 143–163. [CrossRef]
43. Ghazanfar, S.A.; McDaniel, T. Floras of the Middle East: A Quantitative Analysis and Biogeography of the Flora of Iraq. *Edinburgh J. Bot.* **2015**, *73*, 1–24. [CrossRef]
44. Galal, T.M.; Al-Yasi, H.M.; Fadl, M.A. Vegetation Zonation along the Desert-Wetland Ecosystem of Taif Highland, Saudi Arabia. *Saudi J. Biol. Sci.* **2021**, *28*, 3374–3383. [CrossRef]
45. Obaid, W.A.; Boutraa, T.; Abdel-Hameed, U.K. Floristics and Soil Characteristics of Ohud Mountain, Al-Madinah Al-Munawarah, Western Saudi Arabia. *J. Mt. Sci.* **2020**, *17*, 2405–2417. [CrossRef]
46. Osman, A.K.E.; Abdein, M.A.E. Floristic Diversity of Wadi Ar'ar, Saudi Arabia. *J. Taibah Univ. Sci.* **2019**, *13*, 772–789. [CrossRef]
47. Al Shaye, N.A.; Masrahi, Y.S.; Thomas, J. Ecological Significance of Floristic Composition and Life Forms of Riyadh Region, Central Saudi Arabia. *Saudi J. Biol. Sci.* **2020**, *27*, 35–40. [CrossRef] [PubMed]
48. Thakur, A.S. Floristic Composition, Life-Forms and Biological Spectrum of Tropical Dry Deciduous Forest in Sagar District, Madhya Pradesh, India. *Trop. Plant Res.* **2015**, *2*, 112–119.
49. Al-Nafie, A.H. Phytogeography of Saudi Arabia. *Saudi J. Biol. Sci.* **2008**, *15*, 159–176.
50. El-banhawy, A.; Nour, I.H.; Acedo, C.; Elkordy, A.; Faried, A.; Al-Juhani, W.; Gawhari, A.M.H.; Olwey, A.O.; Ellmouni, F.Y. Taxonomic Revisiting and Phylogenetic Placement of Two Endangered Plant Species: *Silene leucophylla* Boiss. and *Silene schimperiana* Boiss. (Caryophyllaceae). *Plants* **2021**, *10*, 740. [CrossRef]
51. Abdelhady, M. A Comparative Study of the Saudi Arabia Flora, Medicinal Plants of Al- Mandaq Province, Al-Baha Region. *Al-Azhar J. Pharm. Sci.* **2021**, *63*, 57–71. [CrossRef]
52. Hamada, F.A.; El-Banhawy, A.; Ellmouni, F.Y.; Al-Juhani, W.; Makharita, R.R.; Nour, I.H. Comparative Taxonomic Study of *Balanites aegyptiaca* (L.) Delile (Zygophyllaceae). *Biol. Life Sci. Forum* **2021**, *11*, 72. [CrossRef]
53. Fakhr, M.A.; Mazrou, Y.S.A.; Ellmouni, F.Y.; Elsaied, A.; Elhady, M.; Elkelish, A.; Nour, I.H. Investigating the Phenotypic Plasticity of the Invasive Weed *Trianthema portulacastrum* L. *Plants* **2022**, *11*, 77. [CrossRef]
54. Marzouk, R.I.; El-Darier, S.M.; Kamal, S.A.; Nour, I.H. Comparative Taxonomic Study of *Launaea* Cass. (Asteraceae, Cichorioideae) in Egypt. *Taxonomy* **2021**, *1*, 192–209. [CrossRef]
55. El-Banhawy, A.; Uluer, D.A.; Fayed, A.A.; Mohamed, M.; Faried, A. DNA Barcoding and Phylogenetic Placement of the *Genus Euphorbia* L. (Euphorbiaceae) in Egypt. *Biol. Life Sci. Forum* **2021**, *4*, 58.

56. Faried, A.; EL-Banhawy, A.; Elqahtani, M. Taxonomic, DNA Barcoding and Phylogenetic Reassessment of the *Egyptian Ephedra* L. (Ephedraceae). *Catrina* **2018**, *17*, 1–13. [CrossRef]
57. Abdel Khalik, K.; Al-Ruzayza, S.; Assiri, A.; Elkordy, A. Pollen Morphology of Malvaceae Genera from Saudi Arabia and Its Taxonomic Significance. *Aust. J. Crop Sci.* **2021**, *15*, 725–742. [CrossRef]
58. El Darier, S.M.; Kamal, S.A.; Marzouk, R.I.; Nour, I.H. Anti-Proliferative Activity of *Launaea fragilis* (Asso) Pau and *Launaea nudicaulis* (L.) Hook.F. Extracts. *Biomed. J. Sci. Tech. Res.* **2021**, *35*, 27492–27496. [CrossRef]
59. Thomas, J. Endemic Plants of Saudi Arabia. Available online: http://www.plantdiversityofsaudiarabia.info/Biodiversity-Saudi-Arabia/Flora/EndemicPlants.htm (accessed on 30 August 2022).
60. Fadl, M.A.; Al-Yasi, H.M.; Alsherif, E.A. Impact of Elevation and Slope Aspect on Floristic Composition in Wadi Elkor, Sarawat Mountain, Saudi Arabia. *Sci. Rep.* **2021**, *11*, 1–10. [CrossRef]
61. Abbas, A.M.; Al-Kahtani, M.A.; Alfaifi, M.Y.; Elbehairi, S.E.I.; Badry, M.O. Floristic Diversity and Phytogeography of Jabal Fayfa: A Subtropical Dry Zone, South-West Saudi Arabia. *Diversity* **2020**, *12*, 345. [CrossRef]
62. Abdel Khalik, K.; El-Sheikh, M.; El-Aidarous, A. Floristic Diversity and Vegetation Analysis of Wadi Al-Noman, Mecca, Saudi Arabia. *Turk. J. Bot.* **2013**, *37*, 894–907. [CrossRef]
63. Erfanzadeh, R.; Pétillon, J.; Maelfait, J.P.; Hoffmann, M. Environmental Determinism versus Biotic Stochasticity in the Appearance of Plant Species in Salt-Marsh Succession. *Plant Ecol. Evol.* **2010**, *143*, 43–50. [CrossRef]
64. Vico, G.; Brunsell, N.A. Tradeoffs between Water Requirements and Yield Stability in Annual vs. Perennial Crops. *Adv. Water Resour.* **2018**, *112*, 189–202. [CrossRef]
65. Al-Eisawi, D.M.; Al Ruzayza, S. The Flora of Holy Mecca District, Saudi Arabia. *Int. J. Biodivers. Conserv.* **2015**, *7*, 173–189. [CrossRef]
66. Alatar, A.; El-Sheikh, M.A.; Thomas, J. Vegetation Analysis of Wadi Al-Jufair, a Hyper-Arid Region in Najd, Saudi Arabia. *Saudi J. Biol. Sci.* **2012**, *19*, 357–368. [CrossRef]
67. Elkordy, A.; Elshikh, O.; Abdallah, N. Floristic Diversity and Vegetation Analysis of Riparian and Aquatic Plants of the Canals in the Sohag Governorate, Egypt. *Phytol. Balc.* **2019**, *25*, 81–95.
68. Das, D.S.; Rawat, D.S.; Maity, D.; Dash, S.S.; Sinha, B.K. Species Richness Patterns of Different Life-Forms along Altitudinal Gradients in the Great Himalayan National Park, Western Himalaya, India. *Taiwania* **2020**, *65*, 154–162. [CrossRef]
69. Al-Yasi, H.M.; Alotaibi, S.S.; Al-Sodany, Y.M.; Galal, T.M. Plant Distribution and Diversity along Altitudinal Gradient of Sarrawat Mountains at Taif Province, Saudi Arabia. *Biosci. Res.* **2019**, *16*, 1198–1213.
70. Al-Sodany, Y.M.; Mosallam, H.A.; Al-Yasi, H.M. Floristic Diversity and Plant Communities Associated with *Juniper* Forests in High Altitudes. *Int. J. Curr. Life Sci.* **2014**, *4*, 118–133.
71. Sorrell, B.K.; Hawes, I. Convective Gas Flow Development and the Maximum Depths Achieved by Helophyte Vegetation in Lakes. *Ann. Bot.* **2010**, *105*, 165–174. [CrossRef]
72. Ling, J.E.; Casanova, M.T.; Shannon, I.; Powell, M. Development of a Wetland Plant Indicator List to Inform the Delineation of Wetlands in New South Wales. *Mar. Freshw. Res.* **2018**, *70*, 322–344. [CrossRef]
73. Kong, Z.L.; Johnson, A.; Ko, F.C.; He, J.L.; Cheng, S.C. Effect of *Cistanche tubulosa* Extracts on Male Reproductive Function in Streptozotocin–Nicotinamide-Induced Diabetic Rats. *Nutrients* **2018**, *10*, 1562. [CrossRef]
74. Zohary, M.; Feinbrun-Dothan, N. *Flora Palaestina*; Academy of Sciences and Humanities: Jerusalem, Israel, 1966.
75. Ramezani, E.; Talebi, T.; Alizadeh, K.; Shirvany, A.; Hamzeh'ee, B.; Behling, H. Long-Term Persistence of Steppe Vegetation in the Highlands of Arasbaran Protected Area, Northwestern Iran, as Inferred from a Pollen Record. *Palynology* **2021**, *45*, 15–26. [CrossRef]
76. Manafzadeh, S.; Staedler, Y.M.; Conti, E. Visions of the Past and Dreams of the Future in the Orient: The Irano-Turanian Region from Classical Botany to Evolutionary Studies. *Biol. Rev.* **2017**, *92*, 1365–1388. [CrossRef]
77. Talebi, K.S.; Sajedi, T.; Pourhashemi, M. *Forests of Iran: A Treasure from the Past, a Hope for the Future*; Springer: London, UK, 2014; ISBN 9789400773707.
78. Jamdade, R.; Upadhyay, M.; Shaer, K.; Al Harthi, E.; Al Sallani, M.; Al Jasmi, M.; Al Ketbi, A. Evaluation of Arabian Vascular Plant Barcodes (RbcL and MatK): Precision of Unsupervised and Supervised Learning Methods towards Accurate Identification. *Plants* **2021**, *10*, 2741. [CrossRef]
79. Mosa, K.A.; Gairola, S.; Jamdade, R.; El-Keblawy, A.; Al Shaer, K.I.; Al Harthi, E.K.; Shabana, H.A.; Mahmoud, T. The Promise of Molecular and Genomic Techniques for Biodiversity Research and DNA Barcoding of the Arabian Peninsula Flora. *Front. Plant Sci.* **2019**, *9*, 1929. [CrossRef] [PubMed]
80. Gostel, M.R.; Kress, W.J. The Expanding Role of DNA Barcodes: Indispensable Tools for Ecology, Evolution, and Conservation. *Diversity* **2022**, *14*, 213. [CrossRef]
81. Kress, W.J.; Erickson, D.L.; Jones, F.A.; Swenson, N.G.; Perez, R.; Sanjur, O.; Bermingham, E. Plant DNA Barcodes and a Community Phylogeny of a Tropical Forest Dynamics Plot in Panama. *Proc. Natl. Acad. Sci. USA* **2009**, *106*, 18621–18626. [CrossRef] [PubMed]
82. Shapcott, A.; Forster, P.I.; Guymer, G.P.; McDonald, W.J.F.; Faith, D.P.; Erickson, D.; Kress, W.J. Mapping Biodiversity and Setting Conservation Priorities for SE Queensland's Rainforests Using DNA Barcoding. *PLoS ONE* **2015**, *10*, e0122164. [CrossRef] [PubMed]

agronomy

Article

Palynological Study of Weed Flora from Potohar Plateau

Anwer Usma [1], Mushtaq Ahmad [1,*], Muhammad Zafar [1], Shazia Sultana [1], Fazal Ullah [2], Saddam Saqib [3,4], Asma Ayaz [5] and Wajid Zaman [6,*]

1 Department of Plant Sciences, Quaid-i-Azam University, Islamabad 45320, Pakistan
2 State Key Laboratory of Grassland Agro-Ecosystems, School of Life Sciences, Lanzhou University, Lanzhou 730000, China
3 State Key Laboratory of Systematic and Evolutionary Botany, Institute of Botany, Chinese Academy of Sciences, Beijing 100093, China
4 University of Chinese Academy of Sciences, Beijing 100049, China
5 State Key Laboratory of Biocatalysis and Enzyme Engineering, School of Life Sciences, Hubei University, Wuhan 434200, China
6 Department of Life Sciences, Yeungnam University, Gyeongsan 38541, Korea
* Correspondence: mushtaqflora@hotmail.com (M.A.); shangla123@gmail.com (W.Z.)

Abstract: The pollen morphology of weeds was investigated by scanning electron microscopy (SEM). A morpho-palynological investigation of 18 species of weeds that belongs to 16 angiosperms families was performed using SEM to document distinguishable microscopic features. The main objective of the present study was to provide basic knowledge about morpho-palynological features of weed species that helps delimit the weed flora of the Potohar Plateau. The results show diversity among the qualitative and quantitative characteristics of pollen shape, equatorial and polar axis diameter, the exine's thickness, and the exine's surface ornamentation. The pollen grains were spherical, prolate-spheroidal, oblate-spheroidal, and sub-oblate. The exine ornamentation in most species was reticulate, scarbate, aerolate, faveolate, reticulate-perforate, and reticulate-scabrate. All the species described possessed tricolpate pollen. The variations found in the thickness of the exine and other characters were helpful at the genus and species-specific levels. In accordance with these variations, a taxonomic key was prepared using these characteristics to identify and differentiate weed plant species. SEM images of pollen grains can help delimit the taxa to the species level. This study provides baseline information to distinguish the species of weeds.

Keywords: exine sculpturing; palynomorph; pollen grains; scanning electron microscopy; Pakistan

Citation: Usma, A.; Ahmad, M.; Zafar, M.; Sultana, S.; Ullah, F.; Saqib, S.; Ayaz, A.; Zaman, W. Palynological Study of Weed Flora from Potohar Plateau. *Agronomy* **2022**, *12*, 2500. https://doi.org/10.3390/agronomy12102500

Academic Editor: Rodolfo Gentili

Received: 5 September 2022
Accepted: 11 October 2022
Published: 13 October 2022

1. Introduction

Weeds are considered undesirable plants competing with crops' growth and development. The growth of weeds in agricultural fields can limit the production of staple crops and causes severe damage to their production [1]. According to Khan et al. [2] weeds grow in cultivated and wild habitats and damage the growing crop plants [3]. Weedy plants threaten human health, welfare, biodiversity, ecosystem services, and food security. Here, weeds are defined as any plants that have detrimental socioeconomic and environmental effects, endanger human health, and jeopardize global food security, biodiversity, and ecosystem services [4]. Weeds play an important role in achieving an ecological balance in a cropping system by supporting different life forms. Throughout the world, major vegetation of weeds belongs to families such as Asteraceae, Poaceae, Amaranthaceae, and Fabaceae. The growth of weeds depends mostly on climatic conditions [5]. In some previously reported studies, the most commonly grown weeds of cotton crops are *Cynodon dactylon*, *Amaranthus viridis*, *Phyla nodiflora*, and *Phragmites australis* [6]. Common weeds of wheat crops are *Phragmites* sp., *Cynodon* sp., *Ranunculus* spp., and *Polypogon* sp. [6]. Similarly, *Achyranthes aspera* is a common weed in sugarcane crops and *Plantago lanceolata* is an important weed in chickpea fields [7]. Weeds belonging to the family Fabaceae can obtain essential nutrients from the soil and compete with crops because they gain nutrition faster than other plants [8]. These plants obtain nitrogen from their surroundings and convert it to nitrogen-rich compounds to enhance the growth of essential microorganisms, thereby increasing soil fertility [9]. Many researchers work on the world's weed flora and documented 50 parasitic weeds from Pakistan [10].

In Pakistan, the region of Potohar is located on the northeast side, with an area of approximately 25,000 km^2. Topographically, hilly and plain areas are found in this region [11]. This area also comprises of river Indus and the Jhelum, stretching from the salt range to the foothills of the Himalayas. The Potohar region is between latitude 32°5′34″ N and longitude 71°30′73″ E. The Jhelum plains lie at the lowest altitude, i.e., 825 ft (250 m). Non-native plant species have been introduced into the study area for greenery, which competes with native plant species. The natural vegetation of the Potohar region includes dry deciduous scrubby forests and few species of trees. This area's agriculture purely depends on the monsoon rainfall [12]. The construction of roads, houses, and industries has affected the natural vegetation, resulting in a change in soil composition.

The study of pollen grains plays a vital role in studying plant taxonomy and biodiversity because it helps identify plant species in a particular area [13]. Many taxonomists use different disciplines of plant systematics to identify plant species, such as morphology, anatomical studies, and palynological studies, to find the precise position of plants within taxa [14–19]. On the other hand, the studies of pollen characters are considered the most important tools for classifying and identifying morphologically similar plants. Considerable research has been done on pollen grains in Pakistan and worldwide. Meo and Khan [19] conducted Morpho-palynological studies on weeds for the first time in Rawalpindi, Pakistan. Meo et al. [20] reported the diverse features of weeds and pollen grains belonging to the family Asteraceae. On the other hand, there is not a single document on the morpho-palynology of weeds from the Potohar region of Pakistan. Current research aims to provide detailed information about weeds in the Potohar region of Pakistan, the diversity of their pollen grains, and their exine structure based on qualitative and quantitative characters using SEM as a basis for future studies. A taxonomic key is made based on microscopic characteristics, which helps to distinguish the micro-morphological characteristics to strengthen the identification of complicated weed species. The purpose of the present study was twofold: (1) to determine the pollen fertility of economically important weed species, which will help identify the freely reproducing species for conservation purposes; (2) to the taxonomic identification of the weed species through pollen morphological characteristics.

2. Materials and Methods

2.1. Specimen Collection and Plant Identification

From different sites in the Potohar region of Pakistan (Figure 1), 18 weeds were collected in a single season between August 2021 and November 2021. Approximately three to five samples were collected from each collection site randomly. Table 1 lists the collection site and voucher numbers. After collection, plants were identified either by comparing their morphological characteristics with the flora of Pakistan or by matching their morphology with specimens placed in the Herbarium of Pakistan (ISL). After mounting, the botanical names and author citations were validated using the International Plant Name Index (IPNI) (www.ipni.org (accessed on 7 July 2022)) and accessioned into the Herbarium of Pakistan (ISL) at Quaid-i-Azam University Islamabad, Pakistan.

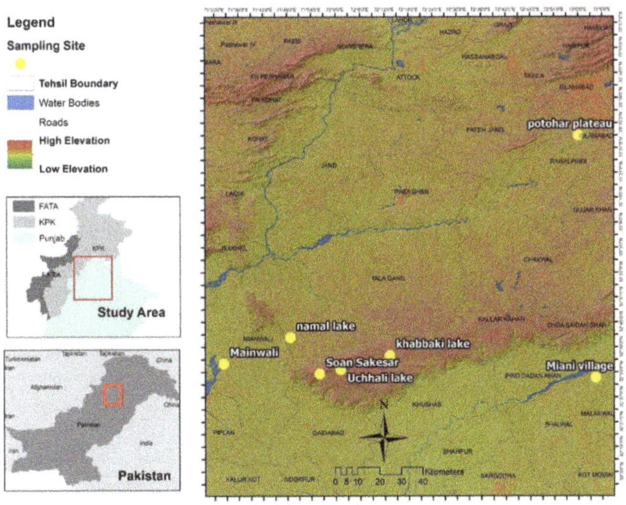

Figure 1. Map of the collection sites of the study area. (Source: Google).

Table 1. Weed samples collected from the Potohar region.

Sr. No	Plant Taxa	Family	Locality	Voucher Specimen Number
1.	*Achyranthus aspera* L.	Amaranthaceae	Salt Range Chakwal	AU-107
2.	*Aerva javanica* (Burm.f.) Juss. ex Schult.	Amaranthaceae	Miani Village Talagang	AU-113
3.	*Amaranthus graecizans* L.	Amaranthaceae	Khewra mines	AU-115
4.	*Asphodelus tenuifolius* Cav.	Asphodelaceae	Uchali Lake	AU-99
5.	*Atriplex stocksii* Boiss.	Amaranthaceae	Uchali Lake	AU-77
6.	*Boerhavia procumbens* Banks ex Roxb.	Nyctaginaceae	Rawalpindi	AU-65
7.	*Brassica furatii* Mouterde	Brassicaceae	Musa khel Mianwali	AU-39
8.	*Capparis decidua* (Forssk.) Edgew.	Capparaceae	Miani Village	AU-55
9.	*Celosia argentea* L.	Amaranthaceae	Kabkbi Lake	AU-110
10.	*Cenchrus ciliaris* L.	Poaceae	Islamabad	AU-205
11.	*Chenopodium murale* L.	Chenopodiaceae	Salt Range	AU-209
12.	*Corchorus depressus* (L.) Stocks	Tiliaceae	Attock	AU-211
13.	*Cissampelos pareira* L.	Menispermaceae	Salt Range	AU-215
14.	*Citharexylum spinosum* L.	Verbenaceae	Islamabad	AU-217
15.	*Cleome brachycarpa* (Forssk.) Vahl ex DC	Capparaceae	Miani village Talagang	AU-31
16.	*Cleome viscosa* L.	Capparaceae	Miani village Talagang	AU-45
17.	*Cucumis melo* subsp. *agrestis* (Naudin) Pangalo.	Cucurbitaceae	Miani village Talagang	AU-53
18.	*Euphorbia granulata* Forssk.	Euphorbiaceae	Thoha Bahadur Chakwal	AU-69

2.2. Study of Pollen under an Optical Microscope

Either dry or fresh plant material extracts pollen grains from anthers. With the help of sharp needles, the anthers were separated from the flowers and placed on glass slides with a single drop of acetic acid, which helps appropriately crush anthers. A slight modification of the method reported by Butt et al. [21] was used to prepare the pollen grains slides. One drop of glycerin jelly was added to the pollen grains, and the extra debris was removed from the glass slide with the help of needles. Glycerin jelly will help stain the pollen grains [21,22]. A cover slip was placed on the pollen grains, and the pollen characteristics were observed by optical microscopy (OM, Leica Dialux 20).

2.3. *Study of Pollen under a Scanning Electron Microscope*

The pollen grains were prepared for SEM using a slight modification of the methodology reported by Erdtmann [23]. In this method, pollen grains were crushed with 45% acetic acid and were placed on double-sided tape fixed to aluminum stubs using a fine pipette and sputter coated with gold to 150Å. Subsequently, these slides were examined by scanning electron microscopy (Model JEOL JSM- 5910) installed in the Central Resource Library (CRL), Department of Physics, University of Peshawar, Pakistan.

2.4. *Data Analysis*

The minimum represented the quantitative features (mean ± standard error) and maximum (e.g., 27.7 (85.2 ± 10.8) 105). Five readings of each character were noted for each pollen grain. The quantitative data were noted and processed using SPSS software (version 16) to determine the minimum, maximum, mean, and standard error. These data are very helpful in identifying species and the nature of the different pollen grains. These indices provide information on pollen diameter, colpi and pore size, exine thickness, and polar-to-equatorial ratio.

2.5. *Percentages of Sterility and Fertility among Pollen*

The pollen fertility was determined using the techniques reported by Khan and Stace [24]. The sterility and fertility percentages were determined using the formula [5,25].

The formula for fertility is given as:

$$\text{Fertility} = (\text{F}/\text{F} + \text{S}) \times 100$$

The formula of sterility is expressed as:

$$\text{Sterility} = (\text{S}/\text{F} + \text{S}) \times 100$$

In the formula, F denotes the number of fertile pollen on a single ocular, while S denotes the number of sterile pollen on one ocular.

2.6. *Taxonomic Key*

After examining the pollen grains under OM and SEM, different microscopic characteristics were observed, such as pollen, exine ornamentations, and colpi. The taxonomic keys have been developed based on these distinguishing characteristics.

3. Results

A palynological study of 18 weed species from the Potohar Plateau of Pakistan belonging to 12 families was analyzed by OM and SEM. The examined species were explained quantitatively and qualitatively in Tables 2–4. The palynological study showed that weedy species of the study area were quite diverse, and their pollen structure helped identify them. Using these qualitative characters, a taxonomic key was prepared to identify and differentiate the weed plant species (Table 5).

Table 2. Qualitative data of the pollen of weeds members from the Potohar region of Pakistan.

Sr. No	Taxa	Shape	Aperture Condition	Mesocolpium	Exine Ornamentation	Colpi Orientation
1.	*Achyranthus aspera*	Oblate-spheroidal	Tricolporate	Scabrate	Scabrate	Sunken, angular
2.	*Aerva javanica*	Prolate-spheroidal	Tricolporate	Aerolate	Reticulate	Sunken, angular, margins distinct, and ends tapering
3.	*Amaranthus graecizans*	Prolate-spheroidal	Trizonocolporate	Reticulate	Scabrate-reticulate	Prominent and rounded at ends
4.	*Asphodelus tenuifolius*	Oblate-spheroidal	Tricolporate	Scabrate-reticulate	Aerolate	Sunken, slit-like margins, and end tapering
5.	*Atriplex stocksii*	Oblate	Tricolporate	Scabrate	Reticulate-Perforate	Prominent and rounded at ends
6.	*Boerhavia procumbens*	Oblate-spheroidal	Tricolporate	Reticulate	Scabrate	Sunken, angular, margins distinct
7.	*Brassica furatii*	Spherical	Tricolporate	Scabrate	Echinate	Sunken, further divided into three slit-like portions, ends are tapering
8.	*Capparis decidua*	Oblate-spheroidal	-	Aerolate	Faveolate	Sunken, slit-like, margins wavy and slightly pointed at ends
9.	*Celosia argentea*	Oblate-spheroidal	Tricolporate	Perforate	Scabrate	Sunken, slit-like, margins distinct, and pointed at ends
10.	*Cenchrus ciliaris*	Prolate-spheroidal	-	Scabrate	Reticulate	Prominent margins
11.	*Chenopodium murale*	Oblate-spheroidal	Tricolporate	Reticulate	Aerolate	Sunken, long and slit-like
12.	*Corchorus depressus*	Oblate-spheroidal	Not visible	Scabrate	Reticulate-scabrate	Prominent and tapering ends
13.	*Cissampelos pareira*	Sub-oblate	Tricolporate	Reticulate	Perforate	Sunken, angular, margins distinct
14.	*Citharexylum spinosum*	Oblate-spheroidal	Tricolporate	Aerolate-Perforate	Scabrate	Long and slit-like, tapering ends
15.	*Cleome brachycarpa*	Oblate-spheroidal	Tricolporate	Reticulate	Reticulate	Sunken and tapering ends
16.	*Cleome viscosa*	Prolate-spheroidal	Tricolporate	Scabrate	Scabrate-reticulate	Long and slit-like margins
17.	*Cucumis melo* subsp. *agrestis*	Oblate-spheroidal	Not Visible	Perforate	Perforate	Sunken, slit-like, wavy margins
18.	*Euphorbia granulata*	Prolate-spheroidal	Tricolporate	Scabrate-reticulate	Aerolate	Sunken, wavy, and margins distinct

Table 3. Quantitative data of pollen of weeds members from Potohar region of Pakistan.

Plant Name	Exine Thickness Mean (Min-Max) S.E µm	Polar Diameter Mean (Min-Max) S.E µm	Equatorial Diameter Mean (Min-Max) S.E µm	P/E Ratio	No of Pores	Colpi Length Mean (Min-Max) S.E µm	Colpi Width Mean (Min-Max) S.E µm	Pores Length Mean (Min-Max) S.E µm	Pores Width Mean (Min-Max) S.E µm
Achyranthus aspera	1.56 (1.35–1.95) ± 0.11	14.5 (12.1–15.6) ± 0.63	15.6 (12.3–18.0) ± 0.96	0.93	–	3.81 (1.65–7.95) ± 3.81	2.55 (1.50–4.50) ± 0.53	–	–
Aerva javanica	1.80 (1.35–2.55) ± 0.23	18.03 (16.8–18.6) ± 0.34	17.8 (15.3–19.6) ± 0.84	1.01	–	6.12 (1.65–10.20) ± 1.40	3.45 (1.95–7.20) ± 0.97	–	–
Amaranthus graecizans	1.50 (1.20–1.80) ± 0.11	28.8 (25.5–33.15) ± 1.65	27.0 (25.6–30.3) ± 1.2	1.07	–	10.05 (4.05–17.55) ± 2.17	6.24 (2.85–13.20) ± 1.92	–	–
Asphodelus tenuifolius	6.12 (1.65–10.20) ± 1.40	3.45 (1.95–7.20) ± 0.97	6.12 (1.6–10.20) ± 1.40	0.97	–	12.2 (10.0–15.1) ± 2.44	5.79 (3.00–8.55) ± 2.31	–	–
Atriplex stocksii	2.01 (1.80–2.10) ± 0.10	11.7 (11.4–11.9) ± 0.15	17.3 (17.2–17.5) ± 0.09	0.67	–	12.30 (9.30–15.10) ± 1.80	7.20 (5.70–9.20) ± 1.03	–	–
Boerhavia procumbens	0.6 (0.5–0.75) ± 0.06	13.3 (12.2–15.25) ± 0.5	13.5 (12.2–15.5) ± 0.54	0.98	–	4.2 (3.2–5.00) ± 0.52	5.4 (4.9–6.3) ± 0.45	–	–
Brassica furatii	1.55 (1.0–2.0) ± 0.16	23.9 (22.75–25.5) ± 0.52	23.9 (22.2–25.5) ± 0.72	1.00	–	5.5 (4.6–6.7) ± 0.62	2.9 (2.3–3.5) ± 0.34	–	–
Capparis decidua	1.05 (0.75–1.25) ± 0.09	13.9 (12.0–15.50) ± 0.63	13.4 (12.0–15.5) ± 0.70	0.99	–	5.2 (4.2–6.1) ± 0.15	6.4 (5.7–7.3) ± 0.15	–	–
Celosia argentea	2.5 (2.25–2.75) ± 0.11	21.0 (15.5–24.75) ± 1.63	22.6 (18.0–25.2) ± 1.3	0.92	–	7.8 (6.8–9.1) ± 0.9	4.1 (3.4–5.1) ± 0.4	–	–
Cenchrus ciliaris	2.65 (2.25–3.0) ± 0.12	15.3 (14.7–15.75) ± 10.16	14.8 (14.0–15.7) ± 0.3	1.03	1	4.8 (4.4–5.4) ± 0.29	2.7 (2.4–3.1) ± 0.2	7.00 (5.25–8.50) ± 0.54	4.05 (3.00–5.00) ± 0.40
Chenopodium murale	1.9 (1.0–2.75) ± 0.34	14.0 (13.0–14.7) ± 0.32	14.5 (13.5–15.2) ± 0.3	0.96	–	6.7 (5.8–8.1) ± 0.6	3.1 (2.4–4.00) ± 0.4	–	–
Corchorus depressus	2.95 (2.75–3.25) ± 0.09	36.9 (33.7–43.00) ± 1.74	38.7 (37.2–40.5) ± 0.65	0.95	–	7.8 (5.8–9.5) ± 1.07	4.3 (3.7–5.00) ± 0.3	–	–
Cissampelos pareira	2.80 (2.25–3.25) ± 0.16	40.4 (35.7–45.50) ± 1.68	47.2 (45.2–48.5) ± 0.65	0.85	–	1.86 (1.35–2.40) ± 0.17	1.23 (.75–1.95) ± 0.21	–	–
Citharexylum spinosum	0.78 (0.6–1.1) ± 0.09	27.3 (22.7–32.7) ± 1.65	30.0 (24.5–33.5) ± 1.77	0.91	–	1.65 (1.05–2.40) ± 0.22	1.68 (1.05–2.10) ± 0.20	–	–
Cleome brachycarpa	2.5 (2.25–2.75) ± 0.11	34.05 (29.0–39.0) ± 1.69	35.1 (32.2–39.2) ± 1.32	0.96	–	2.25 (0.75–3.45) ± 0.47	2.64 (2.10–3.30) ± 0.20	–	–
Cleome viscosa	2.15 (1.5–3.0) ± 0.3	16.2 (13.5–18.0) ± 0.85	15.6 (13.2–18.5) ± 1.0	1.03	–	2.70 (1.95–3.30) ± 0.22	3.03 (2.40–3.60) ± 0.20	–	–
Cucumis melo subsp. *agrestis*	0.95 (0.75–1.25) ± 0.09	13.9 (12.0–15.5) ± 0.63	13.4 (12.0–15.5) ± 0.70	0.99	–	1.14 (0.75–1.50) ± 0.13	1.77 (1.50–2.10) ± 0.12	–	–
Euphorbia granulata	2.0 (1.50–2.25) ± 0.13	19.8 (17.2–21.25) ± 0.69	18.2 (16.2–20.5) ± 0.69	1.08	1	–	–		

Table 4. Pollen Fertility and Sterility percentages of examined taxa.

Sr. No	Taxa	No. of Fertile Pollen	No. of Sterile Pollen	Fertility %	Sterility %
1.	*Achyranthus aspera*	70	15	82	17
2.	*Aerva javanica*	90	40	69	30
3.	*Amaranthus graecizans*	77	10	88	11
4.	*Asphodelus tenuifolius*	99	12	89	10
5.	*Atriplex stocksii*	88	22	80	20
6.	*Boerhavia procumbens*	50	9	84	15
7.	*Brassica furatii*	80	2	97	2
8.	*Capparis decidua*	80	7	91	8
9.	***Celosia argentea***	75	10	88	11
10.	*Cenchrus ciliaris*	90	11	89	10
11.	*Chenopodium murale*	118	7	94	5
12.	*Corchorus depressus*	102	10	91	8
13.	*Cissampelos pareira*	90	5	95	5
14.	*Citharexylum spinosum*	110	14	89	11
15.	*Cleome brachycarpa*	86	11	89	11
16.	*Cleome viscosa*	99	12	89	10
17.	***Cucumis melo*subsp. *agrestis***	143	9	94	5
18.	*Euphorbia granulata*	81	14	85	14

Table 5. Dichotomous key for identifying the Pothor (Pakistan) weed species based on the pollen morphological characteristics.

Link Characters	Present (+) /Absent (−)	Diagnostic Characters	Species Name
1	+	pollen shape oblate-spheroidal, mesocolpium scabrate, colpi orientation sunken, angular	*Achyranthus aspera*
	−	pollen shape prolate-spheroidal, mesocolpium aerolate, colpi orientation Sunken, angular, margins distinct, and ends tapering	2
2	+	aperture tricolporate, exine reticulate	*Aerva javanica*
	−	aperture trizonocolporate, exine scabrate-reticulate	3
3	+	The shape of pollen prolate-spheroidal, mescolopium reticulate	*Amaranthus graecizans*
	−	pollen shape oblate-spheroidal, mescolopium scabrate-reticulate	4
4	+	aperture tricolporate, colpi sunken, slit like margins, and end tapering	*Asphodelus tenuifolius*
	−	aperture tricolporate, colpi prominent, and rounded at ends	5
5	+	shape oblate, mescolopium scabrate	*Atriplex stocksii*
	−	Shape oblate-spheroidal, mescolopium reticulate	6
6	+	exine scabrate, colpi Sunken, angular, margins distinct	*Boerhavia procumbens*
	−	exine echinate, colpi Sunken, further divided into three slit-like portions, ends are tapering	7
7	+	pollen shape spherical, mescolopium scabrate	*Brassica furatii*
	−	pollen shape oblate spheroidal, mescolpium aerolate	8
8	+	exine faveolate, colpi sunken slit-like, margins wavy, and slightly pointed at the end	*Capparis decidua*
	−	pollen shape oblate spheroidal, exine scabrate	9
9	+	aperture tricolporate, mesocolpium perforate	*Celosia argentea*
	−	aperture not clear, mescolpium scabrate	10
10	+	pollen shape prolate-spheroidal, exine reticulate	*Cenchrus ciliaris*
	−	pollen shape oblate-spheroidal, exine aerolate	11
11	+	aperture tricolporate, colpi sunken, long and slit like	*Chenopodium murale*
	−	aperture not visible, colpi prominent, and tapering	12

Table 5. *Cont.*

Link Characters	Present (+) /Absent (−)	Diagnostic Characters	Species Name
12	+	pollen shape oblate-spheroidal, mesocolpium scabrate	*Corchorus depressus*
	−	pollen shape sub-oblate, mescolpium perforate	13
13	+	aperture tricolporate, exine sunken, angular, margins distinct	*Cissampelos pareira*
	−	aperture tricolporate, exine long and slit like, tapering ends	14
14	+	pollen shape oblate-spheroidal, mesocolpium aerolate-perforrate	*Citharexylum spinosum*
	−	pollen shape oblate-spheroidal, mescolpium reticulate	15
15	+	exine reticulate, colpi sunken, and tapering ends	*Cleome brachycarpa*
	−	exine scabrate-reticulate, colpi long and slit-like margins	16
16	+	pollen shape prolate-spheroidal, aperture tricolporate	*Cleome viscosa*
	−	pollen shape oblate-spheroidal, aperture not visible	17
17	+	mesocolpium perforate, exine perforate	*Cucumis melo* subsp. *agrestis*
	−	mescolpium scabrate-reticulate, exine aerolate	*Euphorbia granulata*

3.1. Pollen Size, Shape, and P/E Ratio

Variations in grain shape were observed in equatorial and polar views, size, and pollen shapes. The P/E ratio helps determine the shapes of pollen. Pollen grains of the analyzed species have spherical, prolate-spheroidal, oblate-spheroidal, oblate, and sub-oblate shapes (Figure 2). The maximum polar diameter was 40.40 µm for *Cissampelos parira* and the minimum was 13.95 µm for *Atriplex stoeksii*. *Cissampelos parira* have a maximum equatorial diameter of 47.20 µm, whereas *Cucumis melo* subsp. *Agrestis* have a minimum of 13.45 µm (Figure 3). Similarly, *Euphorbia grandulata* have a maximum P/E ratio of 1.08, and *Cissampelos parira* have a minimum of 0.85 (Figure 4).

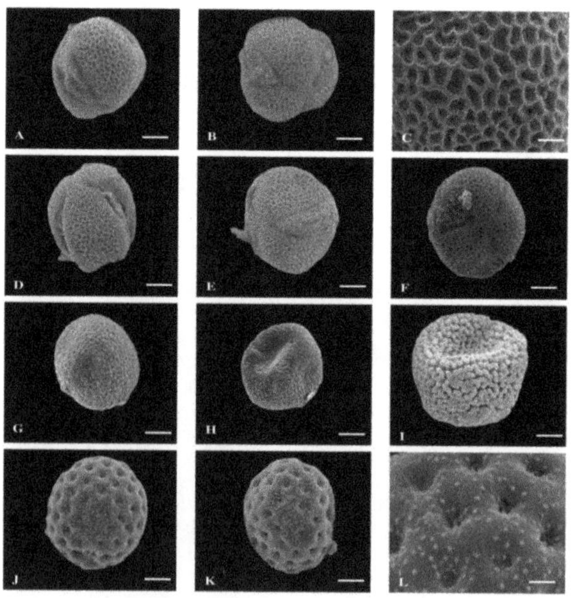

Figure 2. SEM images of pollen grains (Equatorial and Polar axis) and pollen surface ornamentation **(A–C)**, The scale bar represents 5 µm for *Achyranthus aspera* **(D–F)**, 5 µm for *Aerva javanica* **(G–I)**, 10 µm for *Amaranthus graecizans* **(J–L)**, and 5 µm for *Asphodelus tenuifolius*.

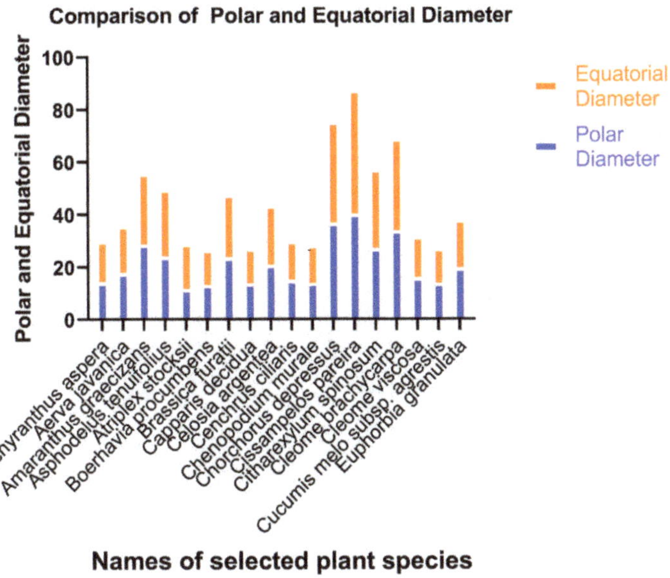

Figure 3. Comparison of the polar and equatorial diameter (μm) of the pollen of selected plant species.

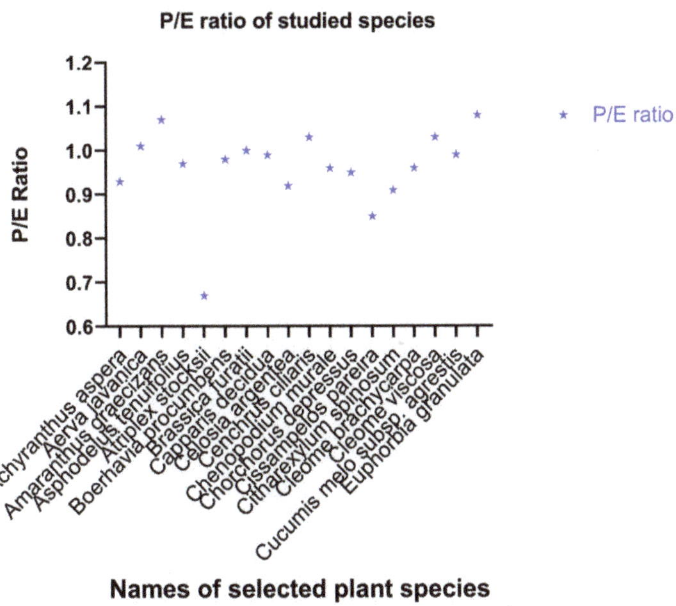

Figure 4. P/E ratio of different species of weeds collected from the Potohar region of Pakistan.

3.2. Quantitative and Qualitative Characters of Colpi

All the documented species contain colpi (Figure 5) in their pollen except for *Euphorbia grandulata*. *Atriplex stoeksii* had a maximum colpi length of 12.30 μm, and *Citharexylum spinosum* had a minimum of 2.25 μm. *Atriplex stoeksii* had a maximum colpi width of 7.20 μm, and *Cissampelos parira* had a minimum of 1.23 μm. All species examined had tricolporate pollen (Figure 6).

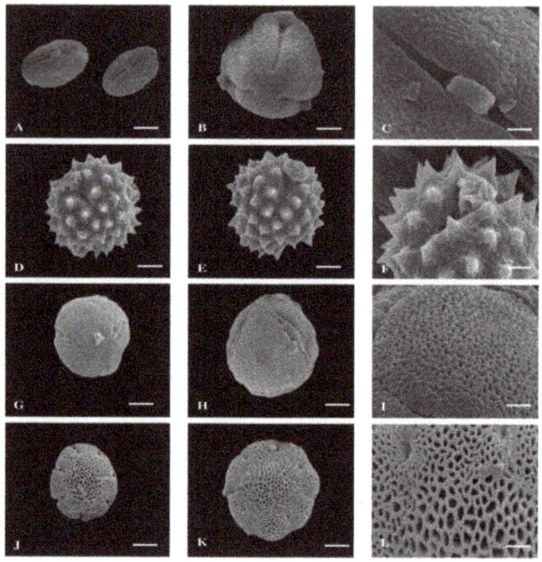

Figure 5. SEM images of pollen grains (Equatorial and Polar axis) and pollen surface ornamentation (**A–C**). The scale bar represents 5 µm for *Atriplex stoeksii* (**D–F**), 5 µm for *Boerhivia procumbense* (**G–I**), 10 µm for *Brassica furatii* (**J–L**), and 10 µm for *Capparis decidua*.

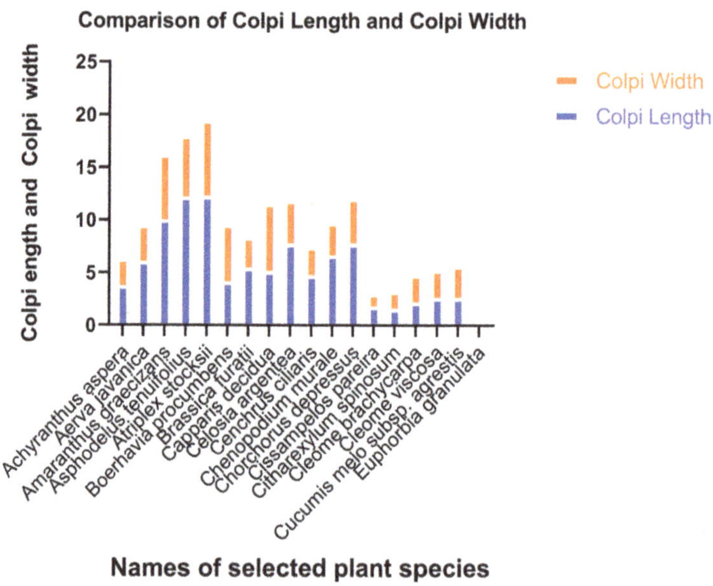

Figure 6. Comparison of the Colpi length and width (µm) of selected plant species.

3.3. Exine Thickness

The quantitative pollen characters were observed by SEM and OM (Figure 7). The maximum exine thickness was 2.95 µm for *Chorchorus depressus*, and the minimum was 0.60 µm for *Boerhivia procumbense* (Figure 8).

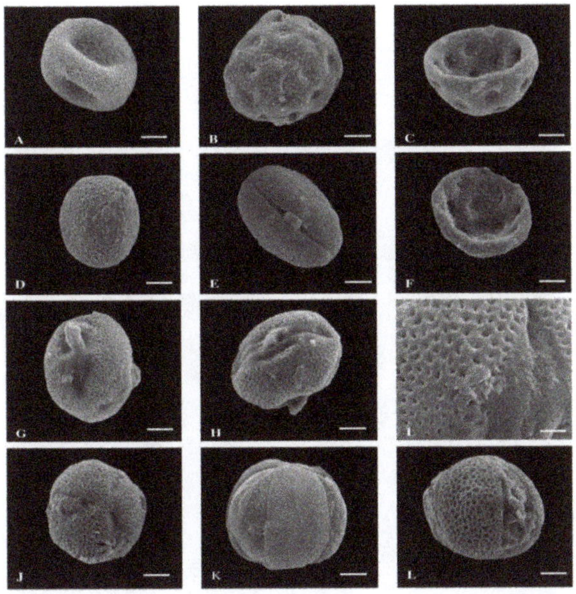

Figure 7. SEM images of pollen grains (Equatorial and Polar axis) and pollen surface ornamentation (**A–C**). The scale bar represents 5 μm for *Celosia aegentae* (**D–F**), 10 μm for *Cenchrus ciliaris* (**G–I**), 5 μm for *Chenopodium murale* (**J–L**), and 5 μm for *Chorchorus depressus*.

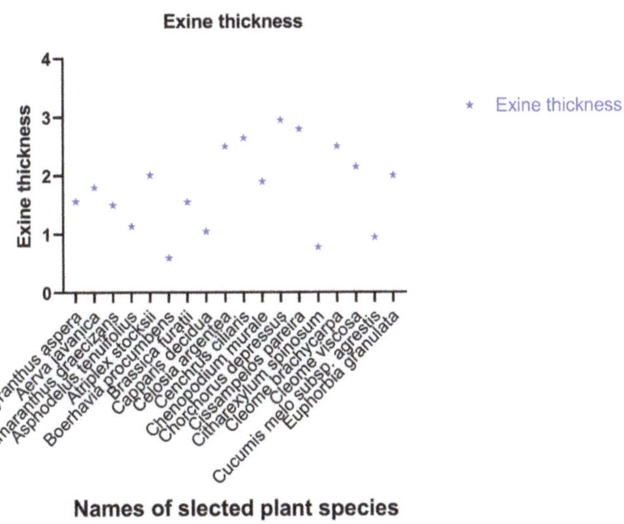

Figure 8. Graph representing the exine thickness of the selected plant species.

3.4. Exine Sculpturing

The exine is the most distinguishing characteristic of the species (Figures 9 and 10). All the pollen grains included in this study had different exine sculpturing. Exine ornamentations are essential features for identifying closely related species and genera with each other. The present study recorded the following ornamentations: reticulate, carbonate, aerolate, faveolate, reticulate-perforate, and reticulate-scabrate. *Achyranthus aspera*, *Celosia aegentae*, and *Citharexylum spinosum* have scarbate exine ornamentations. *Aerva javanica*, *Cleome brachycarpa*, and *Cenchrus ciliaris* have reticulate *ornamentations*. *Amaranthus gracigiense* and *Cleome viscosa* have scarbate-reticulate. *Asphodelus tenuifolius*, *Euphorbia grandulata*, and *Chenopodium murale* have aerolate, while *Cissampelos parira* and *Cucumis melo* subsp. *agrestis* have perforated exine sculpturing.

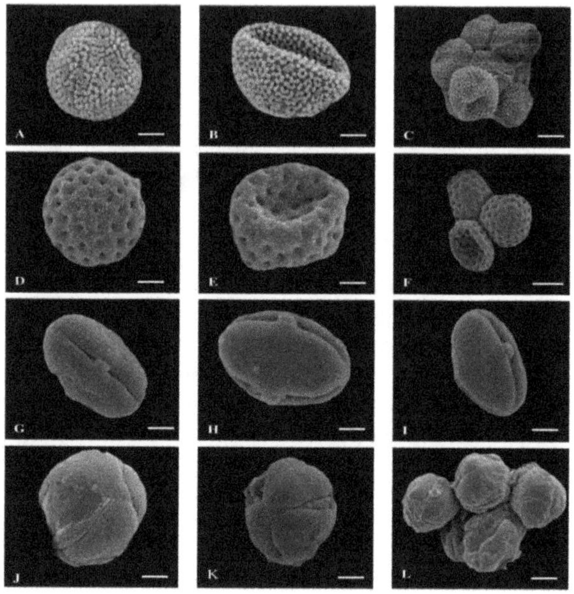

Figure 9. SEM images of pollen grains (Equatorial and Polar axis) and pollen surface ornamentation (**A–C**). The scale bar represents 10 μm for *Cissampelos parira* (**D–F**), 5 μm for *Citharexylum spinosum* (**G–I**), 5 μm for *Cleome brachycarpa* (**J–L**), and 5 μm for *Cleome viscosa*.

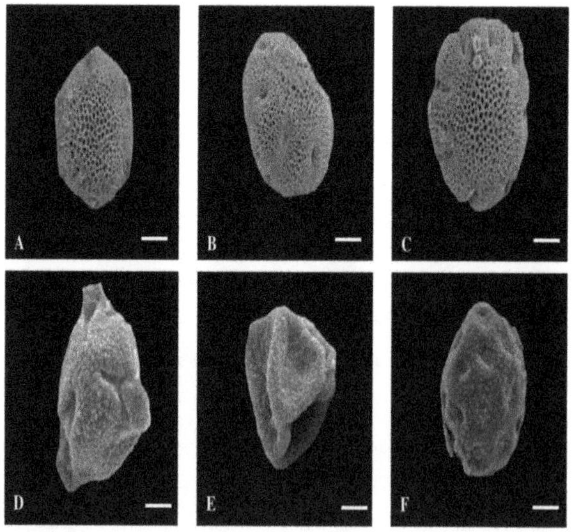

Figure 10. SEM images of pollen grains (Equatorial and Polar axis) and pollen surface ornamentation (**A–C**). The scale bar represents 10 μm for *Cucumis melo* subsp. *agrestis* (**D–F**) and 5 μm for *Euphorbia grandulata*.

3.5. Fertility and Sterility Percentages

Table 4 lists the pollen fertility and sterility results. The highest fertility percentage was observed in *Brassica furatii* (97%) and *Cissampelos pareira* (95%), whereas the lowest percentage of fertility was observed in *Aerva javanica* (69%). In the case of sterility, the highest sterile pollen was observed in *Aerva javanica* (30%), while the lowest sterile pollen grains were found in *Brassica furatii* (2%).

4. Discussion

This study examined the role of palynological studies for species identification from the Potohar regions of Pakistan using microscopic techniques. The pollen grains of the weeds varied in shape, aperture, mesocolpium, colpi orientation, and exine ornamentations. Pollen shape, exine sculpturing, spines, number of colpi, and exine thickness of weeds from district Bannu were described by Ahmad et al. [26]. A wide variation in pollen parameters helps differentiate closely related species and genera. The morphological characteristics of pollen provide additional knowledge in plant taxonomy and systematics. The pollen investigated through LM showed great variability in equatorial diameter, polar axis diameter, width and length of colpi, and exine thickness. An extensive range of variations was observed in the pollen morphology of selected species of weeds from the study area showing that the field of palynology plays a vital role in the evolutionary studies of plant species. Variations in exine sculpturing are the diagnostic characteristic in plant systematic studies.

The current study showed that pollen morphological features, particularly the shape of pollen grains, are the major distinguishing feature for identifying weeds species. The present study varies from the previous studies conducted on weeds in terms of the qualitative and quantitative microscopic characteristics. Amaranthaceae were considered the dominant family of weeds in the present research family, followed by Capparaceae. In contrast, the remainder of each family, i.e., Euphorbiaceae, Cucurbitaceae, Verbenaceae, Menispermaceae, Teliaceae, Chenopodiaceae, Poaceae, Asphodelaceae, Brassicaceae, and Nyctaginaceae, contain single species. Pollen grains were in the form of monad, with most species having scabrate exine ornamentation. Similar findings have been reported [27]. Nazish et al. [28] examined the apolar, spheroidal, and scabrate. Micro Echinate pollen ornamentations and sunken pore ornamentations have several pores, ranging from 11 to 23.

Various researchers working on the palynological study of weeds believed that exine ornamentation is the key feature for differentiating taxonomic characteristics [25–30]. Significant differences were observed in the following: the exine thickness, the diameter of the equatorial and polar axis, the size of the colpus, and the number of pores. Palynological studies are used for species identification in plant taxonomy and systematics. The study is linked directly to environmental sciences, molecular biology, plant ecology, pharmaceutical studies, genetics, and aerobiology [13]. In the present study, some pollen has the same aperture conditions, i.e., tricolporate. Khan et al. [31] worked on the pollen morphological features of 16 plants using scanning electron microscopy, which is somewhat similar to our results [32]. They used OM and SEM and described the pollen micro-morphology of 24 wetland weeds. Bolick et al. [33] recorded the exine sculpturing patterns of the pollen of Asteraceae with particular emphasis on evolution. The morphological pollen characteristics can be valuable in solving complications linked with the systematic study of grass. The study of the pollen provides the base for more structures to properly identify the plant species [34]. A palynological study of Asteraceous taxa from the Potohar region helps differentiate species taxonomically [35]. Research work has been done on many aspects of flora. The study area has diverse species, but no work has been performed on the pollen morphological study of weedy species.

Oblate-spheroidal observations were observed abundantly in this study. Most of the colpi orientations noted were sunken and angular. In addition, many prominent, waxy, long, slit-like, and tapering ends of colpi were observed in the pollen. The pollen examined in this study was reticulate in sculpturing, while some contained faveoelate. *Cenhrus ciliaris* contained a single pore in the center of the pollen. Fertility and sterility percentages were determined for the stability of pollen in the study area. The fertility and sterility percentages in pollen grains indicated the relationship among the species' ancestors. The fertility of pollen grains helps identify the genetic variations in flora [36]. Meo et al. [37] reported different shapes of pollen grains of the genus *Cenchrus*. They observed spherical to perbolate shape in *C. pennisetiformis* and spheroidal to subulate in *C. setigerus*. The pollen of this genus is monoporate with psilate exine ornamentation. The position of pores was ectoporous, but endoporus pores were also observed in some cases. Three species, i.e., *C. ciliaris*, *C. biflorus*, and *C. pennisetiformis* are endoporus, whereas *C. setigerus* is ectoporus. The size and shape of pollen grains of the genus *Cenchrus* are used to distinguish the species. It is a valuable tool for calculating the strength of species in a specific region. The pollen examined in this study were oblate, suboblate, oblate-spheroidal, and prolate-spheroidal in shape with the aperture conditions of tricolporate, mesocolpium scarbate, aerolate, reticulate, and perforated exine ornamentations. By contrast, the colpi orientations were long, slit-like, sunken, angular, margins distinct, waxy, rounded, and tapering ends.

Perveen et al. [38] reported the morphological characteristics of pollen of the family Brassicaceae, which have reticulate exine ornamentation compared to the current study results *Brassica furatii* exhibit echinate type of exine surface. Appel and Al-Shehbaz [39] reported that all pollen grains observed in their research exhibit tricolpate with the reticulate type of exine, similar to the present study results in that *Brassica furatii* has tricolpate pollen grains. Moore et al. [40] reported tricolpate pollen with reticulate sculpturing of the exine. Considerable work has been done on weeds of the family Brassicaceae by [41,42], and it was concluded that the shape of the lumen is the most significant characteristic in the delimitation of different species in the family Brassicaceae.

Weed flora has been documented previously by many researchers. Reference [43] reported 73 weed species that are medicinally important and utilized for treating different diseases of both human beings and animals. These weed species help manufacture different plant yields and the indigenous sellers of crude drugs of plants. The pattern of distribution, richness, and abundance of weed species (e.g., *A. viridis*, *P. lanceolata*, *D. annulatum*, *V. thapsus*, and *C. dactylon*) depend on the different environmental conditions, such as $CaCO_3$, pH, organic matter, soil texture, phosphorous, electrical conductivity, and concentration of nitrogen. Rafay et al. (2014) reported that weeds play a significant role in the agroecosystem by increasing the cropping system's environmental heterogeneity and floral diversity. Williams and Kremen [44] reported that non-crop plants support different types of invertebrates because many insects depend upon specific weeds for survival.

On the other hand, weeds play a serious role in crop production and threaten cultivated lands because they use soil nutrients. Weeds adversely affect the yield and growth of crop plants because of the nutrient demand. Weed vegetation is eliminated using different control measures, e.g., chemical approaches, cultural procedures, and biological mechanisms. In cultural operations, weeds are thrown away from the rice field [43]. Identification of different weed species and farming practices plan with revised management has been suggested [45].

The results of the present study of this family were compared with previously published data showing that these species are quite similar to the previously reported species, with few variations. Butt et al. [32] described the palynological study of weeds for pollen features, i.e., pore number, the diameter of the equatorial and polar axis, length, width of colpi, and its outline, which are very helpful for classifying species in plant taxonomy. According to the results, these characteristics are insufficient for species identifications; sometimes, many species hold similar features. The most vital characteristics are exine thickness, mesocolpium, and aperture conditions. The study also provides information on the delimitation of species from the genus to the family level. Nevertheless, further research will be needed to achieve more conclusive results using more cosmopolitan species.

5. Conclusions

A microscopic study of weeds and pollen grains collected from the Potohar region of Pakistan has been documented. The studied pollen helps identify the weed plant species, genus, and family. Eighteen plant species belonging to 16 families were investigated in the present research; the morpho-palynological characters were investigated quantitatively and qualitatively. In the case of palynomorph, major variations were observed in size, appearance, and diameter of the equatorial and polar axis, the wideness of the exine, and its ornamentation. The pollen traits were proven to have taxonomic potential to support and strengthen the systematics of this subfamily. The dominant family of weeds in the present study is Capparaceae followed by Amaranthaceae, and the oblate shape of pollen is commonly observed in weeds. All the pollen grains were tricolpate with scabrate exine sculpturing. The current study was compared with previous studies, confirming that the pollen morphology under OM and SEM provides important information for identifying weeds species. The microscopic features provide considerable knowledge and have important information for correctly identifying flora. This study also contributed to collecting the data on weeds plants, i.e., preventive, therapeutic, and diagnostic potential. Further phylogenetic, pharmacognosy, and molecular studies are recommended for weed plants in the future to support and strengthen the systematics of weed flora.

Author Contributions: A.U., M.A., M.Z. and S.S. (Shazia Sultana) designed the research, performed the experiments, and analyzed the data., W.Z., F.U., A.A. and S.S. (Saddam Saqib) visualization, methodology, writing—review and editing, investigation, re-sources, data curation. All authors have read and agreed to the published version of the manuscript.

Funding: This research received no external funding.

Data Availability Statement: Not applicable.

Acknowledgments: All the authors are thankful to (CRL) Central resource library, Department of the Physics University of Peshawar, for providing the facility of Scanning Electron microscopy.

Conflicts of Interest: The authors declare no conflict of interest.

References

1. Oerke, E.C.; Dehne, H.W. Safeguarding production—Losses in major crops and the role of crop protection. *Crop Prot.* **2004**, *23*, 275–285. [CrossRef]
2. Khan, F.; Muhammad, Z.; Khan, K.; Ahmad, S.; Khan, M.J.; Bakht, T.; Kamal, A. Palynological investigation of allergenic and invasive weeds plants for biodiversity in district Lakki Marwat using scanning electron microscopy. *Pak. J. Weed Sci. Res.* **2020**, *26*, 349–365. [CrossRef]
3. Qureshi, R.; Waheed, A.; Arshad, M. Weed communities of wheat crop in district Toba Tek Singh, Pakistan. *Pak. J. Bot* **2009**, *41*, 239–245.
4. Neve, P.; Barney, J.N.; Buckley, Y.; Cousens, R.D.; Graham, S.; Jordan, N.R.; Lawton-Rauh, A.; Liebman, M.; Mesgaran, M.B.; Schut, M. Reviewing research priorities in weed ecology, evolution and management: A horizon scan. *Weed Res.* **2018**, *58*, 250–258. [CrossRef]
5. Ullah, F.; Zafar, M.; Ahmad, M.; Dilbar, S.; Shah, S.N.; Sohail, A.; Zaman, W.; Iqbal, M.; Bahadur, S.; Tariq, A. Pollen morphology of subfamily Caryophylloideae (Caryophyllaceae) and its taxonomic significance. *Microsc. Res. Tech.* **2018**, *81*, 704–715. [CrossRef] [PubMed]
6. Mamoona, M.; Mir, A.K.; Mushtaq, A.; Nighat, S.; Sidra, N.A.; Kanwal, T.; Saira, T.; Tehmeena, M.; Madhia, A.; Shazia, B. Foliar epidermal anatomy of some ethnobotanically important species of wild edible fruits of northern Pakistan. *J. Med. Plants Res.* **2011**, *5*, 5873–5880.
7. Jogi, Q.; Hajano, G.A.; Kandharo, M.N.; Shah, A.N.; Soomro, A.A.; Abbasi, Z.A.; Burioro, M.; Baloch, S.K.; Kalwar, Z.A.; Abbasi, N.A. Examine different weed management techniques in sugarcane (*Saccharum officinarum* L.). *Pure Appl. Biol. (PAB)* **2019**, *8*, 151–159.
8. Mehsud, A.; Mehmood, S.; Muhammad, A.; Khan, R.U.; Khan, S.U.; Khan, H.U.; Wazir, R.; Hussain, Z. Morphology and anatomy of some weeds from flora of district Bannu, Pakistan. *Pak. J. Weed Sci. Res.* **2013**, *19*, 437–445.
9. Amarger, N. Rhizobia in the field. *Adv. Agron.* **2001**, *23*, 741–748.
10. Marwat, K. A checklist and key to parasitic weeds of Pakistan. *Pak. J. Weed Sci. Res.* **1933**, *6*, 1–15.
11. Naz, J.; Ahmad, M.; Sultana, S.; Zafar, M.; Naz, S.; Zaman, W.; Ullah, F. Invasive alien species of Potohar Plateau, Northern Pakistan. *Ann. Bot.* **2021**, *11*, 171–179.

12. Rashid, K.; Rasul, G. Rainfall variability and maize production over the Potohar Plateau of Pakistan. *Pak. J. Meteorol.* **2011**, *8*, 63–74.
13. Muhammad, Z.; Khan, M.; Mushtaq, A.; Shazia, S. Palynological and taxonomic studies of some weeds from flora of Rawalpindi. *Pak. J. Weed Sci. Res.* **2006**, *12*, 99–109.
14. Hameed, A.; Zafar, M.; Ahmad, M.; Sultana, S.; Akhter, M.S.; Zaman, W.; Saqib, S.; Ullah, F. Micromorphology, phytochemical and pharmacological evaluation of *Isodon rugosus* (wall. Ex benth.) Codd. *J. Anim. Plant Sci.* **2022**, *32*, 736–745.
15. Jehanzeb, S.; Zafar, M.; Ahmad, M.; Sultana, S.; Zaman, W.; Ullah, F. Comparative petioler anatomy of tribe Mentheae subfamily Nepetoideae, Lamiaceae from Pakistan. *Feddes Repert.* **2020**, *131*, 163–174. [CrossRef]
16. Shah, S.N.; Ahmad, M.; Zafar, M.; Ullah, F.; Zaman, W.; Malik, K.; Rashid, N.; Gul, S. Taxonomic importance of spore morphology in Thelypteridaceae from Northern Pakistan. *Microsc. Res. Tech.* **2019**, *82*, 1326–1333. [CrossRef]
17. Ullah, F.; Ayaz, A.; Saqib, S.; Parmar, G.; Bahadur, S.; Zaman, W. Taxonomic implication of leaf epidermal anatomy of selected taxa of Scrophulariaceae from Pakistan. *Microsc. Res. Tech.* **2020**, *84*, 521–530. [CrossRef]
18. Usma, A.; Ahmad, M.; Zafar, M.; Sultana, S.; Lubna; Kalsoom, N.; Zaman, W.; Ullah, F. Micromorphological variations and taxonomic implications of caryopses of some grasses from Pakistan. *Wulfenia* **2020**, *27*, 86–96.
19. Zaman, W.; Ullah, F.; Parmar, G.; Saqib, S.; Ayaz, A.; Park, S. Foliar micromorphology of selected medicinal Lamiaceae taxa and their taxonomic implication using scanning electron microscopy. *Microsc. Res. Tech.* **2022**, *2022*, 1–20. [CrossRef]
20. Meo, A.; Khan, M. Palynological studies of some weeds of asteraceae from Pakistan. *Pak. J. Weed Sci. Res. (Pak.)* **2004**, *10*, 87–92.
21. Ronald, O.K. *Pollen and Spores*, 2nd ed.; American Association of Stratographic Palynologists Foundation: College Station, TX, USA, 2000; pp. 13–21.
22. Ashfaq, S.; Ahmad, M.; Zafar, M.; Sultana, S.; Nazish, M.; Khan, A.N. Systematics of medicinally important weeds of genus *Convolvulus*: Convolvulaceae. *Planta Daninha* **2020**, *38*. [CrossRef]
23. Erdtmann, G. Pollen morphology and plant taxonomy III, Morina L. with an addition on pollen morphological terminology. *Sven. Bot. Tidskr.* **1952**, *39*, 187–191.
24. Khan, M.A.; Stace, C.A. Breeding relationships in the genus *Brachypodium* (Poaceae: Pooideae). *Nord. J. Bot.* **1999**, *19*, 257–269. [CrossRef]
25. Ahmad, S.; Zafar, M.; Ahmad, M.; Yaseen, G.; Sultana, S. Microscopic investigation of palyno-morphological features of melliferous flora of Lakki Marwat district, Khyber Pakhtunkhwa, Pakistan. *Microsc. Res. Tech.* **2019**, *82*, 720–730. [CrossRef]
26. Ahmad, S.; Zafar, M.; Ahmad, M.; Ozdemir, F.; Yaseen, G.; Sultana, S.; Kutlu, M. Palynological studies of winter weeds melliferous flora of district Bannu, Khyber Pakhtunkhwa, Pakistan. *Ann. Bot.* **2020**, *10*, 77–86.
27. Raees, K.; Abidin, S.Z.U.; Mumtaz, A.S.; Jamsheed, S.; Ullah, H. Comparative leaf and pollen micromorphology on some Grasses taxa (Poaceae) distributed in Pakistan. *Int. J. Nat. Life Sci.* **2017**, *1*, 72–82.
28. Nazish, M.; Zafar, M.; Ahmad, M.; Sultana, S.; Ullah, R.; Alqahtani, A.S.; Ullah, F.; Ahmad, S.; Ashfaq, S.; Ullah, F. Palyno-morphological investigations of halophytic taxa of Amaranthaceae through SEM from Salt range of Northern Punjab, Pakistan. *Microsc. Res. Tech.* **2019**, *82*, 304–316. [CrossRef]
29. Amina, H.; Ahmad, M.; Bhatti, G.R.; Zafar, M.; Sultana, S.; Butt, M.A.; Bahadur, S.; Haq, I.U.; Ghufran, M.A.; Ahmad, S. Microscopic investigation of pollen morphology of Brassicaceae from Central Punjab-Pakistan. *Microsc. Res. Tech.* **2020**, *83*, 446–454. [CrossRef]
30. Perveen, A.; Qaiser, M. Pollen flora of Pakistan-LXIX. Poaceae. *Pak. J. Bot.* **2012**, *44*, 747–756.
31. Khan, K.; Ahmad, M.; Zafar, M.; Malik, K.; Sultana, S.; Ahmad, S.; Khan, F.; Ullah, K. Palyno-Morphological Study of Weedy Melliferous (Bee Visited) Plants Using Light Microscopic Techniques From Southern Khyber Pakhtunkhwa, Pakistan. *Pak. J. Weed Sci. Res.* **2021**, *27*, 163–172. [CrossRef]
32. Butt, M.A.; Zafar, M.; Ahmad, M.; Kayani, S.; Bahadur, S.; Ullah, F.; Khatoon, S. The use of taxonomic studies to the identification of wetlands weeds. *Adv. Weed Sci.* **2021**, *39*, 1–13. [CrossRef] [PubMed]
33. Bolick, M.R. Taxonomic, evolutionary, and functional considerations of Compositae pollen ultrastructure and sculpture. *Plant Syst. Evol.* **1978**, *130*, 209–218. [CrossRef]
34. Aftab, R.; Perveen, A. A palynological study of some cultivated trees from Karachi. *Pak. J. Bot.* **2006**, *38*, 15–28.
35. Usma, A.; Ahmad, M.; Ramadan, M.F.; Khan, A.M.; Zafar, M.; Hamza, M.; Sultana, S.; Yaseen, G. Micro-morphological diversity of pollen among Asteraceous taxa from Potohar Plateau-Pakistan. *Microsc. Res. Tech.* **2022**, *85*, 2467–2485. [CrossRef] [PubMed]
36. Ullah, F.; Ullah, A.; Sohail, A.; Khan, R. Diversity, distribution and ecological importance of weeds in the maize crop at Maidan Valley, Dir (l), Khyber Pakhtunkhwa, Pakistan. *Pak. J. Weed Sci. Res.* **2015**, *21*, 543–553.
37. Meo, A.; Hafiz, H.; Baig, I. Pollen grain description of 13 non allergic and four allergic Graminaceous species. *J. Pure Appl. Sci.* **1989**, *8*, 19–26.
38. Perveen, A.; Qaiser, M.; Khan, R. Pollen Flora of Pakistan–XLII. Brassicaceae. *Pak. J. Bot.* **2004**, *36*, 683–700.
39. Appel, O.; Al-Shehbaz, I. Cruciferae. In *Flowering Plants Dicotyledons*; Springer: Berlin/Heidelberg, Germany, 2003; pp. 75–174.
40. Moore, P.D.; Webb, J.A.; Collison, M.E. *Pollen Analysis*; Blackwell Scientific Publications: Hoboken, NJ, USA, 1991.
41. Brochmann, C. Polyploid evolution in arctic-alpine (Brassicaceae). *Sommerfeltia* **1992**, *4*, 1–44. [CrossRef]
42. Al-Shehbaz, I.; Beilstein, M.; Kellogg, E. Systematics and phylogeny of the Brassicaceae (Cruciferae): An overview. *Plant Syst. Evol.* **2006**, *259*, 89–120. [CrossRef]

43. Khan, M.; Kakar, S.; Marwat, K.; Khan, I. Differential response of Zea mays L. in relation to weed control and different macronutrient combinations. *Sains Malays.* **2013**, *42*, 1405–1411.
44. Williams, N.M.; Kremen, C. Resource distributions among habitats determine solitary bee offspring production in a mosaic landscape. *Ecol. Appl.* **2007**, *17*, 910–921. [CrossRef] [PubMed]
45. Iqbal, M.; Khan, S.M.; Khan, M.A.; Ahmad, Z.; Abbas, Z.; Khan, S.M.; Khan, M.S. Distribution pattern and species richness of natural weeds of wheat in varying habitat conditions of district Malakand, Pakistan. *Pak. J. Bot.* **2017**, *49*, 2371–2382.

agronomy

MDPI

Article

Species Composition, Diversity, and Biomass Estimation in Coastal and Marine Protected Areas of Terengganu, Peninsular Malaysia

Elizabeth Pesiu [1,*], Gaik Ee Lee [1,2,*], Muhammad Razali Salam [1], Jamilah Mohd Salim [1], Kah Hoo Lau [3], Jean Wan Hong Yong [4] and Mohd Tajuddin Abdullah [5,6]

1 Faculty of Science and Marine Environment, Universiti Malaysia Terengganu, Kuala Nerus 21030, Terengganu, Malaysia
2 Institute of Tropical Biodiversity and Sustainable Development, Universiti Malaysia Terengganu, Kuala Nerus 21030, Terengganu, Malaysia
3 Forest Biodiversity Division, Forest Research Institute Malaysia, Kepong 52109, Selangor, Malaysia
4 Department of Biosystems and Technology, Swedish University of Agricultural Sciences, 23053 Alnarp, Sweden
5 Faculty of Fisheries and Food Science, Universiti Malaysia Terengganu, Kuala Nerus 21030, Terengganu, Malaysia
6 West Wing, Menara Matrade, Jalan Sultan Haji Ahmad Shah, Kuala Lumpur 50480, Selangor, Malaysia
* Correspondence: elizabethpesiu@gmail.com (E.P.); gaik.ee@umt.edu.my (G.E.L.)

Citation: Pesiu, E.; Lee, G.E.; Salam, M.R.; Salim, J.M.; Lau, K.H.; Yong, J.W.H.; Abdullah, M.T. Species Composition, Diversity, and Biomass Estimation in Coastal and Marine Protected Areas of Terengganu, Peninsular Malaysia. *Agronomy* **2022**, *12*, 2380. https://doi.org/10.3390/agronomy12102380

Academic Editors: Alessio Papini, Mushtaq Ahmad, Fazal Ullah, Wajid Zaman and Youssef Rouphael

Received: 10 August 2022
Accepted: 18 September 2022
Published: 1 October 2022

Publisher's Note: MDPI stays neutral with regard to jurisdictional claims in published maps and institutional affiliations.

Abstract: We investigated and compared the tree species composition and diversity of different forest types in Setiu Wetlands and on the three major islands of Terengganu. A total of 24 plots of 25 m × 25 m with four plots in each study site were established, viz. *Melaleuca* swamp forest in Kampung Fikri, freshwater swamp forest in Kampung Gong Batu, mangrove forest in UMT Setiu research station, and the islands, namely Pulau Bidong, Pulau Redang, and Pulau Perhentian. We calculated the basal area, stand density, Importance Value Index, species diversity, and above-ground biomass in the designated study areas. We assessed 139 tree species from 96 genera and 50 families based on a total of 2608 tree samples of 5 cm DBH and above. The freshwater swamp forest harbored the highest number of species with 20 species in Setiu Wetlands, and among the islands, Pulau Redang had the highest with 56 species. *Melaleuca cajuputi* was the most dominant species in the *Melaleuca* swamp forest, while *Alstonia spatulata* and *Rhizophora apiculata* are expected in the freshwater swamp and mangrove forest, respectively. Pulau Bidong, Pulau Redang, and Pulau Perhentian are mostly represented by *Licania splendens, Shorea glauca*, and *Vatica* sp., respectively. All the dominant species but *Licania splendens* contributed to the highest amount of above-ground biomass. Our current study indicated that different forest types vary in composition and structure, which may contribute to their unique ecological roles within their specific environment.

Keywords: coastal and insular vegetation; protected areas; coastal wetland; freshwater swamp; mangrove; lowland dipterocarp forest

1. Introduction

The forest cover of Terengganu on the east coast of Peninsular Malaysia is still relatively extensive, with ca. 654,000 ha of forested areas [1]. Within these areas, 30,000 ha have been gazetted as a state park in Kenyir and ca. 1600 ha in Setiu Wetlands, including the freshwater wetland of Tasik Berombak [2,3]. National or state parks are thus far considered the best protected areas that do not allow logging or monoculture plantations. On the other hand, ca. 544,000 ha have been gazetted as Permanent Reserved Forests, which are still partly subjected to state forest management for timber production [1]. There are 45 forest reserves in Terengganu, consisting mainly of dipterocarp forests. Among these are Besul, Bukit Bauk, Bukit Kesing, Bukit Terendak, Cerul, Gunung Tebu, Hulu Besut, Hulu

Telemong, Hulu Terengganu, Jengai, Jerangau, Merchang, Pasir Raja Barat and Selatan, Pelagat, Petuang, Rasau Kerteh, Sungai Nipah, and Tembat, each with above 5000 ha [1]. Furthermore, there are 13 marine parks in Terengganu with 36 sea turtle nesting grounds, the largest amount in Peninsular Malaysia [4]. Thus, Terengganu's islands and the long coastline in Setiu Wetlands have been proposed as important marine ecological corridors: the Northern Terengganu Marine Park Islands-Setiu Wetlands Ecological Corridor of Peninsular Malaysia (Figure 1).

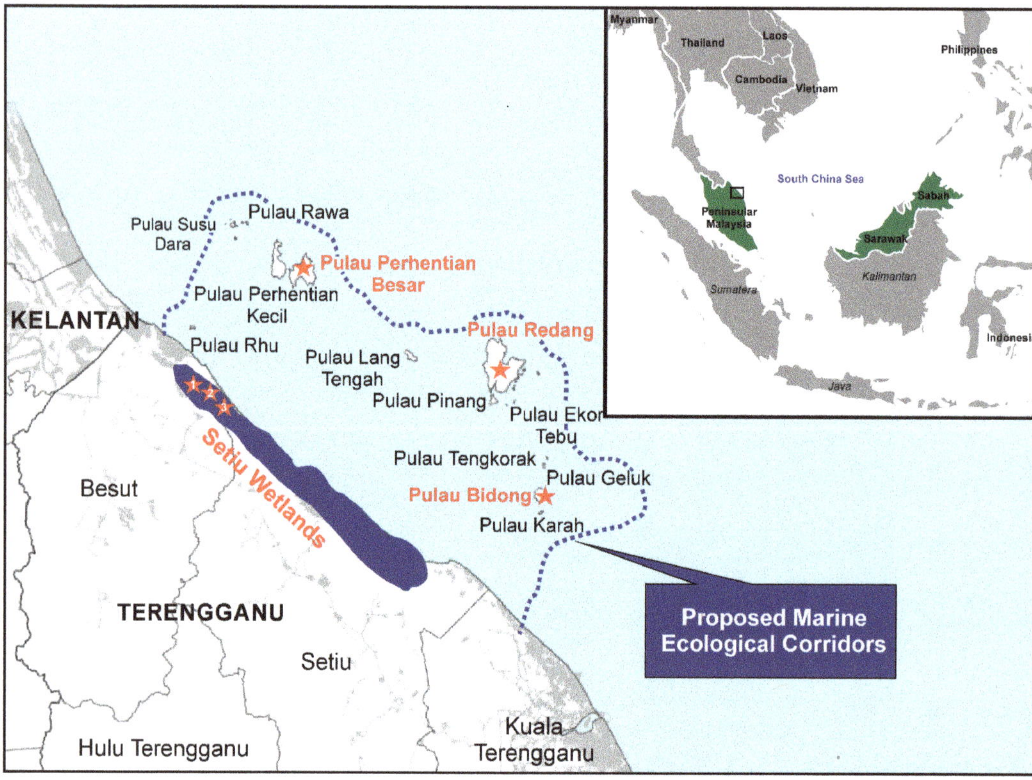

Figure 1. Map showing the Setiu Wetlands and islands of Terengganu. Red stars and bold red texts indicate the location of the study areas. Red star 1: mangrove in UMT research station, 2: freshwater swamp in Gong Batu, 3: *Melaleuca* swamp in Kampung Fikri. The dotted blue line is the proposed Northern Terengganu Marine Park Islands-Setiu Wetlands Ecological Corridor. Map modified from the 6th National Report of Malaysia to the Convention on Biological Diversity.

The Setiu Wetlands are a complex and heterogeneous landscape containing several different habitat types ranging from coastal forests to mangrove estuaries and freshwater swamps to hill dipterocarp forests [5,6]. Wetlands provide basic ecosystem regulatory services such as coastal protection [7,8], flood mitigation and erosion control [9,10], and nurseries for juvenile marine fishes and also support traditional livelihoods for the local population [6]. The wetlands are located northeast of Peninsular Malaysia in Terengganu and form part of the Setiu river basin. Much of the natural vegetation in Setiu Wetlands is still well-represented. These wetlands are considered the largest and possibly the most intact coastal wetland complex on the east coast of Peninsular Malaysia [2,6]. Due to their geographic isolation from the source population on the mainland, islands have long been known to support a set of unique populations, communities, and ecosystems. These resistance entities have adapted through a geographical limitation with vital processes, properties, and interactions that occur in a simpler way [11]. The forests in coastal areas bring about ecological and socio-economic importance in terms of goods and services (such as forestry and fisheries resources; recreation and ecotourism) [12].

Assessment and monitoring of forest biological diversity in coastal areas can provide insight into conservation value and are essential for sustainable forest management. According to Noss [13], the challenge lies in defining sound and practical biodiversity monitoring systems that deliver scientific data to inform sustainable forest management. The use of inventories to determine biodiversity (composition, diversity of tree species, and frequency) is a common way to gather information for forest management operations [14]. On the other hand, the different types of forest types occurring in Terengganu may act as carbon sources and carbon sinks. Biologically, estimates of above-ground biomass are essential for studies of carbon stocks and the effects of deforestation and carbon sequestration on the global carbon balance [15,16]. However, such data are lacking for coastal areas in Terengganu.

Currently, published data analyses of various forest types in Setiu Wetlands and major islands in Terengganu are still lacking. Although much research has been conducted independently to record plant species, e.g., [17–19], the forest structure and floristic composition are still unknown. Most of the available literature only focused on qualitative measures attempting to document the absence or presence of tree species but not the physical structures, such as quantitative measures of species diversity within a plot, the numbers of individuals within a species, and comparison among forest types in each species. Therefore, we aimed to investigate and compare the tree species composition, diversity, and above-ground biomass of three different forest types in Setiu Wetlands: *Melaleuca* swamp, freshwater swamp, and mangrove forests. Furthermore, three marine parks, namely Pulau Bidong, Pulau Perhentian, and Pulau Redang, were also included in the present study. We hypothesize that the tree species diversity in different habitats in Setiu Wetlands varies according to soil and environment factors and changes along the coastline towards the inland of Setiu.

2. Materials and Methods

The study sites are in Setiu Wetlands State Park and on three major islands of Terengganu, i.e., Pulau Bidong, Pulau Redang, and Pulau Perhentian (Figure 1). Three different forest types, i.e., *Melaleuca* swamp, freshwater swamp, and mangrove forests in Setiu Wetlands, were selected in the present study (Figures 2 and 3). The islands of Terengganu are generally made up of lowland dipterocarp forests.

We used plot-based tree inventories to test the hypothesis. A total of 24 plots of 25 m × 25 m were established in all the study sites, with four plots in each of the six study sites. All trees with 5 cm diameter breast height (DBH) and above were surveyed, measured, and tagged with flagging ribbons. Trees were identified in-situ, while voucher specimens were collected for further identification and verification purposes. Specimens were deposited at the herbarium of Universiti Malaysia Terengganu (UMTP). Tree species were identified by E. Pesiu and M.R. Salam using the collection at UMTP and herbarium of Forest Research Institute Malaysia (KEP).

Figure 2. Different habitats at the study areas. (**A**,**B**) Setiu River, one of the four main rivers in Setiu Wetlands; (**C**) freshwater swamp forest; (**D**) mangrove forest; (**E**) *Melaleuca* swamp forest, (**F**) lowland dipterocarp forest in Pulau Redang.

Figure 3. Some of the plants found in the study areas. (**A**) Top canopy of *Melaleuca cajuputi;* (**B**) leaves and inflorescence of *Melaleuca cajuputi;* (**C**) top canopy of *Rhizophora apiculata;* (**D**) leaves and propagules of *Rhizophora apiculata;* (**E**) bark of *Vatica* sp. (inset top: white resin, bottom: leaf arrangement); (**F**) habitat of *Talipariti tiliaceus* (inset top: fruit, bottom: flower).

We calculated the basal area, stand density, Importance Value Index, species diversity, and above-ground biomass in the designated study areas. The sample-based rarefaction curve [20] was used to compare the species abundance in different forest types at each study site. The 95% confidence intervals were constructed where samples were drawn without replacement and calculated as ±2 standard deviations from the expected values [21]. Data was computed by using PAST Version 3.0 software [22]. Community composition within the plots was carried out with several abundance parameters such as basal area (m^2/ha), stand density (Ind/ha) and Importance Value index (IV_i). The IV_i was calculated by summing the values of relative density (RD), relative dominance (based on basal area) (RB), and relative frequency (RF) of each species or family (IV_i = RD + RB + RF) [23]. Species diversity was determined using the Shannon Diversity Index (H′) [24] as follows:

$$H = \sum_{i=1}^{s} -(Pi * \ln Pi)$$

where s is the number of species; Pi is the proportion of individuals or the abundance of the ith species expressed as a proportion of total abundance. The estimation of above-ground biomass was calculated following Kato et al. [25]. The equation takes into account of the stem biomass (W_s), branches biomass (W_b) and leaves biomass (W_l). The above-ground biomass is estimated by adding up the value of stem, branches, and leaves biomass. The equations are as follows:

$$\text{Stem biomass } (W_s) = 0.313\,(DBH^2\,H)^{0.9733}$$

$$\text{Branches biomass } (W_b) = 0.136\,W_s^{1.041}$$

$$\text{Leaves biomass } (W_l) = \frac{125 \times 0.124\,W_s^{0.794}}{0.124\,W_s^{0.794} + 125}$$

where,

$$\text{Tree height, } H = \frac{(122 \times DBH)}{(2\,DBH + 61)}$$

$$\text{Above-ground biomass (kg)} = W_s + W_b + W_l$$

3. Results

3.1. Species Richness and Diversity

We assessed 139 tree species from 96 genera and 50 families. This sampling was based on a total of 2608 tree samples of 5 cm DBH and above (Tables 1 and 2). The detail of the tree species is shown in Table 2. In Setiu Wetlands, the freshwater swamp forest showed the highest number of species with 20 species compared to the *Melaleuca* swamp and mangrove forest with 13 species each. On the islands, Pulau Redang recorded the highest number of species with 56 species, followed by Pulau Bidong and Pulau Perhentian, with 55 and 20 species, respectively. In terms of density, freshwater swamp forest had the highest density and largest basal area with 1992 ind/ha and 17.38 m^2/ha; among the islands, Pulau Bidong had the highest density with 2700 ind/ha and Pulau Redang had the largest basal area with 52.09 m^2/ha (Table 1). The rarefaction curves differed significantly, and 95% confidence intervals did not overlap among the six study areas (Figures S1 and S2). All the curves are still in the exponential phase and are non-asymptotic with currently available sample sizes.

Table 1. Comparison of basic tree inventory metrics and the total above-ground biomass in each study site.

	Setiu Wetlands			Islands		
	***Melaleuca* Swamp**	**Freshwater Swamp**	**Mangrove**	**Pulau Bidong**	**Pulau Redang**	**Pulau Perhentian**
Number of families	13	17	10	28	27	13
Number of genera	14	20	13	43	42	14
Number of species	13	20	13	55	56	20
Stem density (trees DBH ≥ 5 cm DBH, trees/ha)	1432	1992	1208	2700	1088	1500
Basal area of trees (m^2/ha)	15.22	17.38	8.84	25.04	52.09	24.64
Total above-ground biomass (t/ha)	130.92	147.98	69.47	216.73	728.04	244.05

Table 2. Species list based on plot-based tree inventories. MS = *Melaleuca* swamp, FS = freshwater swamp, M = mangrove in Setiu Wetlands; PB = Pulau Bidong, PR = Pulau Redang, PP = Pulau Perhentian.

Family	**Species**	**MS**	**FS**	**M**	**PB**	**PR**	**PP**
Anacardiaceae	*Bouea oppositifolia*	−	−	−	+	+	−
	Bouea sp.	−	−	−	−	+	−
	Buchanania arborescens	−	−	−	+	−	−
	Campnosperma coriaceum	−	+	−	−	−	−
	Dracontomelon dao	−	−	−	+	−	−
	Mangifera macrocarpa	−	−	−	−	+	−
	Mangifera odorata	−	−	−	−	+	−
	Parishia insignis	−	−	−	−	+	−
	Swintonia floribunda	−	−	−	−	+	−
	Swintonia schwenkii	−	−	−	+	−	−
Annonaceae	*Goniothalamus tenuifolius*	−	−	−	−	+	−
	Polyalthia sumatrana	−	−	−	−	+	−
Apocynaceae	*Alstonia spatulata*	−	+	−	−	−	−
Aquifoliaceae	*Ilex cymosa*	+	+	+	−	−	−
Asparagaceae	*Dracaena maingayi*	−	−	−	−	+	−
Bignoniaceae	*Dolichandrone spathacea*	−	−	+	−	−	−
Burseraceae	*Dacryodes rostrata*	−	−	−	−	+	−
	Santiria rubiginosa	−	−	−	−	+	−
	Santiria sp.	−	−	−	+	−	−
Calophyllaceae	*Calophyllum rupicola*	−	−	−	+	−	+
	Mesua ferrea	−	−	−	−	+	−
	Mesua lepidota	−	−	−	+	+	−
Cannabaceae	*Gironniera* sp.	−	−	−	+	−	−
Celastraceae	*Euonymus cochinchinensis*	−	−	−	−	−	−
	Kokoona sessilis	−	−	−	−	+	−
Chrysobalanaceae	*Licania splendens*	−	−	−	+	−	−
	Maranthes corymbosa	−	−	−	−	+	−
	Maranthes sp.	−	−	−	+	−	−
Clusiaceae	*Garcinia eugenifolia*	−	+	−	+	−	−
	Garcinia hombroniana	+	+	−	+	−	−
	Garcinia nigrolineata	−	−	−	+	−	−
	Garcinia sp.	−	−	−	−	+	−
Dipterocarpaceae	*Dipterocarpus chartaceus*	−	−	−	+	−	−
	Shorea glauca	−	−	−	−	+	−
	Shorea materialis	−	−	−	+	−	−
	Vatica sp.	−	−	−	+	+	+
Ebenaceae	*Diospyros pilosanthera*	−	−	−	−	+	−
	Diospyros sp. 1	−	−	−	+	−	−

Table 2. *Cont.*

Family	Species	MS	FS	M	PB	PR	PP
	Diospyros sp. 2	−	−	−	−	+	−
	Diospyros sp. 3	−	−	−	−	+	−
Elaeocarpaceae	*Elaeocarpus macrocerus*	+	+	−	−	−	−
	Elaeocarpus mastersii	−	−	−	−	−	−
	Elaeocarpus sp.	−	−	−	−	−	+
Erythroxylaceae	*Erythroxylum cuneatum*	−	−	−	+	−	−
Euphorbiaceae	*Croton oblongus*	−	−	−	−	+	−
	Drypetes sp.	−	−	−	+	−	−
	Excoecaria agallocha	−	−	+	−	−	−
	Koilodepas sp.	−	−	−	−	+	−
	Macaranga hypoleuca	−	+	−	−	−	−
	Suregada multiflora	+	+	−	−	+	−
Fabaceae	*Acacia mangium*	−	−	+	−	−	−
	Archidendron contortum	−	−	−	+	−	−
	Callerya atropurpurea	−	−	−	−	−	−
	Intsia bijuga	−	−	+	−	−	−
	Ormosia sumatrana	−	−	−	−	+	−
	Sindora cochinchinensis	−	−	−	−	−	−
	Lithocarpus rassa	−	−	−	+	−	+
Flacourtiaceae	*Hydnocarpus* sp.	−	−	−	−	+	−
Hypericaceae	*Cratoxylum arborescens*	−	+	−	−	−	−
	Cratoxylum formosum	−	−	−	−	−	+
Lamiaceae	*Teijsmanniodendron coriaceum*	−	−	−	+	+	−
Lauraceae	*Cinnamomum sintoc*	−	−	−	−	+	−
	Litsea sp.	−	−	−	−	+	−
	Neolitsea zeylanica	+	+	−	−	−	−
Lecythidaceae	*Barringtonia macrostachya*	−	−	−	−	−	−
	Barringtonia scortechinii	−	−	−	−	+	−
Loganiaceae	*Norrisia malaccensis*	−	−	−	+	−	−
Malvaceae	*Heritiera littoralis*	−	−	+	−	−	−
	Heritiera simplicifolia	−	−	−	−	+	−
	Heritiera sp. 1	−	−	−	+	−	−
	Heritiera sp. 2	−	−	−	+	−	−
	Talipariti tiliaceus	+	+	+	−	−	−
Melastomataceae	*Memecylon edule*	+	−	−	−	−	−
	Memecylon sp.	−	−	−	+	+	−
Meliaceae	*Sandoricum koetjape*	−	−	−	−	−	−
	Xylocarpus rumphii	−	−	+	−	−	−
Moraceae	*Artocarpus kemando*	−	−	−	−	+	−
	Artocarpus lanceifolius	−	−	−	−	+	−
	Artocarpus scortechinii	−	−	−	−	−	−
	Ficus sp.	−	+	−	−	−	−
	Streblus taxoides	+	−	−	−	−	−
Myricaceae	*Myrica esculenta*	+	+	−	−	−	−
Myristicaceae	*Knema glauca*	−	−	−	+	+	−
	Knema globularia	−	−	−	−	+	−
	Knema laurina	−	−	−	+	−	−
Myrtaceae	*Melaleuca cajuputi*	+	+	−	−	−	−
	Rhodamnia cinerea	−	−	−	+	−	+
	Syzygium cerinum	−	−	−	+	−	+
	Syzygium cinereum	−	−	−	+	−	−
	Syzygium grande	−	−	−	+	+	−
	Syzygium sp. 1	−	−	−	+	+	+
	Syzygium sp. 2	−	−	−	+	−	+
	Syzygium sp. 3	−	−	−	+	−	+
	Syzygium sp. 4	−	−	−	+	−	+
	Syzygium sp. 5	−	−	−	−	−	+

Table 2. *Cont.*

Family	Species	MS	FS	M	PB	PR	PP
	Syzygium sp. 6	–	–	–	–	–	+
	Syzygium syzygioides	–	–	–	+	–	–
	Syzygium zeylanicum	–	+	–	+	+	–
Ochnaceae	*Brackenridgea hookeri*	–	+	–	+	–	–
	Campylospermum serratum	–	–	–	+	–	–
Pentaphylacaceae	*Adinandra dumosa*	–	–	–	–	–	–
Peraceae	*Chaetocarpus castanocarpus*	–	–	–	+	–	+
Phyllanthaceae	*Baccaurea parviflora*	–	–	–	–	+	–
	Cleistanthus sp. 1	–	–	–	–	+	–
	Cleistanthus sp. 2	–	–	–	–	+	–
	Cleistanthus sumatranus	–	–	–	–	+	–
	Glochidion sp.	–	–	+	–	–	–
Picrodendraceae	*Austrobuxus nitidus*	–	–	–	+	–	–
Pittosporaceae	*Pittosporum ferrugineum*	+	+	–	–	–	–
Polygalaceae	*Xanthophyllum* sp.	–	–	–	+	–	–
Primulaceae	*Rapanea porteriana*	–	–	–	–	–	+
Rhizophoraceae	*Bruguiera gymnorrhiza*	–	–	+	–	–	–
	Gynotroches axillaris	–	+	–	–	–	–
	Rhizophora apiculata	–	–	+	–	–	–
Rubiaceae	*Aidia wallichiana*	–	–	–	–	+	–
	Canthium glabrum	–	–	–	–	+	–
	Canthium nitidum	–	–	–	+	–	–
	Canthium sp.	–	–	–	–	+	–
	Diplospora malaccensis	–	–	–	+	+	–
	Ixora pendula	–	–	–	–	+	–
	Morinda elliptica	–	–	–	+	–	–
	Psydrax sp.	–	–	–	+	–	+
	Timonius wallichianus	–	–	–	+	–	–
Rutaceae	*Acronychia pedunculata*	–	–	–	+	+	–
	Atalantia monophylla	–	–	–	–	+	–
Sapindaceae	*Guioa bijuga*	–	–	+	+	–	+
	Nephelium sp.	–	–	–	–	–	–
Sapotaceae	*Madhuca longifolia*	–	–	–	+	–	–
	Madhuca sericea	–	–	–	–	+	–
	Madhuca tubulosa	–	–	–	–	+	–
	Palaquium obovatum	–	–	–	+	–	+
	Pouteria malaccensis	–	–	–	–	+	–
	Pouteria obovata	+	+	+	+	+	–
Simaroubaceae	*Eurycoma longifolia*	–	–	–	+	–	–
Sterculiaceae	*Sterculia parviflora*	–	–	–	–	+	–
Symplocaceae	*Symplocos adenophylla*	–	–	–	+	–	+
Verbenaceae	*Vitex pinnata*	–	–	–	–	–	+
	Vitex trifolia	+	+	–	–	–	–
Violaceae	*Rinorea* sp.	–	–	–	–	+	–
Total: 50 families	139 species	13	20	13	55	56	20

In Setiu Wetlands, the Shannon–Wiener Diversity Index (H′) shows that freshwater swamp has the highest tree diversity (Table 3). On the other hand, the diversity index on islands were slightly higher, of which Pulau Bidong recorded the highest with 3.22 (Table 3). As for species evenness, the forests indicated low evenness. Pulau Perhentian had the highest evenness with 0.68, while Pulau Bidong and Pulau Redang indicated moderate evenness.

Regarding DBH size classification, the study areas in Setiu Wetlands showed a low number of diameter class distributions with only 2–3 main diameter classes (Figure 4). Most of the trees within the plot were small to medium in size, and the majority were in the 5–14.99 cm DBH class, followed by 15–24.9 cm, and only a few in the 25–34.9 cm. In the

Melaleuca swamp forest, *Melaleuca cajuputi* from the family Myrtaceae dominated the forest plot, while in the mangrove and freshwater swamp, *Rhizophora apiculata* (Rhizophoraceae) and *Alstonia spatulata* (Apocynaceae) recorded the highest, respectively.

Table 3. The Shannon–Wiener Diversity Index (H′) and species evenness (E) in Setiu Wetlands and three major islands of Terengganu.

Study Site	Shannon Index (H′)	Evenness (E)
Setiu Wetlands:		
Freshwater swamp forest	1.85	0.38
Melaleuca swamp forest	1.67	0.30
Mangrove forest	0.98	0.20
Islands:		
Pulau Bidong	3.22	0.46
Pulau Redang	3.18	0.43
Pulau Perhentian	1.95	0.68

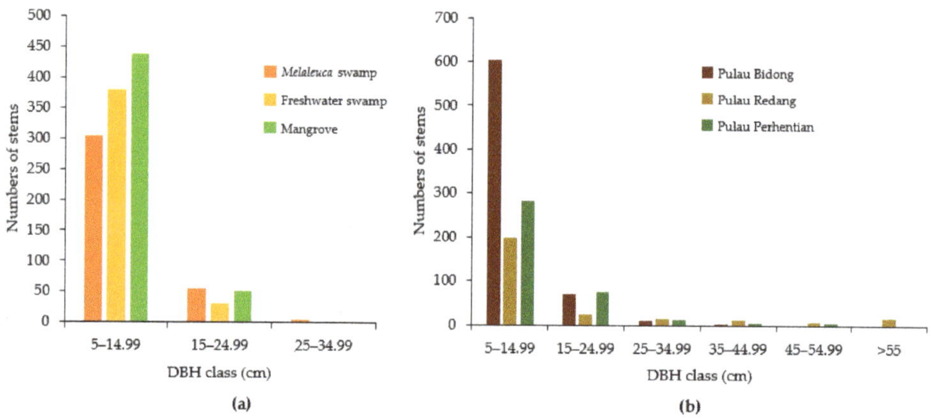

Figure 4. Diameter class distribution. (**a**) The *Melaleuca* swamp, freshwater swamp and mangrove forests in Setiu Wetlands; (**b**) the three major islands of Terengganu, Pulau Bidong, Pulau Redang, and Pulau Perhentian.

The other families were scattered, such as Aquifoliaceae, Malvaceae, Sapotaceae, Pittosporaceae, Euphorbiaceae, Verbenaceae, Anacardiaceae, and Fabaceae. Myrtaceae had the largest basal area, particularly in the *Melaleuca* swamp, followed by the Aquifoliaceae, Sapotaceae, and Malvaceae. The abundance of the family Rhizophoraceae with the most dominant species, *Rhizophora apiculata*, formed the major structure of the mangrove forest with a total basal area of 7.08 m^2 ha^{-1}. In the freshwater swamp, the family Apocynaceae (*Alstonia spatulata*) recorded the largest basal area with a total of 9.77 m^2 ha^{-1}.

From the islands, the study areas demonstrated more DBH classes of at least 4–6 classes of diameter distributions (Figure 4). Most of the trees within the plot were small to medium in size, and the majority were in the 5–14.99 cm DBH class, followed by 15–24.9 cm and 25–34.9 cm. In Pulau Bidong and Pulau Perhentian, the family Myrtaceae and genus *Syzygium* showed the highest number of individuals in diameter class distribution. However, the species varies accordingly based on the diameter size. In Pulau Bidong, *Licania splendens* tends to be abundant at small sizes within the forest, while *Lithocarpus rassa* and *Syzygium cinereum* were found growing in medium to bigger sizes. In Pulau Perhentian, *Vatica* sp. was dominant in small and larger diameter sizes, while *Palaquium obovatum* and *Buchanania arborescens* were in more medium to large sizes. In Pulau Redang, *Diospyros pilosanthera* from the family Ebenaceae recorded the highest at 5–14.99 diameter. Meanwhile, *Madhuca sericea* from the family Sapotaceae was dominant at 15–24.99 diameter class. For the diameter class of 25 cm DBH and above, *Shorea glauca* from the family Dipterocarpaceae recorded the highest.

In terms of density, the family Myrtaceae showed the highest in Pulau Bidong and Pulau Perhentian, with 572 individual/ha and 444 individual/ha, respectively. The family Dipterocarpaceae was the second highest and thrived well on all three islands. Each site had 384 ind/ha, 172 ind/ha, and 280 ind/ha in Pulau Bidong, Pulau Redang, and Pulau Perhentian, respectively. Furthermore, in the basal area, the family Dipterocarpaceae showed the highest in Pulau Redang with 35.49 m^2/ha, and Myrtaceae in Pulau Perhentian with 9.58 m^2/ha and Pulau Bidong with 7.09 m^2/ha. *Shorea glauca* from the family Dipterocarpaceae in Pulau Redang showed the largest basal area with 35.28 m^2/ha while *Vatica* sp. in Pulau Perhentian and *Syzygium cinereum* in Pulau Bidong, each with 5.04 m^2/ha and 2.68 m^2/ha respectively.

3.2. Importance Value Index

In the *Melaleuca* swamp forest, *Melaleuca cajuputi* had the highest importance value (IV_i = 112.52) (Table 4). The other dominant species within the same forest type includes *Ilex cymosa* (IV_i = 62.72), *Talipariti tiliaceus* (IV_i = 23.18), and *Pouteria obovata* (IV_i = 19.09). For the mangrove forest, the most dominant species were the *Rhizophora apiculata* (IV_i = 173.67), followed by *Excoecaria agallocha* (IV_i = 22.60), *Pouteria obovata* (IV_i = 20.84), *Talipariti tiliaceus* (IV_i = 15.20), *Intsia bijuga* (IV_i = 14.97), and *Heritiera littoralis* (IV_i = 18.82). The least dominant species was *Dolichandrone spathaceae* with IV_i of 3.75. The most dominant species in freshwater swamp forest was *Alstonia spatulata* with IV_i of 105.96. The other dominant species were *Gynotroches axillaris* (IV_i = 49.99), *Macaranga hypoleuca* (IV_i = 33.91), *Vitex trifolia* (IV_i = 17.40), and *Campnosperma coriaceum* (IV_i = 13.49).

Table 4. Relative dominance (RD_i), relative coverage (RC_i), relative frequency (Rf_i), and importance value (IV_i) of tree species with DBH of 5 cm in Setiu Wetlands and three major islands of Terengganu.

Species	RD_i	RC_i	Rf_i	IV_i
Setiu Wetlands: *Melaleuca* swamp forest				
Melaleuca cajuputi	56.72	44.69	11.11	112.52
Ilex cymosa	25.07	26.54	11.11	62.72
Talipariti tiliaceus	4.53	7.54	11.11	23.18
Pouteria obovata	3.23	4.75	11.11	19.09
Neolitsea zeylanica	3.15	3.63	11.11	17.89
Suregada multiflora	1.98	2.23	11.11	15.33
Pittosporum ferrigeneum	2.48	4.19	5.56	12.23
Streblus taxoides	0.79	0.84	5.56	7.18
Myrica esculenta	0.32	0.56	5.56	6.43
Vitex trifolia	0.61	1.96	2.78	5.34
Suregada multiflora	1.98	2.23	11.11	15.33
Mangrove forest				
Rhizophora apiculata	81.83	78.05	13.79	173.67
Excoecaria agallocha	6.89	5.37	10.34	22.6
Pouteria obovata	2.41	4.63	13.79	20.84
Talipariti tiliaceus	1.68	3.17	10.34	15.2
Intsia bijuga	1.94	2.68	10.34	14.97
Heritiera littoralis	2.04	2.44	10.34	14.82
Guioa bijuga	0.70	1.22	6.90	8.82
Ilex cymosa	1.13	0.49	6.90	8.51
Bruguiera gymnorhiza	0.55	0.73	3.45	4.73
Acacia mangium	0.47	0.24	3.45	4.16

Table 4. *Cont.*

Species	RD_i	RC_i	Rf_i	IV_i
Freshwater swamp forest				
Alstonia spatulata	56.1	41.16	8.70	105.96
Gynotroches axillaris	17.0	24.30	8.70	49.99
Macaranga hypoleuca	11.96	13.25	8.70	33.91
Vitex trifolia	3.28	5.42	8.70	17.40
Campnosperma coriaceum	1.99	2.81	8.70	13.49
Cratoxylum arborescens	1.07	1.41	6.52	8.99
Melaleuca cajuputi	2.32	1.20	4.35	7.88
Ilex cymosa	1.28	1.61	4.35	7.24
Pouteria obovata	0.72	0.80	4.35	5.87
Brackenridgea hookeri	0.61	0.80	4.35	5.76
Islands: Pulau Bidong				
Licania splendens	10.58	3.17	12.89	26.64
Calophyllum rupicola	9.75	3.17	9.63	22.55
Vatica sp.	8.32	3.17	10.96	22.46
Syzygium cinereum	10.78	3.17	7.26	21.21
Lithocarpus rassa	10.09	3.17	3.56	16.82
Syzygium cerinum	4.71	3.17	4.44	12.33
Syzygium sp. 2	5.94	2.38	3.85	12.17
Austrobuxus nitidus	3.80	3.17	4.74	11.72
Symplocos adenophylla	2.32	3.17	5.33	10.83
Buchanania arborescens	3.22	3.17	3.11	9.51
Pulau Redang				
Shorea glauca	67.77	15.19	4.49	87.45
Madhuca sericea	4.53	12.22	4.49	21.25
Diospyros pilosanthera	0.67	9.26	4.49	14.42
Diospyros sp.	0.91	10	3.37	14.28
Litsea sp.	0.90	7.41	4.49	12.81
Barringtonia scortechinii	0.98	4.44	3.37	8.80
Artocarpus lanceifolius	3.69	1.85	2.25	7.79
Cleistanthus sumatranus	0.96	3.33	3.37	7.66
Koilodepas sp.	0.21	3.33	3.37	6.91
Swintonia floribunda	4.01	0.37	1.12	5.50
Pulau Perhentian				
Vatica sp.	20.46	18.93	6.78	46.18
Syzygium cerina	13.28	16.80	6.78	36.86
Buchanania arborescens	12.91	12.00	6.78	31.69
Symplocos adenophylla	7.41	9.87	6.78	24.06
Syzygium sp. 1	5.46	6.67	6.78	18.91
Psydrax sp.	3.28	8.80	5.08	17.16
Chaetocarpus castanocarpus	3.66	5.87	6.78	16.31
Calophyllum rupicola	5.43	5.60	5.08	16.12
Syzygium sp. 3	4.54	2.40	5.08	12.03
Syzygium sp. 4	5.67	1.07	3.39	10.13

In Pulau Bidong, *Licania splendens* was the most dominant species within the plot, having the highest importance value ($IV_i = 26.64$). The least dominant species were *Timonius wallichianus*, *Memecylon* sp., *Diplospora malaccensis*, and *Garcinia eugenifolia* with IV_i of 0.98. The most dominant species in Pulau Redang was *Shorea glauca* ($IV_i = 87.85$), followed by *Madhuca sericea* ($IV_i = 21.25$), *Diospyros pilosanthera* ($IV_i = 14.42$), *Diospyros* sp. 1 ($IV_i = 14.28$), and *Litsea* sp. ($IV_i = 12.81$). Meanwhile, the least dominant species were *Mangifera odorata*, *Cleistanthus* sp. 1, *Diospyros* sp. 2, *Suregada multiflora*, *Madhuca tubulosa*, *Dracaena maingayi*, *Memecylon* sp., *Ormosia sumatrana*, and *Syzygium zeylanicum* with IV_i of 1.51. In Pulau

Perhentian, the most dominant species was *Vatica* sp. (IV_i = 46.18), and the co-dominant species were *Syzygium cerina* (IV_i = 36.86) and *Buchanania arborescens* (IV_i = 31.69) (Table 4).

3.3. Above-Ground Biomass Estimation

The family Myrtaceae in Pulau Bidong and Pulau Perhentian recorded the highest above-ground biomass with a total of 67.07 t/ha and 100.32 t/ha, respectively, contributing to 30.95% and 41.11%. Meanwhile, in Pulau Redang, the family Dipterocarpaceae was the highest, with a total of 534.35 t/ha (73.40%). In Pulau Bidong, *Syzygium cinereum* contributed the most with a total of 26.49 t/ha (11.99%), while the *Shorea glauca* was the highest in Pulau Redang, 532.37 t/ha (73.10%). In Pulau Perhentian, *Vatica* sp. contributed the most, with an estimated total above-ground biomass of 49.71 t/ha (20.36%).

The total amount of above-ground biomass concerning the family differs according to each study site that follows the types of forest formation and vegetation. The family Myrtaceae was the highest for the Melaleuca swamp forest with 84.88 t/ha (65.01%) (Table 5). Meanwhile, in the mangrove forest, the family Rhizophoraceae contributed 55.70 t/ha (80.18%). The family Apocynaceae were the highest in the freshwater swamp forest with 87.40 t/ha or 58.06% of the total above-ground biomass. Generally, the families Myrtaceae and Dipterocarpaceae usually contributed a high amount of above-ground biomass in islands. In Pulau Bidong and Pulau Perhentian, the Myrtaceae was the highest, with a total amount of 67.07 t/ha and 100.32 t/ha contributing to 30.95% and 41.11% of the total above-ground biomass. Meanwhile, the family Dipterocarpaceae was the highest in Pulau Redang, with a total of 534.35 t/ha (73.40%).

Table 5. List of five species that contribute to the highest amount of above-ground biomass (AGB).

Study Area	Species	Family	AGB (t/ha)	%
Setiu Wetlands: *Melaleuca* swamp forest	*Melaleuca cajuputi*	Myrtaceae	84.88	64.83
	Ilex cymosa	Aquifoliaceae	28.10	21.46
	Pouteria obovata	Sapotaceae	4.17	3.18
	Pittosporum ferrigenium	Pittosporaceae	3.59	2.74
	Talipariti tiliaceus	Malvaceae	3.51	2.68
Mangrove forest	*Rhizophora apiculata*	Rhizophoraceae	55.70	80.17
	Excoecaria agallocha	Euphorbiaceae	5.65	8.13
	Heritiera littoralis	Malvaceae	2.01	2.89
	Pouteria obovata	Sapotaceae	1.79	2.58
	Intsia bijuga	Fabaceae	1.47	2.12
Freshwater swamp forest	*Alstonia spatulata*	Apocynaceae	87.27	58.97
	Gynotroches axillaris	Rhizophoraceae	22.79	15.40
	Macaranga hypoleuca	Euphorbiaceae	17.7	11.96
	Vitex trifolia	Verbenaceae	4.37	2.95
	Melaleuca cajuputi	Myrtaceae	3.55	2.40
Islands: Pulau Bidong	*Syzygium cinereum*	Myrtaceae	26.49	11.99
	Lithocarpus rassa	Fagaceae	25.18	11.62
	Licania splendens	Chrysobalanaceae	21.22	9.79
	Calophyllum rupicola	Calophyllaceae	20.09	9.27
	Vatica sp.	Dipterocarpaceae	16.21	7.48
Pulau Redang	*Shorea glauca*	Dipterocarpaceae	532.27	73.10
	Swintonia floribunda	Anacardiaceae	34.20	4.70
	Mangifera macrocarpa	Anacardiaceae	29.67	4.08
	Artocarpus lanceifolius	Moraceae	24.11	3.31
	Madhuca sericea	Sapotaceae	23.12	3.18
Pulau Perhentian	*Vatica* sp.	Dipterocarpaceae	49.71	20.36
	Buchanania arborescens	Anacardiaceae	31.79	13.02
	Syzygium cerina	Myrtaceae	29.34	12.02
	Syzygium sp.5	Myrtaceae	18.15	7.44
	Symplocos adenophylla	Symplocaceae	17.48	7.16

The list of the five species that contributed to the highest amount of above-ground biomass is shown in Table 5. In Setiu Wetlands, *Melaleuca cajuputi*, *Rhizophora apiculata*, and *Alstonia spatulata* showed the highest amount of total above-ground biomass, with each contributing 84.88 t/ha (64.83%), 55.70 t/ha (80.17%) and 87.27 t/ha (58.97%) due to their abundance and dominant in each forest types formation. In Pulau Bidong, *Syzygium cinereum* contributed the most, comprising 26.49 t/ha (11.99%). *Shorea glauca* was the highest in Pulau Redang, with 532.27 t/ha (73.10%). In Pulau Perhentian, *Vatica* sp. contributed the most with an estimated total above-ground biomass of 49.71 t/ha (20.36%).

4. Discussion

Generally, the distribution of plants and vegetation in Malaysia is influenced by the climate, soil, and soil water [26], as well as habitat. Our study shows that the diversity of tree species in the lowland dipterocarp forests in the three major islands of Terengganu was higher than that of the edaphic forest types in Setiu Wetlands. The *Melaleuca* swamp, freshwater swamp and mangrove forests are among the various edaphic vegetation formations found in Malaysia, in which species composition is greatly influenced by adaptation to soil conditions [27]. Setiu Wetlands shows disparities in vegetation types, tree abundance, and diversity due to different BRIS soil characteristics and series that may explain the plant distribution. The BRIS soil is classified into two orders, namely entisol and spodosol; the former can be described as young and sandy soil, which is found in areas close to the sea while the latter is acidic soil combined with a mor-humus (acidic humus), which can be found more to the inland areas [28]. Several studies have been carried out on the physical characteristics of BRIS soil in the coastal area of Terengganu, which further characterized the soil order into four series: Baging, Rhu Tapai, Rudua, and Jambu [29,30]. These soil series are generally lacking in selected mineral nutrients, have low water retention capacity and are poorly structured, that limit the ability of plants to grow [22,23,29,30]. Therefore, only certain locally adaptable plants with specific soil tolerance can survive and live within these sites [3,6].

The tree density, basal area and H′ value recorded in mangrove forest for this study was slightly lower but within a similar range when compared to the other regional studies, e.g., Ayer Hangat Forest Reserve, Matang Forest Reserve, Tok Bali Forest Reserve, and Kisap Forest Reserve in Peninsular Malaysia [31–34], Bahile Mangrove Forest, the Philippines [35] and in Trang Province, Southern Thailand [36] (Table S1). Our study indicated that the mangrove forest around the area was subjected to higher disturbance and experiencing low regeneration, as inferred from the low tree density and small basal area. Various studies have reported that mangrove forests are threatened by land use changes worldwide, e.g., [37–39]. In Setiu Wetlands, mangrove forests are subjected to anthropogenic disturbances such as land reclamation for development, deforestation mainly for charcoal and firewood and aquaculture activities such as shrimp farming and fish ponds [3]. These observations highlighted the urgent need for effective conservation management in the Setiu Wetlands. According to Ashton et al. [40], mangroves can regenerate productively and sustainably when given conducive conditions for stock trees to produce propagules or seeds. Meanwhile, onsite protection by law and continuous monitoring revealed higher mangrove tree regeneration survival after disturbances at another Malaysian mangrove site [34]. In terms of species diversity, the low H′ value was expected due to the lack of species variation within the mangrove stands. Various comparative studies also concluded that mangrove forests had lower diversity because of their unique abiotic adaptive requirements and stand formation compared to the other tropical terrestrial forest ecosystems [32,34,41].

The freshwater swamp forest types are generally less studied in Southeast Asia when compared to mangroves [36,42,43]. A study in the freshwater swamp forest of Otuwe, Nigeria demonstrated similar findings where the H′ value ranged from 0.98 to 2.13 [44]; this study recorded the H′ value of 1.85. The tree density and basal area recorded for this study are slightly lower when compared to Igu [44]. This was attributed to the low tree diameter class distribution as most of the Setiu freshwater swamp trees belonged to the small-to-medium diameter class distribution, while Igu [44] recorded more trees with large diameter sizes. The species important value (IV_i) shows that swamp forests are usually dominated by a few tree species [45], which agrees with the results of other studies, e.g., [32,35,36,44,46].

Our study found that most of the trees enumerated in Setiu Wetlands were in 6–25 cm DBH, and large trees greater than 25 cm DBH were rarely seen. Similarly, on the dune landscape in Jambu Bongkok, Terengganu, 433 out of 451 individuals were in the lowest diameter class of 5–15 cm DBH [18]. Poor nutrient soil properties significantly affect tree growth and diameter and may have contributed to fewer trees with large diameters greater than 25 cm [47]. On the other hand, among all the study sites, only in Pulau Redang and Pulau Perhentian, trees were seen in more than 45 cm DBH and above, e.g., *Shorea glauca* from the family Dipterocarpaceae. While in Pulau Bidong, none of the Dipterocarpaceae family was found, due partly to the impact of Vietnamese refugees that settled on the island in the 1980s and harvested most of the trees for building materials [48]. The highest numbers of trees recorded in the islands were below 24.9 cm DBH, indicating one of the characteristics of dipterocarp forest [49,50]. The distribution pattern resembled the inverted 'J' that represents the decrease in the number of individuals as the DBH of trees increases. The finding suggests that the forests are in regenerating phase due to the abundance of the understory layer [51]. Similar tree distribution is also observed in previous studies, e.g., the primary lowland dipterocarp forest in Pasoh Forest Reserve, Negeri Sembilan [52], and the lowland dipterocarp forest in Bukit Belata, Selangor [53].

Our study shows that the stem density varies according to the forest types (Table 1) and corroborates previous studies carried out in Malaysia, with a range of 800–2200 Ind/ha. For example, a total of 796 Ind/ha were recorded in the swamp forest in Sugut Forest Reserve, Sabah, 1113 Ind/ha in Tanjung Tuan Forest Reserve, Negeri Sembilan, 1576 Ind/ha in Timun Island, Pulau Langkawi, 1670 Ind/ha in the coastal forest of Terengganu, 1840 Ind/ha in Pulau Singa Besar, Pulau Langkawi, and 2200 Ind/ha in the inland forest of Pulau Redang [12,54,55]. Overall, the total basal area recorded in the three islands from 24–52 m^2/ha is consistent with data obtained in other studies focusing on the lowland dipterocarp forest. Generally, the total basal area is within the range of 28–52 m^2/ha [56], e.g., Nizam et al. [57] reported a total basal area of two forest plots established in Kenong Forest Park were 26.91 m^2/ha and 29.23 m^2/ha. Moreover, the lowland dipterocarp forest harbors a higher amount of above-ground biomass than the other forest types. Similar studies reported a range of 107.5–955.61 t/ha for total above-ground biomass [12,15,16,25,55,58–60].

Towards this point, our studies successfully assessed the forest stand structure and evaluated the level of species composition, distribution and diversity within the covered area. However, long-term monitoring of the forests may be initiated to understand more of the forest dynamics that attempt to relate present community patterns to past events and predict the outcomes of future ones concerning suitable environmental parameters such as climate and soil properties. The study's findings are much needed to enhance management practices, concurrently conserve forest resources, and guide and inform forest management activities and assessment. Furthermore, this information may contribute to the conservation, biodiversity assessment, and sustainable forest management strategies for Setiu Wetlands and islands of Terengganu.

Supplementary Materials: The following supporting information can be downloaded at: https://www.mdpi.com/article/10.3390/agronomy12102380/s1, Figure S1: Sample-based rarefaction curve for *Melaleuca* swamp forest, freshwater swamp forest and mangrove forest in Setiu Wetlands. Figure S2: Sample-based rarefaction curve for Pulau Redang, Pulau Perhentian, and Pulau Bidong. Table S1: Stem density, basal area, and diversity in mangrove forest for this study compared with other studies in Southeast Asia.

Author Contributions: Planning and designing of the research and draft of the manuscript: E.P., M.R.S. and G.E.L.; fieldwork and data curation: E.P. and M.R.S.; identification and statistical analyses: E.P. and M.R.S.; data analyses: E.P. and G.E.L.; funding acquisition: M.T.A.; writing—original draft:

E.P. and G.E.L.; writing—review and editing: M.R.S., J.M.S., K.H.L., J.W.H.Y. and M.T.A. All authors have read and agreed to the published version of the manuscript.

Funding: This research was funded by Trans-disciplinary Research Grant Scheme (TRGS, 2015/59373), Niche Research Grant Scheme (NRGS/2015/5313/2), Geran Galakan Penyelidikan (GGP/68007/2014/127), and Dana Pembangunan Geopark Kenyir (GEOPARK/2015/53167/3) led by M.T. Abdullah and colleagues. The APC was funded by the Research Management and Innovation Centre (RMIC), Universiti Malaysia Terengganu.

Institutional Review Board Statement: Not applicable.

Informed Consent Statement: Not applicable.

Data Availability Statement: Not applicable.

Acknowledgments: We would like to thank the Universiti Malaysia Terengganu for the permission and administrative and logistic support in conducting this study since 2015, the Forestry Department of Peninsular Malaysia, and the Setiu District Office. The second author is grateful to Jarina Mohd Jani for her insightful discussion on the local ecological knowledge about sustainable harvesting in Setiu Wetlands.

Conflicts of Interest: The authors declare no conflict of interest.

References

1. Terengganu State Foresty Department. Available online: http://trgforestry.terengganu.gov.my (accessed on 3 August 2022).
2. Jarina, M.J.; Rohani, S.; Wan Saefullah, W.N.; Sathiamurthy, E.; Lee, G.E.; Jamilah, M.S.; Aqilah, M.; Nur Omaima, H.; Nor Zalipah, M.; Wan Bayani, W.A.; et al. Setiu Wetlands. In *A Celebration of Nature and Culture*; Majlis Pengurusan Taman Negeri Terengganu: Kuala Terengganu, Malaysia, 2021; pp. 1–108.
3. Jamilah, M.S.; Siti Mariam, M.N.; Lee, G.E.; Razali, M.S.; Salwa, S.; Pesiu, E.; Chong, J.L.; Jarina, M.J.; Horsali, N.A.I.; Faridah, M.; et al. A checklist of vascular plants and its uses for livelihood-making in Setiu Wetlands, Terengganu, Malaysia. *Phytokeys* **2020**, *160*, 7–43. [CrossRef]
4. Asean Centre for Biodiversity: Sixth National Report of Malaysia. Available online: https://asean.chm-cbd.net/documents/sixth-national-report-malaysia (accessed on 3 August 2022).
5. Tamblyn, A.; Turner, C.; Rainess, P. *Malaysia Tropical Forest Conservation Project Report of the Setiu Wetlands Phase*; Coral Cay Conservation: London, UK, 2006; pp. 1–89.
6. Faridah, M.; Jamilah, M.S.; Jarina, M.J.; Rohani, S. *Setiu Wetlands: Species, Eosystems and Livelihood*; Penerbit Universiti Malaysia Terengganu: Kuala Nerus, Malaysia, 2015; pp. 1–228.
7. Gedan, L.B.; Kirwan, M.L.; Wolanski, E.; Barbier, E.B.; Silliman, B.R. The present and future role of coastal wetland vegetation in protecting shorelines: Answering recent challenges to the paradigm. *Clim. Change* **2011**, *106*, 7–29. [CrossRef]
8. Temmerman, S.; Meire, P.; Bouma, T.J.; Herman, P.M.J.; Ysebaert, T.; De Vriend, H.J. Ecosystem-based coastal defence in the face of global change. *Nature* **2013**, *504*, 79–83. [CrossRef] [PubMed]
9. De Groot, R.S.; Wilson, M.A.; Boumans, R.M.J. A typology for the classification, description and valuation of ecosystem functions goods and services. *Ecol. Econ.* **2002**, *41*, 393–408. [CrossRef]
10. Acremen, M.; Holden, J. How wetlands affect floods. *Wetlands* **2013**, *33*, 773–786. [CrossRef]
11. Vitousek, P.M.; Loope, L.L.; Adsersen, H.; D' Antonio, C.M. Island ecosystems: Do they represent natural experiments in biological diversity and ecosystems functions. In *Functional Role of Biodiversity: A Global Perspective*; Mooney, H.A., Cushman, J.H., Medina, E., Sala, O.E., Schulze, E.D., Eds.; John Wiley & Sons Ltd.: Chichester, UK, 1996; pp. 245–259.
12. Khairil, M.; Nashriyah, M.; Norhayati, N.; Shahril, A.; Fatihah, N. Tree species composition, diversity and above ground biomass of two forest types at Redang Island, Peninsular Malaysia. *Walailak J. Sci. Technol.* **2012**, *10*, 77–90.
13. Noss, R.F. Assessing and monitoring forest biodiversity: A suggested framework and indicators. *For. Ecol. Manag.* **1999**, *115*, 135–146. [CrossRef]
14. Suratman, M.N. Tree species diversity and forest stand structure of Pahang National Park, Malaysia. In *Biodiversity Enrichment in a Diverse World*; Lameed, G.A., Ed.; Intech Open: London, UK, 2012; pp. 473–492.
15. Kueh, J.H.R.; Lim, M.T. An estimate of forest biomass in Ayer Hitam forest reserve. *Pertanika J. Trop. Agric. Sci.* **1999**, *22*, 117–122.
16. Hikmat, A. Biomass estimation, carbon storage and energy content of three virgin jungle reserves in Peninsular Malaysia. *Media Konserv.* **2005**, *10*, 1–8.
17. Jamilah, M.S.; Nur-Faiezah, A.G.; Siti Kehirah, A.; Siti Mariam, M.N.; Razali, M.S. Woody plants on dune landscape of Terengganu, Peninsular Malaysia. *J. Trop. For. Sci.* **2014**, *26*, 267–274.
18. Razali, M.S.; Muhammad Khairul, F.Z.; Wan Juliana, W.A. Distribution and rarity of mangrove and coastal plants in developing indicators of hotspots in Setiu Wetlands. In *Prosiding Seminar Ekspedisi Saintifik Tanah Bencah Setiu*; Abdullah, D.M., Nasir, M.Z., Mohd Mazlan, N., Wan Faridah, A.W.K., Eds.; WWF Malaysia: Petaling Jaya, Malaysia, 2017; pp. 55–66.

19. Pesiu, E.; Abdullah, M.T.; Jamilah, M.S.; Razali, M.S. Tree species composition in Pulau Bidong and Pulau Redang. *J. Sustain. Sci. Manag.* **2016**, *1*, 48–60.
20. Gotelli, N.J.; Colwell, R.K. Quantifying biodiversity: Procedures and pitfalls in the measurement and comparison of species richness. *Ecol. Lett.* **2001**, *4*, 379–391. [CrossRef]
21. Devries, P.J.; Alexander, L.G.; Chacon, I.A.; Fordyce, J.A. Similarity and differences among rainforest fruit-feeding butterfly communities in Central and South America. *J. Anim. Ecol.* **2012**, *81*, 472–482. [CrossRef]
22. Hammer, O.; Harper, D.A.T.; Ryan, P.D. PAST: Paleontological statistics of software package for education and data analysis. *Palaeontol. Electron.* **2001**, *4*, 9.
23. Brower, J.E.; Zar, J.H.; Von Ende, C. *Field and Laboratory Methods for General Ecology*, 4th ed.; WCB McGraw-Hill: New York, NY, USA, 1997; pp. 1–176.
24. Shannon, C.E.; Weaver, W. *The Mathematical Theory of Communication*; University of Illinois Press: Champaign, IL, USA; Urbana, IL, USA, 1949; pp. 1–296.
25. Kato, R.; Tadaki, Y.; Ogawa, H. Plant biomass and growth increment studies in Pasoh Forest. *Malay. Nat. J.* **1978**, *30*, 21–122.
26. Wyatt-Smith, J. *Manual of Malayan Silviculture for Inland Forest. Malayan Forest Records No. 23*; Forest Research Institute Malaysia: Kuala Lumpur, Malaysia, 1963; Volume 1, pp. 1–465.
27. Saw, L.G. Vegetation of Peninsular Malaysia. In *Flora of Peninsular Malaysia. Series II: Seed Plants*; Kiew, R., Chung, R.C.K., Saw, L.G., Soepadmo, E., Eds.; Forest Research Institute Malaysia: Kuala Lumpur, Malaysia, 1963; pp. 21–45.
28. Roslan, I.; Shamshuddin, J.; Fauziah, C.I.; Anuar, A.R. Fertility and suitability of the Spodosols formed on sandy beach ridges interspersed with swales in the Kelantan-Terengganu plains of Malaysia for Kenaf production. *Malays. J. Soil Sci.* **2011**, *15*, 472–482.
29. Mohd Ekhwan, T.; Mazlin, M.; Muhammad Barzani, G.; Nor Azlina, A.A. Analysis of the physical characteristics of BRIS soil in coastal Kuala Kemaman, Terengganu. *Earth Sci. Res. J.* **2009**, *1*, 1–6.
30. Usman, M.I.; Umara, B.; Edi Armantho, H.M.; Adzemi, M.A. Assessment and evaluation of BRIS soil and its implication on maize crop in Merang, Terengganu, Region of Malaysia. *J. Biol. Agr. Health.* **2014**, *4*, 69–76.
31. Muktar, A.; Awang Noor, A.G.; Nobilly, F.; Alerk, R.J. Stand structure and species composition of mangrove forests in Kuala Trong range, Matang, Perak. In *A Sustainable Management of Matang Mangroves: 100 Years and Beyond*; Shaharuddin, M.I., Azahar, M., Rozani, U., Kamaruzaman, A.B., Lim, K.I., Suhaili, R., Jalil, M.S., Latiff, A., Eds.; Forest Department of Peninsular Malaysia: Kuala Lumpur, Malaysia, 2005; pp. 253–262.
32. Norhayati, A.; Nizam, M.S.; Wan Juliana, W.A.; Shukor, M.N.; Fera Fizani, A.F.; Hafizah, S.; Sariah, H.; Jamaliah, J.; Intan Nordira, A.K.; Latiff, A. Comparison of mangrove tree species composition at selected location in Peninsular Malaysia. In *A Sustainable Management of Matang Mangroves: 100 Years and Beyond*; Shaharuddin, M.I., Azahar, M., Rozani, U., Kamaruzaman, A.B., Lim, K.I., Suhaili, R., Jalil, M.S., Latiff, A., Eds.; Forest Department of Peninsular Malaysia: Kuala Lumpur, Malaysia, 2005; pp. 263–272.
33. Kasawani, I.; Kamaruzaman, J.; Nurul-Nadhirah, M.I. Biological diversity assessment of Tok Bali Mangrove Forest, Kelantan, Malaysia. *WSEAS Trans. Environ. Dev.* **2007**, *2*, 37–44.
34. Wan Norilani, W.I.; Wan Juliana, W.A.; Razali, M.S.; Latiff, A. Structural and floristic pattern in a disturbed mangrove tropical swamp forest: A case study from the Langkawi UNESCO Global Geopark Forest, Peninsular Malaysia. *Sains Malays.* **2018**, *47*, 861–869.
35. Abino, A.C.; Castillo, J.A.A.; Lee, Y.J. Assessment of species diversity, biomass and carbon sequestration potential of a natural mangrove stand in Samar, the Philippines. *Forest Sci. Technol.* **2014**, *10*, 2–8. [CrossRef]
36. Sudtongkong, C.; Webb, E.L. Outcomes of state versus community-based mangrove management in Southern Thailand. *Ecol. Soc.* **2008**, *13*, 27–51. [CrossRef]
37. Valiela, I.; Bowen, J.L.; York, J.K. Mangrove forests: One of the world's threatened major tropical environments. *J. Bio. Sci.* **2001**, *51*, 807–815.
38. Alongi, D.M. Present state and future of the world's mangrove forests. *Environ. Conserv.* **2002**, *29*, 331–349. [CrossRef]
39. Polidoro, B.A.; Carpenter, K.E.; Collins, L.; Duke, N.C.; Ellison, A.M.; Ellison, J.C.; Farnsworth, E.J.; Fernando, E.S.; Kathiresan, K.; Koedem, N.E.; et al. The loss of species: Mangrove extinction risk and geographic areas of global concern. *PLoS ONE* **2010**, *5*, e10095. [CrossRef]
40. Ashton, E.C.; Macintosh, D.J. Preliminary assessment of the plant diversity and community ecology of Sematan mangrove forest, Sarawak, Malaysia. *For. Ecol. Manag.* **2002**, *166*, 111–129. [CrossRef]
41. Lu, C.Y.; Wong, Y.S.; Tam, N.F.Y.; Berry, R. Vegetation analysis of a typical mangrove swamp- Lai Chi Wo coast of Hong Kong. *Chin. J. Oceanol. Limnol.* **1998**, *16*, 72–77.
42. Yamada, I. *Tropical Rainforest of Southeast Asia—A Forest Ecologist's View*; University of Hawaii Press: Honolulu, HI, USA, 1997; pp. 1–440.
43. Dudgeon, D. Riverine biodiversity in Asia—A challenge for conservation biology. *Hydrobiologia* **2000**, *418*, 1–13. [CrossRef]
44. Igu, N.I. Species diversity and structure of an intact freshwater swamp forest in the Niger Delta. *Open. J. For.* **2017**, *7*, 242–254. [CrossRef]
45. Corlett, R.T. *The Ecology of Tropical East Asia*; Oxford University Press: New York, NY, USA, 2009; pp. 1–333.
46. Rozainah, M.Z.; Mohamad, M.R. Mangrove forest species composition and density in Balok River, Pahang, Malaysia. *Ecoprint* **2006**, *13*, 23–28. [CrossRef]

47. Richards, P.W. *The Tropical Rain Forests*; Cambridge University Press: Cambridge, UK, 1952. [CrossRef]
48. Vu, Q.G.N. Journey of the abandoned: Endless refugees' camp and incurable traumas. *Signs* **2007**, *32*, 580–584. [CrossRef]
49. Whitmore, T.C.; Burnham, C.P. *Tropical Rain Forests of the Far East*; Clarendon Press: Oxford, UK, 1975.
50. Wan Norilani, W.I.; Wan Juliana, W.A.; Latiff, A. Floristic composition, stand structure, above ground biomass and conservation value of a coastal lowland forest at Bukit Bauk Urban Forest, Dungun, Terengganu, Malaysia. *Malays. For.* **2013**, *76*, 17–26.
51. Newberry, D.M.; Campbell, E.J.F.; Lee, Y.F.; Ridsdale, C.E.; Still, M.J. Primary lowland dipterocarp forest at Danum Valley, Sabah, Malaysia: Structure, relative abundance and family composition. *Phil. Trans. R. Soc. Lond.* **1992**, *335*, 341–356. [CrossRef]
52. Appanah, S.; Weinland, G. Research Pamphlet No. 112. In *A Preliminary Analysis of the 50-Hectare Pasoh Demography Plot: I. Dipterocarpaceae*; Forest Research Institute of Malaysia: Kuala Lumpur, Malaysia, 1993.
53. Nizam, M.S.; Nur Maisarah, J.; Jildred, J.; Wan, J.W.A. *Komposisi, Kepelbagaian Dan Biojisim Komuniti Pokok Di Hutan Simpan Bukit Belata*; Azahar, M., Koh, H.L., Mohd Shah, N.M.N., Samsul, A.N., Latiff, A., Eds.; Forest Department of Peninsular Malaysia: Kuala Lumpur, Malaysia, 2008.
54. Nilus, R.; Ong, R.C. Floristic composition of the Perupok Swamp Forest of the Sugut Forest Reserve, Sabah. *Mal. Nat. J.* **2001**, *55*, 95–101.
55. Mat-Salleh, K.; Tami, R.; Latiff, A. Ecology and conservation value of Tanjung Tuan, the Myrtaceae-dominant coastal forest reserve of Malaysia. *J. Trop. For. Sci.* **2003**, *15*, 59–73.
56. Swaine, M.D.; Lieberman, D.; Putz, F.E. The dynamics of tree populations in tropical forest: A review. *J. Trop. Ecol.* **1987**, *3*, 359–366. [CrossRef]
57. Nizam, M.S.; Rohani, S.; Wan Juliana, W.A. Floristic variation of tree communities in two distinct habitats within a forest park in Pahang, Peninsular Malaysia. *Sains Malay.* **2012**, *41*, 1–10.
58. Soepadmo, E. Structure, aboveground biomass and floristic composition at Gunung Janin Barat, Ulu Endau, Johore, Malaysia. *Mal. Nat. J.* **1987**, *41*, 275–290.
59. Lajuni, J.J.; Latiff, A. Biomass and floristic composition of Bangi Permanent Forest Reserve, a twice-logged lowland dipterocarp forest in Peninsular Malaysia. *Sains Malay.* **2013**, *42*, 1517–1521.
60. Syuharni, A.; Hakeem, K.; FaridahHanum, I.; Ozturk, M. Ecology of the coastal heath forest flora—A case study from Terengganu, Malaysia. *Emir. J. Food Agric.* **2014**, *26*, 1114. [CrossRef]

agronomy

MDPI

Article

Floristic Association of Moist Temperate Forests of Shangla District, Delineated by a Multivariate Approach

Javed Iqbal [1], Nasiruddin Shaikh [1], Moinuddin Ahmed [2], Wajid Zaman [3,*], Adam Khan [4], Asma Ayaz [5], Diaa O. El-Ansary [6], Hanoor Sharma [7], Hosam O. Elansary [8] and SeonJoo Park [3,*]

1 Department of Botany, Government College University Hyderabad, Sindh 71000, Pakistan; javedkhattak76@yahoo.com (J.I.); shaikhdrnasir@gmail.com (N.S.)
2 Department of Earth and Environmental Systems, Indiana State University, Terre Haute, IN 47809, USA; drmoingeolinks@gmail.com
3 Department of Life Sciences, Yeungnam University, Gyeongsan 38541, Korea
4 Department of Botany, University of Lakki Marwat, Lakki Marwat 28420, Pakistan; adam.khan345@yahoo.com
5 State Key Laboratory of Biocatalysis and Enzyme Engineering, School of Life Sciences, Hubei University, Wuhan 430062, China; asmaayaz@bs.qau.edu.pk
6 Precision Agriculture Laboratory, Department of Pomology, Faculty of Agriculture (El-Shatby), Alexandria University, Alexandria 21545, Egypt; diaa.elansary@alexu.edu.eg
7 School of Microbiology and Immunology, Wright State University, Dayton, OH 45435, USA; hanoor.sharma@clearlabs.com
8 Plant Production Department, College of Food and Agriculture Sciences, King Saud University, Riyadh 11451, Saudi Arabia; helansary@ksu.edu.sa
* Correspondence: shangla123@gmail.com (W.Z.); sjpark01@ynu.ac.kr (S.P.)

Abstract: Multivariate analysis was conducted to explore the moist temperate forests of the Shangla district, Khyber Pakhtunkhwa. The prime objective was to quantitatively describe and differentiate the vegetation groups and the factors that determine the boundaries and composition of plant communities in the Shangla district. This was achieved by sampling all common species in a complex vegetation mosaic coinciding with local gradients in topography and soil distribution. Ward's clustering dendrogram demonstrated four significant vegetation clusters with respect to environmental effects. These four major groups of the tree vegetation were superimposed on the ordination plane: 1. *Pinus wallichiana*, the dominant group associated with *Abies pindrow*; 2. *Abies pindrow* and the *Picea smithiana* group; 3. Dominant *Cedrus deodara* associated with the *Pinus wallichiana, Abies pindrow, Picea smithiana*, and *Quercus baloot* group; 4. *Pinus roxberghii* pure group. The key controlling factors for each group were the environmental characteristics (i.e., edaphic factors, topographic factors, soil physical properties, and soil nutrients). The results revealed elevation (p <0.001) to be the prominent factor in the composition of plant communities. Furthermore, pH, soil moisture, maximum water holding capacity, and soil physical properties (sand, silt, and clay) also showed a significant ($p < 0.05$) relationship with vegetation. The other environmental factor did not show a significant relationship with vegetation. Ward's cluster dendrogram of understory species also demonstrated four groups. Group 1 comprises two subgroups, a and b, with the highest number of species, i.e., *Digeteria sanguinalis, Fragaria nubicola, Verbascum Thapsus, Pinus wallichiana seedlings*, and *Polygonatum multiflorium*, respectively. The second large group contains twenty-five species out of eight stands, and the dominant species was *Tagetis minuta*. Eighteen species out of six stands were found in group 3, which was considered the smallest group. Group 4 consisted of seven stands containing twenty-four species of ground flora, with *Anaphalis scopulosa* followed by *Adiantum venustum* as the dominant species. The environmental characteristics of the understory vegetation showed a resemblance with the tree communities. With the exception of elevation, the other factors did not show a significant correlation.

Keywords: flora; environmental variable; multivariate analysis; soil moisture; Shangla district; Pakistan

Citation: Iqbal, J.; Shaikh, N.; Ahmed, M.; Zaman, W.; Khan, A.; Ayaz, A.; El-Ansary, D.O.; Sharma, H.; Elansary, H.O.; Park, S. Floristic Association of Moist Temperate Forests of Shangla District, Delineated by a Multivariate Approach. *Agronomy* **2022**, *12*, 1723. https://doi.org/10.3390/agronomy12071723

Academic Editor: Marco Scortichini

Received: 6 June 2022
Accepted: 20 July 2022
Published: 21 July 2022

1. Introduction

In an ecological study, multivariate analysis is one of the most important approaches, demonstrating the relationship between species and their local communities [1]. This technique has been used widely to abstract and simplify the massive ecological data set available and explore the possible relationship between various environmental variables and vegetation. Therefore, for understanding the species composition and distribution with respect to environmental variables, multivariate analysis techniques are reliable approaches [2]. In Pakistan, earlier ecological studies were based on observational approaches. Though, for the determination of the floristic composition of vegetation, few quantitative studies have been conducted, they were mostly based on primitive techniques. Advanced multivariate ordination and cluster analysis techniques have been used routinely in Europe and North America for several decades. In Pakistan, Shaukat and Qadir [3] and Ahmed [4,5] applied the multivariate techniques to the vegetation of the calcareous hills around Karachi, the industrial area of Karachi, and the Skardu District for the first time. Shaukat et al. [6] used these techniques to show significant correlations between the local environmental variables and vegetation. Various physical and environmental characteristics, including biotic and abiotic stresses, and particularly anthropogenic disturbance, have been linked to the distribution of plant species and communities [6–11].

Most of the vegetation of Pakistan has been analyzed by various researchers using multivariate techniques. Using the multivariate technique, Shaukat and Uddin [12] investigated the *Achyranthes aspera* tree composition and pattern. Ahmed et al. [13] and Hussain et al. [14] illustrated the vegetation of Chiltan in Baluchistan and the Swabi area of Khyber Pakhtunkhwa, respectively, using multivariate analysis. Ahmad et al. [15] investigated the phytosociological and structural characteristics of the Himalayan forests in several climatic zones of Pakistan. They found that specific communities had similar floristic compositions but differing quantitative values and provided a description of understory species. Ilyas et al. [16] analyzed the anthropogenic pressure on existing temperate forests in the Swat district, Khyber Pakhtunkhwa, including logging, deforestation, overgrazing, and forest removal for terrace farming. Ahmad, Fazal, Valeem, Khan, Sarwar and Iqbal [15] evaluated ecological aspects of roadside vegetation around Havalian city using multivariate techniques and vegetation along the motorway (M-2) in Pakistan. Siddiqui et al. [17] analyzed Pakistan's major moist temperate area vegetation quantitatively using multivariate agglomerative cluster analysis. Siddiqui et al. [18] conducted detailed research on several forests in Pakistan's moist temperate areas, whereas Rashid et al. [19] carried out a phytoecological evaluation with a detailed floristic appraisal of the vegetation around Malam Jabba's forests. Wahab et al. [20] explored the population dynamics of pine tree species in the Dir District, and Khan [21] investigated the vegetation ecology of Chitral Gol National Park.

The understory vegetation plays a very important role in a functioning forest ecosystem and structure. These species make strong association with trees and maintain the nutrient cycle as well as canopy succession [22]. The impact of trees on the understory vegetation and its relationship is essential because ground flora plays a significant role in the functioning of forest ecosystems. Natural and diverse understory vegetation may be significant to plant communities beyond any effect on growth or nutrients. Huo, Feng, and Su [22] suggested that coniferous forests are less favorable to biodiversity than mixed- or hardwood forests. On the other hand, few investigations have compared the vegetation of coniferous species [23,24]. Natural forest patches, mainly inhabited by *Pinus wallichiana*, are commonly distributed in the Hindu Kush Himalayan mountains of the northern area of Pakistan. These vegetated areas, composed mainly of forests, are a critical source of timber, firewood, and water conservation, and prevent the erosion of the local fertile soils in the region. Owing to the rigid and harsh mountainous system and less accessible part of the Khyber Pakhtunkhwa, quantitative and multivariate analysis of the vegetation of this area had been ignored in the past. Therefore, this study examined the vegetation–environmental relationship using multivariate techniques. The proposed study is the first multivariate analysis study from the Shangla district. This study is expected to enhance the understanding of the vegetation and environmental complex of these forests.

2. Materials and Methods

2.1. Vegetation Sampling

Sampling was conducted in forests of the Shangla district of Khyber Pakhtunkhwa (Figure 1). The study sites were selected on the basis of the maturity of forests determined by (1) no sign of recent disturbances and (2) tress that have at least 60 cm dbh. Forty mature sites of conifer tree species were sampled using the PCQ (point-centered quarter) method reported by Cottam and Curtis [25]. At each site, trees under 10 cm dbh and understory species were sampled with a circular plot, 1.5 m in diameter. In addition, the coordination and altitude of the sampling sites were calculated by GPS. At each sampling site, three soil samples were collected from a 10 cm depth up to 500 gm, and a composite of the samples was made for analysis. The slope angle and aspect were recorded using a clinometer. The phytosociological attributes, including the relative and absolute density and basal area of the tree species, were calculated using standard ecological approaches [25–27].

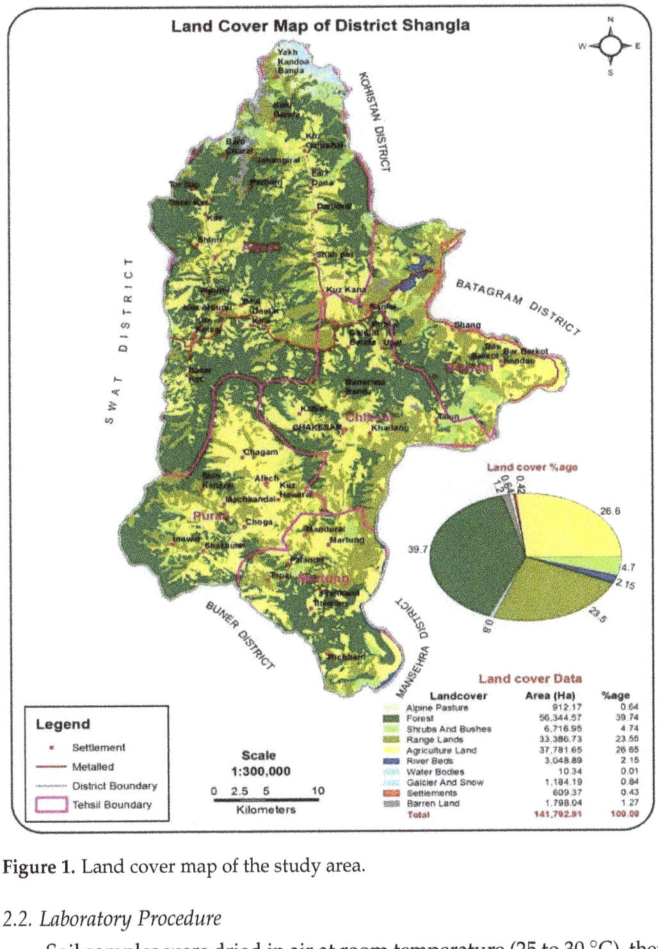

Figure 1. Land cover map of the study area.

2.2. Laboratory Procedure

Soil samples were dried in air at room temperature (25 to 30 °C), then sieved through a 2 mm sieve. The soil characteristics, including (TDS) total dissolved solids, soil pH, salinity, and conductivity of this soil were obtained and analyzed by making a suspension of 20 gm soil and filtering through a Whatman filter paper no. 42. These filtered samples were taken in a beaker, and their results were determined using a multiparameter meter (HANNA Model Sension Tm^{105}). The maximum water holding capacity (MWHC) of the collected soil samples was determined using the approach reported by Eaton [28]. The soil organic matter was determined by applying weight loss following the loss-on-ignition method [29]. The phosphorus (P) level was estimated using the method of Vanado-Molybdate-Yellow (Barton's Reagent) described by [30]. The total phosphorus was calculated as a percentage. The soil texture was then assessed. Soil has particles of different sizes called sand, silt, and clay. Sands are the largest particles in the soil. Silt has midsized particles of soil. Clay is the smallest-sized soil particles. The percentages of sand, silt, and clay were determined using the pipette method reported by Cornell Nutrient Analysis Laboratory (CNAL), website: www.css.cornell.edu/soiltest, accessed on 1 June 2022.

2.3. *Statistical Analysis*

For the statistical analysis, the obtained data were subjected to a computer program (PC-ord version 5). The tree vegetation and understory species were classified by cluster analysis [31]. Two techniques, i.e., principal component analysis (PCA) and two-dimensional nonmetric multidimensional scaling (NMS), described by [32], were used for ordination. The frequency of the understory vegetation and the importance value of the trees were taken to categorize the vegetation into groups. The differences between the local environmental variables among the groups were investigated using the single-way analysis (ANOVA). The understory vegetation was divided into different groups according to the method of Tansley and Chipp [33,34]: (I) 10–20% Rare, (II) 21–30% Occasional, (III) 31–40% Frequent, (IV) 41–50% Abundant, and (V) 51–60% Very abundant.

3. Results

3.1. *Classification Based on Ward's Cluster Analysis (Tree Vegetation Data)*

The cluster dendrogram (Figure 2) separated four major groups of vegetation. Table 1 lists the characteristic features of these groups, whereas Table 2 presents the environmental variables.

Table 1. Four groups derived from Ward's cluster analysis of 40 stands and their average tree species composition (average importance value for each group).

Tree Species	Group 1 (a)	Group 1 (b)	Group 1 (c)	Group 2	Group 3	Group 4
Pinus wallichiana	100 ± 0	98.33 ± 1.67	80.9 ± 2.2	*	26.3 ± 9.3	*
Abies pindrow	*	1.67 ± 1.67	10 ± 6.1	92.5 ± 4.6	9.75 ± 9.75	*
Cedrus deodara	*	*	*	*	44 ± 20.7	*
Picea smithiana	*	*	*	7.5 ± 4.6	16.25 ± 16.25	*
Pinus roxburghii	*	*	*	*	*	100 ± 0
Quercus baloot	*	*	9.0 ± 5.2	*	3.75 ± 2.5	*

Note: (*) = Absent, (±) = Standard error.

3.1.1. Group I *Pinus wallichiana* Dominant Group

Twenty-eight stands comprised of three subgroups were obtained: group 1 (a), composed of 21 stands of pure *Pinus wallichiana* tree species; group 1 (b), comprised of three stands of *Pinus wallichiana* and *Abies pindrow* association; group 1 (c), four stands comprised of two gymnosperm and one angiosperm species (Figure 2).

3.1.2. Group I (a) Pure *Pinus wallichiana*

Among the groups, this was the largest group, which is composed of 21 stands of *Pinus wallichiana*. As this group was based on monospecific stands of Pinus wallichiana, their importance value index (IVI) was 100. The average elevation of this group was 1953.1 ± 69 m, whereas the slope angle was 39.05 ± 1.41°. The total dissolved salts (TDS) and water-holding capacity were 68.0 ± 6.8 g/L and 12.24. ± 0.06%, respectively. The salinity, conductivity, and soil moisture of this group were 0.06 ± 0.01%, 136.6 ± 13.7 µS/cm, and 24.5 ± 1.2%, respectively. The soil nature of this group was alkaline, having a mean pH of 7.94 ± 0.04. Regarding the soil nutrients, this group had mean values of organic matter of 0.6 ± 0.0% and mean phosphorus of 0.43 ± 0.09%. The mean sand, silt, and clay were 53.37 ± 1.90%, 32.37 ± 1.63%, and 13.9 ± 1.53%, respectively.

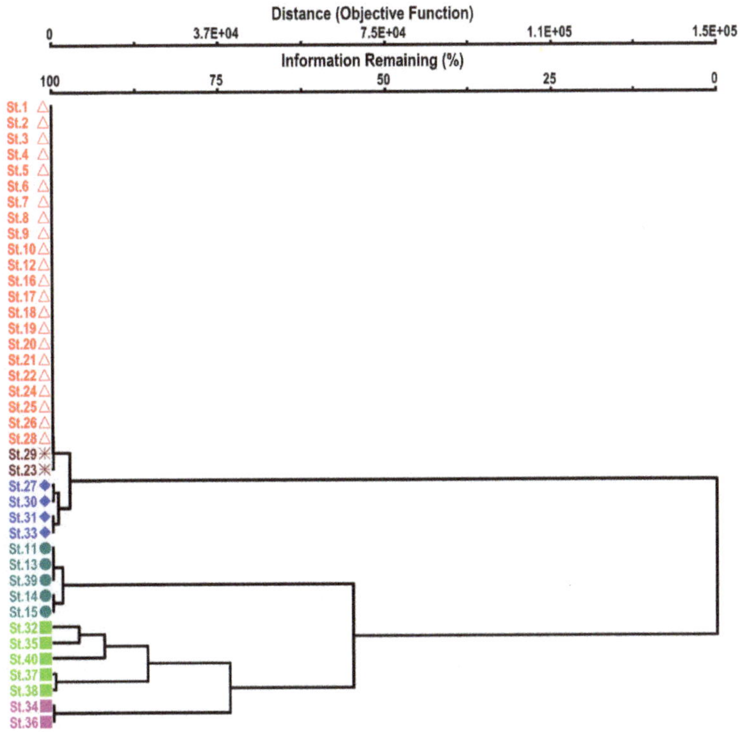

Figure 2. Dendrogram derived from Ward's cluster analysis using the importance value of tree species from the Shangla district. The different colors indicate different floristic composition with respect to their environmental variables. (St indicates different stands in groups of vegetation.)

3.1.3. Group I (b) *Pinus wallichiana* and *Abies pindrow* Association

This subgroup contained three stands showing two gymnosperm tree species: *Pinus wallichiana* and *Abies pindrow*. *Pinus wallichiana* is the leading dominant species with a 98.33 ± 1.67 importance value, and codominant species *Abies pindrow* contributed a very low 1.67 ± 1.67 importance value. The mean elevation of this group was 2203.3 ± 29.6 m with a 38.3 ± 4.4 mean slope angle. The edaphic condition of this group showed that the TDS value, maximum water-holding capacity, salinity, conductivity, and soil moisture were 48.67 ± 5.36, 11.70 ± 0.9, 0.04 ± 0.0, 96.67 ± 10.09%, and 23.4 ± 1.8%, respectively. The pH was alkaline at 8.06 ± 0.03. The soil nutrient organic matter content was 43 ± 0.030 and the phosphorus level was 0.23 ± 0.03%. The sand, silt, and clay contents were 51.73 ± 7.66, 36.9 ± 7.36, and 11.3 ± 1.13%, respectively.

3.1.4. Group I (c) *Pinus wallichiana* Mix Group

This subgroup comprised four stands of trees showing two coniferous species, *Pinus wallichiana* and *Abies pindrow,* with high importance values of 80.9 ± 2.2 and 10 ± 6.1, respectively. By contrast, the angiospermic species *Quercus baloot* contributed a 9.0 ± 5.2%

importance value to this group. This group is generally associated with an elevation of 2171.5 ± 35 m with a 30.8 ± 3.3° steep slope. The edaphic features of this group showed a TDS, maximum water-holding capacity, salinity, conductivity, and soil moisture of 71.3 ± 19.1, 9.5 ± 0.6, 0.07 ± 0.02, 143.8 ± 37.6, and 18.9 ± 1.2%, respectively. The pH was 7.8 ± 0.2. The soil nutrients of this group showed an organic matter level of 0.7 ± 0.2 and a phosphorus level of 0.33 ± 0.11%. The physical characteristics of the soil of this group were composed of sand 49.4 ± 5.0%, silt 42.0 ± 6.0%, and clay 8.7 ± 2.6%.

Table 2. Mean values ± SE of the environmental variables (topographic, edaphic, and soil nutrient) based on three groups derived from Ward's cluster analysis using the tree vegetation data. (Mean ± SE).

Variables		Group 1		Group 2	Group 3	Group 4
	1(a)	1(b)	1(c)			
1. Topographic variables						
1.Elevation(m)	1953.1 ± 69.7	2203.3 ± 29.6	2171.5 ± 35.0	2691.2 ± 47.6	2188 ± 76.25	1374.5 ± 76.5
2. Slope	39.05 ± 1.4	38.33 ± 4.41	30.8 ± 3.3	34.0 ± 7.48	38.2 ± 4.04	35.0 ± 5
2. Edaphic variables						
1.pH	7.94 ± 0.04	8.06 ± 0.03	7.8 ± 0.2	7.78 ± 0.09	7.60 ± 0.06	7.40 ± 0.02
2.WHC	12.24 ± 0.06	11.7 ± 0.9	9.5 ± 0.6	15.4 ± 1.81	12.05 ± 0.83	8.07 ± 4.23
3.Salinity	0.06 ± 0.01	0.04 ± 0.0	0.07 ± 0.02	0.04 ± 0.01	0.05 ± 0.0	0.04 ± 0.02
4.Cond	136.5 ± 13.7	96.67 ± 10.0	143.8 ± 37.6	82.6 ± 7.59	108 ± 12.73	89.5 ± 34.5
5.TDS	68.0 ± 6.8	48.67± 5.3	71.3 ± 19.1	44.4 ± 7.01	51.2 ± 4.95	45.0 ± 17.0
6. Soil Moisture	24.5 ± 1.2	23.4 ± 1.8	18.9 ± 1.2	30.8 ± 3.61	24.1 ± 1.7	16.14 ± 8.46
3. Soil Texture						
1. Sand	53.37 ± 1.9	51.73 ± 7.66	49.4 ± 5.0	45.88 ± 2.59	47.56 ± 5.85	27 ± 0.0
2. Silt	32.37 ± 1.6	36.93 ± 7.36	42.0 ± 6.0	34.48 ± 3.11	43.44 ± 5.61	53.8 ± 11.0
3. Clay	13.9 ± 1.5	11.33 ± 1.13	8.7 ± 2.6	19.64 ± 4.22	9 ± 0.6	19.2 ± 11.0
4. Soil nutrients						
1. OM	0.62 ± 0.07	0.43 ± 0.03	0.7 ± 0.2	0.42 ± 0.07	0.48 ± 0.05	0.4 ± 0.2
2. Phos	0.43 ± 0.09	0.23 ± 0.03	0.33 ± 0.11	0.53 ± 0.17	0.46 ± 0.09	0.3 ± 0.0

SE = standard error; WHC = water-holding capacity; OM = organic matter of soil in%; TDS = total dissolved salt; Cond = conductivity; Phos = phosphorus.

3.1.5. Group II *Abies pindrow* and *Picea smithiana* Association

This group includes five stands; *Abies pindrow* dominates with a 92.5 ± 4.6% average importance value, while *Picea smithiana* attained a 7.5 ± 4.6% average importance value. Compared to the other groups, this group was commonly growing on the highest elevation (2691.2 ± 47.6 m), with a 34.0 ± 7.48° slope angle. In the edaphic variables, the TDS, maximum water-holding capacity, salinity, conductivity, and soil moisture were 44.4 ± 7.01; 15.4 ± 1.81, 0.04 ± 0.01, 82.6 ± 7.59, and 30.8 ± 3.61%, respectively. The pH was 7.78 ± 0.09. Regarding the soil nutrients, this group showed 0.42 ± 0.07% of organic matter and 0.53 ± 0.17% phosphorus. This group contained 45.88 ± 2.59% sand, 34.48 ± 3.11% silt, and 19.64 ± 4.22% clay.

3.1.6. Group III Mixed Group of Conifer Dominating Species

This group contains five stands of 32, 35, 37, 38, and 40, respectively, and included four conifers and one angiospermic tree species. The dominant species was *Cedrus deodara*, with an average importance value of 44 ± 20.7, followed in order by *Pinus wallichiana*, *Picea smithian*, *Abies pindrow*, and *Quercus baloot* at average importance values of 26.3 ± 9.3, 16.25 ± 16.2, 9.75 ± 9.75, and 3.75 ± 2.5%, respectively. The elevation value means of this group were slightly lower at 2188 = 76.2 than that found in the previous group, while the mean slope was 38.2 ± 4.04, higher than the previous group. The edaphic feature of this group showed a mean TDS, maximum water-holding capacity, salinity, conductivity, and soil moisture of 51.2 ± 4.9, 12.05 ± 0.83, 0.05 ± 0.0, 108 ± 12.73, and 24.1 ± 1.7%, respectively. The pH of this group was 7.60 ± 0.06. The soil nutrients of this group were 0.48 ± 0.05 for organic matter and 0.46± 0.09 for phosphorus. The sand, silt, and clay contents were 47.56 ± 5.85, 43.44 ± 5.61, and 9 ± 0.6%, respectively.

3.1.7. Group IV Pure *Pinus roxberghii* Association

Among the groups, group IV was the smallest group, composed of two pure stands of *Pinus roxberghii* tress with 100 ± 00% importance values. This group belongs to the subtropical area. The topographic appearances of this group revealed a comparatively low elevation of 1374.5 ± 76.5 m with a 35.0 ± 5° slope angle. The edaphic feature of this group showed a mean TDS, maximum water-holding capacity, salinity, conductivity, and soil moisture of 45.0 ± 17.0, 8.07 ± 4.23, 0.04 ± 0.02, 89.5 ± 34.5, and 16.14 ± 8.46, respectively. The soil pH was 7.40 ± 0.02. The soil texture was composed of 27 ± 0.0% sand, 53.8 ± 11.0% silt, and 19.2 ± 11.0% clay. Soil organic matter was 0.4 ± 0.2, and the phosphorus level was 0.3 ± 0.0.

3.2. *Univariate Analysis of Variance (ANOVA)*

ANOVA was used to determine the relationship between vegetation and environmental characteristics. The results showed that in the topographic factors, elevation was highly significant ($p < 0.001$), whereas the slope was nonsignificant. Among the edaphic factors, pH was significantly correlated ($p < 0.01$) (Table 3). Soil moisture and maximum water holding capacity also showed a significant $p >$ correlation with vegetation. Salinity, conductivity, and TDS were nonsignificant. The soil texture (sands, silt, and clay) also showed a significant correlation ($p < 0.05$). Regarding the soil nutrients, soil organic matter was correlated significantly ($p < 0.001$), but phosphorus was nonsignificant.

3.3. *Ordination*

Principal Component Analysis (PCA) Ordination of Tree Vegetation Data

PCA was used to determine the different soil factors, i.e., topographic variables (elevation and slope), edaphic factors, soil physical properties, and soil nutrients, and the importance value index of the tree species (Table 4). Four major groups were isolated by Ward's cluster analysis and superimposed on the PCA ordination with axis 1 and 2, 1 and 3, and 2 and 3 (Figure 3). The analyses showed no overlapping among the groups in the two axes of 1,2 and 1,3, whereas axis 2,3 showed little overlap. The largest group among all four groups was group 1, showing 28 stands. This group contained three subgroups, i.e., group I (a), group I (b), and group I (c), showing different cospecies compositions. *Pinus wallichiana* was the main dominant species in the three subgroups. Group I was separated on all axes. This group was present on the three axes of 1–2, 1–3, and 2–3.

Group I (a) consisted of 21 stands mainly containing *Pinus wallichiana*, which was the only species there. The elevation and slope angles of this group were 1953.1 ± 69.7 m and 39.05 ± 1.41, respectively. Group I (b) was composed of three stands containing *Abies pindrow* as the codominant species. The average elevation of this group was 2203.3 ± 29 m, with a 38.33 ± 4.4° slope angle. Group I (c) was composed of four stands. This subgroup contained two species: *Abies pindrow* and *Quercus baloot*. This group was found on a 2171.5 ± 35 m average elevation and a 30.8 ± 3.3° average slope angle. Group II was composed of five stands with *Abies pindrow* the dominant tree, while *Picea smithiana* was codominant. This group was found on the highest elevation (2691.2 ± 47) with a slope angle of 34.0 ± 7.4. Group III was composed of five stands dominated by *Cedrus deodara*. The codominant species were *Pinus wallichiana*, *Picea smithiana*, and *Abies pindrow*. In this group, one angiosperm, *Quercus baloot*, was also recorded. This group was located on an average elevation of 2188.0 ± 76, with a 38.2 ± 4° slope angle. Group IV was the smallest group, which was separated based on their monospecific position. In this group, two stands, i.e., 34 and 36, were recorded as pure *Pinus roxburghii* stands. The lowest average elevation was recorded as 1374.5 ± 76.5 m, with a 35 ± 5° average slope angle. This group was represented only in the subtropical condition.

Table 3. Analysis of the variance of individual environmental variables (topographic, edaphic, soil texture, and soil nutrients) by Ward's cluster analysis using the tree vegetation data of 40 stands.

Source of Variation		SS	df	MS	F	*p*-Level
1. Topographic Variables						
1	**Elevation**					
	Between Groups	3,337,292.82	5	667,458.57	10.15	$p < 0.001$
	Within Groups	2,236,902.77	34	65,791.26		
	Total	5,574,195.6	39			
2	**Slope**					
	Between Groups	307.81	5	61.56	0.814	NS
	Within Groups	2571.17	34	75.62		
	Total	2878.98	39			
2. Edaphic Variables						
1	**PH**					
	Between Groups	1.043235	5	0.208647	5.73	$p < 0.01$
	Within Groups	1.240102	34	0.036474		
	Total	2.2833375	39			
2	**WHC**					
	Between Groups	115.4212	5	23.08424	2.766	$p < 0.05$
	Within Groups	283.7654	34	8.346041		
	Total	399.1866	39			
3	**Salinity**					
	Between Groups	0.003552	5	0.00071	0.981	NS
	Within Groups	0.024625	34	0.000724		
	Total	0.028178	39			
4	**Conductivity**					
	Between Groups	19,125.115	5	3825.023	1.251	NS
	Within Groups	103,938.26	34	3057.008		
	Total	123,063.38	39			
5	**TDS**					
	Between Groups	4271.35833	5	854.2717	1.116	NS
	Within Groups	26,021.4167	34	765.3358		
	Total	30,292.775	39			
6	**Soil Moisture**					
	Between Groups	461.68485	5	92.33697	2.766	$p < 0.05$
	Within Groups	1135.0615	34	33.38416		
	Total	1596.7464	39			
3. Soil Texture						
1	**Sands**					
	Between Groups	1421.31148	5	284.2623	3.241	$p < 0.05$
	Within Groups	2981.75952	34	87.69881		
	Total	4403.071	39			
2	**Silt**					
	Between Groups	1278.40024	5	255.68005	2.967	$p < 0.05$
	Within Groups	2930.09476	34	86.179258		
	Total		39			
3	**Clay**					
	Between Groups	716.5828095	5	143.3166	3.384	$p < 0.05$
	Within Groups	1440.14819	34	42.3573		
	Total	2156.731	39			
4. Soil Nutrients						
1	**Organic matter**					
	Between Groups	9564.9254	5	1912.985	7.120	$p < 0.001$
	Within Groups	9135.4484	34	268.6897		
	Total	18,700.374	39			
2	**Phosphorus**					
	Between Groups	0.24110417	5	0.048221	0.406	NS
	Within Groups	4.04083333	34	0.118848		
	Total	4.28193750	39			

Note: SS = sum of square; MS = mean square; F = F ratio, df = degree of freedom; P level = probability level; ns = nonsignificant.

Table 4. The correlation coefficients of environmental variables (topographic variables, edaphic variables, soil texture, and soil nutrients) with 3 PCA ordination axes obtained by tree vegetation data.

		Axis 1		Axis 2		Axis 3	
S. No.	**Variables**	**r**	**Prob. Level**	**R**	**Prob. Level**	**r**	**Prob. Level**
1. Topographic variables							
1	Elevation	0.316	$p < 0.05$	−0.309	NS	0.509	$p < 0.001$
2	Slope	−0.079	NS	0.122	NS	−0.044	NS
2. Edaphic variables							
1	Ph	−0.511	NS	−0.321	NS	0.059	NS
2	MWHC	0.144	NS	−0.319	NS	0.235	$p< 0.05$
3	Salinity	−0.240	NS	0.008	NS	0.127	NS
4	Conductivity	−0.262	NS	−0.017	NS	0.094	NS
5	TDS	−0.263	NS	−0.013	NS	0.093	NS
6	Soil moisture	0.144	NS	−0.329	NS	0.245	$p < 0.05$
3. Soil Texture							
1	Sand	−0.346	NS	−0.309		0.224	NS
2	Silt	0.255	$p < 0.05$	0.423	$p < 0.005$	−0.088	NS
3	Clay	0.228	NS	−0.149	NS	−0.198	NS
4. Soil nutrients							
1	OM	−0.240	NS	0.008	NS	0.124	NS
2	Phosphorus	0.136	NS	0.017	NS	0.128	NS

Key to abbreviations: r = correlation coefficient; NS = nonsignificant; Prob. Level = probability level; OM = organic matter of soil in %; TDS = total dissolved salt.

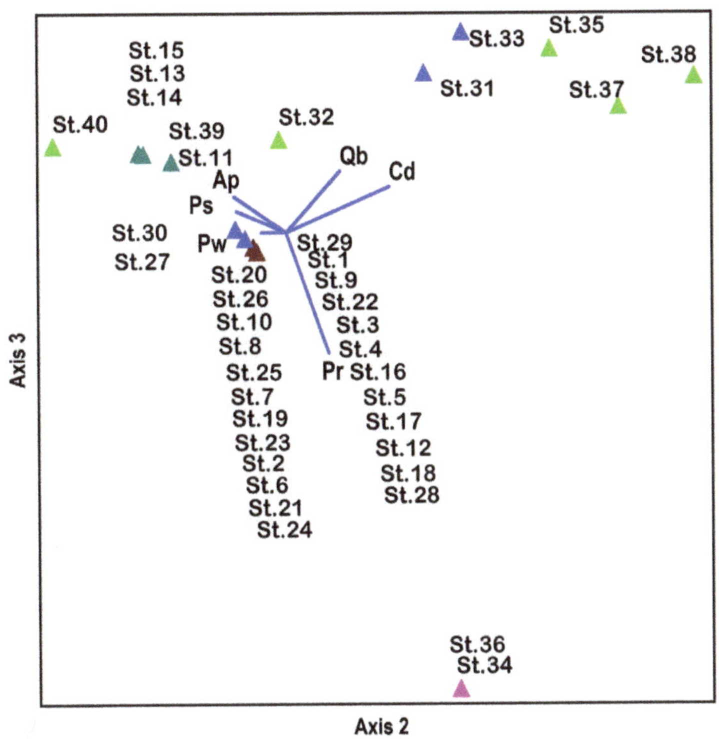

Figure 3. PCA stand ordination based on IVI of tree species from the Shangla district, Pakistan. The floristic composition/group obtained from Ward's cluster analysis were super imposed on ordination axes. The different colors in the above ordination plan indicate different groups. (St indicates different stands in groups of vegetation.)

3.4. Relationship (Correlation Coefficients) of Three Ordination Axes with Environmental Variables

The correlation between the PCA ordination of the three axes with different environmental variables are shown in (Table 4). Axis 1 was significantly correlated with elevation ($p < 0.05$) and silt ($p < 0.05$), while the other environmental factors did not show a significant correlation. Axis 2 ordination showed a significant correlation ($p < 0.005$) with silt, whereas no correlation was observed with the other environmental factors. Ordination on axis 3 was a highly significant correlation ($p <0.001$) on the elevation and a significant correlation between the maximum water-holding capacity and soil moisture, while the other environmental variables did not show a significant correlation.

3.5. Understory Vegetation Data

Ward's Cluster Analysis of Stands

Based on frequency, Ward's cluster dendrogram distributed the understory vegetation into four groups (Figure 4). Table 5 lists the average frequency of these four groups, while Table 6 presents their environmental factors.

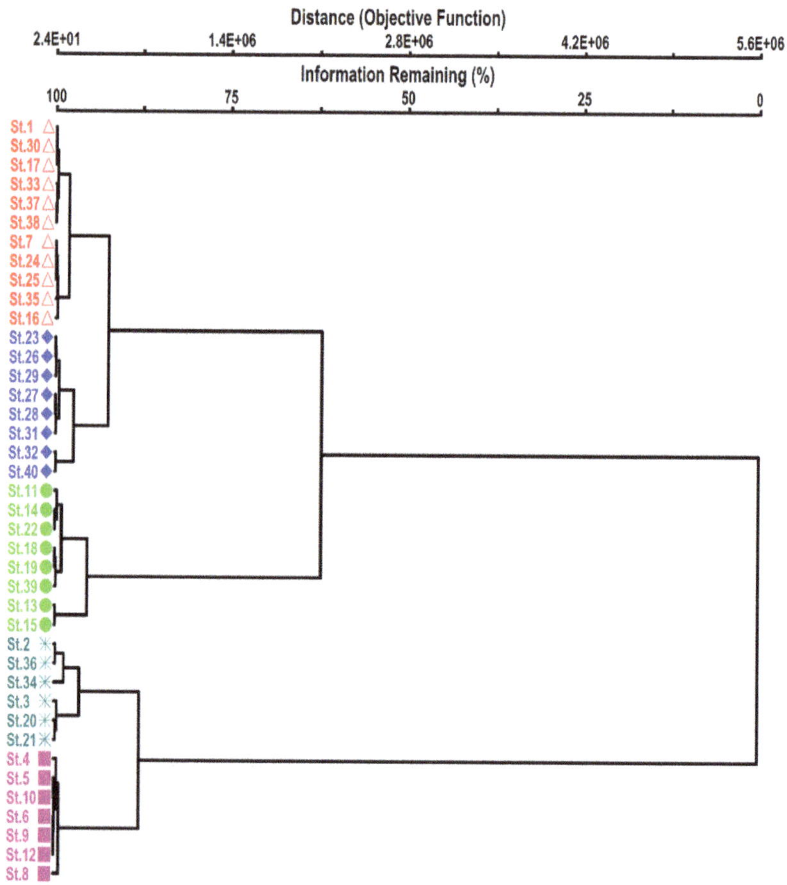

Figure 4. Ward's cluster analysis of the understory vegetation based on frequency. The understory species indicates five distinct groups. (St indicates different stands in groups of vegetation.)

Table 5. Average frequency of the understory vegetation of four main groups was derived from Ward's cluster analysis.

S No.	Species Name	Group I (A)	Group I (B)	Group II	Group III	Group IV
1	*Adiantum venustum* D.Don	22.5 ± 3.2	27.5 ± 2.5	*	30	45
2	*Amaranthus tricolor* L.	25 ± 5	20	*	33.3 ± 4.4	25
3	*Ammannia baccifera* L.	27.5 ± 2.5	*	30 ± 5	32.5 ± 7.5	*
4	*Anaphalis scopulosa* Boriss	*	33.3 ± 9.3	31.7 ± 8.3	*	47.5 ± 2.5
5	*Asplenium ceterach* L.	37.5 ± 3.10	31.25 ± 2.3	30 ± 2.04	36 ± 1.25	35 ± 5
6	*Berberis lycium* L.	*	40	*	*	*
7	*Bistorta amplixiculis* D.Don	*	*	40 ± 5	28.7 ± 4.3	30 ± 2.9
8	*Cannabis sativa* L.	21.7 ± 1.7	27.5 ± 7.5	33.7 ± 3.15	*	37.5 ± 2.5
9	*Cenchrus penusaliformis* L.	30	28.33 ± 8.3	27.5 ± 7.5	30 ± 3.54	20
10	*Cicota virosa* L.	35	*	*	30	*
11	*Conyza bonarensis* L.	33.3 ± 6.01	20 ± 5	22.5 ± 7.5	20	*
12	*Corbichonia decumbers* (Forssk.).	30	35 ± 5	22.5 ± 2.5	*	20
13	*Digiteria sanguinalis* L.	40	25	40	17.5 ± 2.5	27.5 ± 7.5
14	*Droypteris stewartii* L.	20	*	31.7 ± 4.4	*	15
15	*Fragaria nubicola* L.	40	40 ± 10	30 ± 5	45 ± 10	25± 10
16	*Fragaria orientalis* Los.	22.5 ± 12.5	25	30 ± 5	26.7 ± 8.3	36.7 ± 1.7
17	*Hedera nepalensis* K.Koch	33.3 ± 2.1	28.7 ± 5.15	32 ± 4.36	*	29 ± 3.32
18	*Impatiens braclycenera* L.	33.3 ± 8.8	36.6 ± 1.7	*	32.5 ± 12.5	*
19	*Indigofera gerardiana* Wall.	28.3 ± 2.5	*	*	*	25
20	*Launaea procumbens* (Roxb.)	20	15	26.7 ± 4.4	*	15
21	*Morchella esculenta* L.	*	32.5 ± 2.5	23.3 ± 1.7	*	*
22	*Ocimum bacilicum* L.	31.25 ± 4.7	17.5 ± 2.5	25	*	30
23	*Panicum miliaceum* L.	26.7 ± 2.8	32.5 ± 1.7	*	27.5 ± 2.5	31.7 ± 3.3
24	*Persicaria punctata* (Elliott.)	25 ± 5	31.7 ± 3.3	*	*	*
25	*Pinus wallichiana* seedling	40 ± 5	30 ± 5	*	*	30
26	*Pteridium aquilinium* L.	23.5 ± 3.1	*	33.7 ± 5.5	25	25
27	*Phragmites karka* (Retz.)	33.7 ± 2.4	*	50	*	*
28	*Polygonatum multiflorium* L.	22.5 ± 4.8	45	35	*	*
29	*Rubus fruticosus* L.	23.3 ± 1.7	26.7 ± 1.7	30	25	27.5 ± 2.5
30	*Rumex hastatus* D.Don	*	*	40	28.3 ±6.01	32.5 ± 7.5
31	*Solanum nigrum* L.	28.3 ± 4.4	35 ± 20	20	32.5 ± 7.5	25
32	*Tagetis minuta* L.	26.7 ± 6.01	27.5 ± 5.2	55	*	27.5 ± 12.5
33	*Urtica dioica.* L.	15	30	21.7 ± 3.3	*	23.3 ± 3.3
34	*Verbascum Thapsus* L.	40	*	25	26.2 ± 8.3	*

* Shows the absence of this species in a group.

Table 6. Mean values ± SE of environmental variables based on four groups derived from Ward's cluster analysis using understory vegetation data of 40 stands.

Variables	Group 1 (A)	Group 1 (B)	Group 2	Group 3	Group 4
1. Topographic variables					
Elevation	2062.73 ± 17	2249.5 ± 28.9	2645.75 ± 37.01	1509.83 ± 52	1792.14 ± 19
Slope	38.18 ± 2.6	36.13 ± 2.5	36.25 ± 4.6	34.17 ± 2.01	40.71 ± 2.5
2. Edaphic variables					
MWHC	11.49 ± 0.6	10.38 ± 1.3	14.24 ± 1.24	11.68 ± 1.7	12.86 ± 0.7
Salinity	0.05 ± 0.0	0.05 ± 0.01	0.05 ± 0.01	0.06 ± 0.02	0.07 ± 0.01
OM	0.5 ± 0.04	0.58 ± 0.09	0.49 ± 0.07	0.55 ± 0.2	0.71 ± 0.15
3. Soil Texture					
Sand	51.33 ± 3.5	51.5 ± 3.6	48.78 ± 2.7	44.73 ± 5.9	51.34 ± 3.9
Silt	37.2 ± 3.8	37.23 ± 3.5	35.3 ± 2.6	40.93 ± 5.3	32.57 ± 3.2
Clay	11.47 ± 2.2	11.28 ± 1.2	15.93 ± 3.1	14.33 ± 3.3	16.09 ± 3.4

Group I is the largest group among the groups and was divided into two subgroups, Group I (A) and Group I (B), as follows.

Group I (A) contained a large understory vegetation group composed of eleven stands and twenty-nine tree species. Among these species, *Digeteria sanguinalis, Fragaria nubicola, Verbascum Thapsus*, and *Pinus wallichiana* seedlings showed a 40% average frequency, which are the dominant species recorded in these stands. *Asplenium ceterach* was the second dominant species and showed a 37.5% average frequency. *Urtica dioica*, with a 15% average frequency, was recorded as a rare species in this group. Based on the topographic characteristics, this group was recorded at 2062.73 ± 17 m mean elevation with a 38.18 ± 2.6° mean slope angle. In the edaphic factors, the water-holding capacity was highest (11.49 ± 0.6). The recorded mean salinity and organic matter were 0.5 ± 0.0 and 0.5 ± 0.04, respectively. The sand, silt, and clay contents were 51.33 ± 3.5%, 37.2 ± 3.8%, and 11.47 ± 2.2%, respectively.

Group I (B) contained eight stands composed of 25 species. Among these species, 22 were common, and the leading species was *Polygonatum multiflorium*, with a 45% average frequency. *Fragaria nubicola* was the second dominant species, with a 40% average frequency. Other species, such as *Impaliense braclycenera* (36.6%), *Corbichonia decumbers* (35%), and *Solanum nigrum* (35%), were occasional species, while *Launaea procum* (15%) was a rare species. According to the environmental variables, this group has a 2249.5 ± 28.9 m elevation and a 36.13 ± 2.5° mean steep slope. The sand, silt, and clay contents were 51.5 ± 3.6, 37.23 ± 3.5, and 11.28 ± 1.2%, respectively.

Group II was the second largest group of understory vegetation and contained eight stands with 25 different species of ground flora. Sixteen of these species were common in group I (a) and group I (b). *Tagetis minuta* was the dominant species in this group, with a 55% average frequency. *Phragmites karka* had a 50% average frequency and ranked second. *Solanum nigrum* was a rare species with a 20% average frequency. This group showed the highest elevation (2645.75 ± 37.01 m), with a mean slope angle of 36.25 ± 4.6°. Among the edaphic variables of this group, this group had the highest water-holding capacity (14.24 ± 1.24). The salinity and organic matter of this group were 0.05 ± 0.01 and 0.49 ± 0.07%, respectively. The sand, silt, and clay contents were 48.78 ± 2.7, 35.3 ± 2.6, and 15.93 ± 3.1, respectively.

Group III consisted of six stands and 18 species and was considered the smallest group. The group was dominated by *Fragaria nubicola*, representing a 45% average frequency. Twelve species were common with group I (a) and group I (b), whereas thirteen species that occurred in this group were also recorded in group II. *Asplenium ceterach* was the second dominant species in this group, with a 36% average frequency. *Digeteria sanguinalis* was a rare species, with a 17.5% average frequency in this group. With respect to the environmental variables, this group was characterized by a very low elevation (1509.83 ± 52 m) and mean slope (34.17 ± 2.01°). The edaphic variables showed a water-holding capacity of 11.68 ± 1.7, which was similar to group I (a), while the salinity and organic matter mean values were 0.06 ± 0.02 and 0.55 ± 0.2, respectively, which was similar to group I (a) and group I (b).

Compared to the other groups, sand was the lowest (44.73 ± 5.9), whereas silt had the highest value (40.93 ± 5.3). Clay was calculated as 14.33 ± 3.3, which was different from the other groups.

Group IV consisted of seven stands and 24 species of ground flora. Among these species, only five were common in all four groups from the cluster analysis. *Anaphalis scopulosa* was the dominant species in this group, with a 47.5% average frequency. The codominant species was *Adiantum venustum*, with a 45% frequency, while Cannabis *sativa* had a 37.5% average frequency. *Drypteris stewartii* and *Launaea procum* were rare species, with a 15% average frequency. The results of the environmental variables group showed that this group was characterized by an average elevation (1792.14 ± 19 m) with the highest mean steep slope of 40.71 ± 2.5°. The water-holding capacity was 12.86 ± 0.7, whereas the salinity and organic matter were 0.07 ± 0.01 and 0.71 ± 0.15, respectively. The sand of this group was 51.34 ± 3.9, similar to group I (A and B). The recorded silt was 32.57 ± 3.2%, while the clay was 16.09 ± 3.4%, the highest among all groups.

3.6. Univariate Analysis of Variance (ANOVA)

The results of the different environmental variables of the understory vegetation of the four main groups derived from Ward's cluster analysis were examined by ANOVA (Table 7). The results showed that, of the topographic variables, elevation showed a significant ($p < 0.001$) correlation, while slope showed no relationship. The other environmental variables did not significantly correlate with the understory species.

Table 7. Analysis of variance of individual environmental variables (topographic, edaphic, and soil texture) derived by Ward's cluster analysis using understory vegetation data.

Source of Variance	SS	Df	MS	F	p-Level
1. Topographic variables					
1 Elevation					
Between Groups	532,2642	4	1,330,661	185.14	$p < 0.001$
Within Groups	251,553.4	35	7187.239		
Total	5,574,196	39			
2 Slope					
Between Groups	168.701732	4	42.1754329	0.545	Nonsignificant
Within Groups	2710.27327	35	77.4363791		
Total	2878.975	39			
2. Edaphic variables					
1 Water holding capacity					
Between Groups	69.5759929	4	17.394	1.847	Nonsignificant
Within Groups	329.6106046	35	9.417446		
Total	399.1865975	39			
2 Salinity					
Between Groups	0.002504286	4	0.000626	0.854	Nonsignificant
Within Groups	0.025673214	35	0.000734		
Total	0.0281775	39			
3 Organic Matter					
Between Groups	0.250428571	4	0.062607	0.854	Nonsignificant
Within Groups	2.567321429	35	0.073352		
Total	2.81775	39			
3. Soil Texture					
1 Sand					
Between Groups	227.7037056	4	56.92593	0.477	Nonsignificant
Within Groups	4175.367294	35	119.2962		
Total	4403.071	39			
2 Silt					
Between Groups	246.852381	4	61.7131	0.545	Nonsignificant
Within Groups	3961.642619	35	113.1898		
Total	4208.495	39			
3 Clay					
Between Groups	182.6822771	4	45.67057	0.814	Nonsignificant
Within Groups	1963.733723	35	56.10668		
Total	2146.416	39			

Note: SS = sum of square; MS = mean square; F = F ration; df = degree of freedom; p-level = probability level.

3.7. Stand Ordination of the Understory Vegetation Data

Two-dimensional nonmetric multidimensional scaling (NMS) ordination divided the understory species into four distinct groups (Figure 5). A continuous pattern appeared to exist between axes 1 and 2. Groups 1 (A) and 1 (B) were located on the top of the ordination axes, whereas group 2 was slightly lower. Groups 3 and 4 were located on the extreme lower middle side of the ordination plan, indicating differences in species composition and environmental variables.

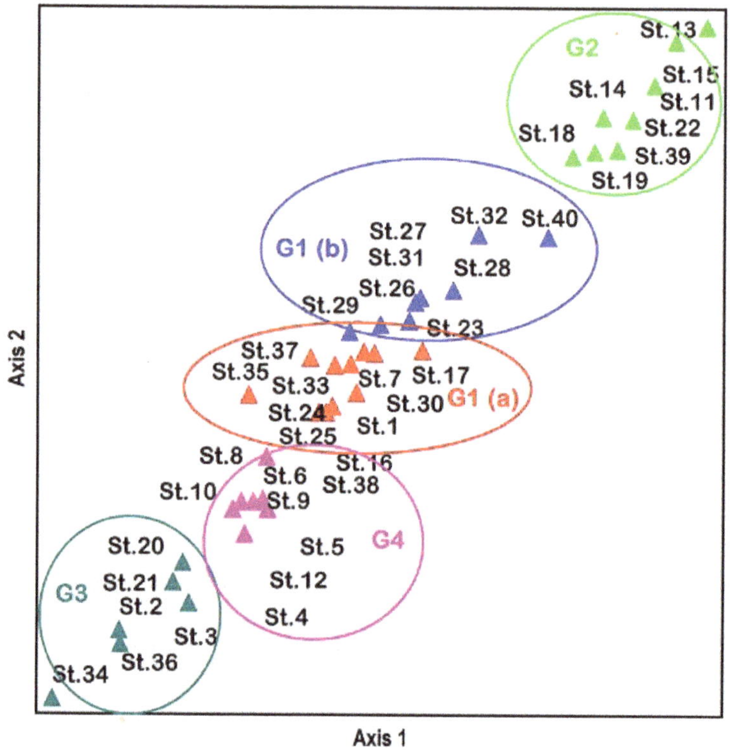

Figure 5. NMS stand ordination of the understory species, based on the frequency from the Shangla district, Pakistan. The four distinct groups (G1–G4) and the subgroup (G1a) obtained from Ward's cluster analysis of understory species were clearly imposed on ordination plan. (St indicates different stands in groups of vegetation.)

Relationship (correlation coefficient) between the NMS ordination axis and the environmental variables of the understory vegetation data.

Table 8 lists the results of the environmental factors with the ordination axis. The topographic and edaphic factors of elevation, with a significant correlation with axis 1 ($p < 0.001$) and axis 2 ($p < 0.001$), of the NMS ordination are shown. The other environmental factors did not exhibit a significant relationship.

Table 8. Relationship (correlation coefficient) of the environmental variables with two axes of NMS ordination based on understory vegetation.

S. No	Variables	Axis 1 R	Axis 1 Prob. Level	Axis 2 R	Axis 2 Prob. Level
1. Topographic variables					
1	Elevation	0.7801	$p < 0.001$	0.9417	$p < 0.001$
2	Slope	−0.1411	NS	−0.0221	NS
2. Edaphic variables					
3	WHC	0.1517	NS	0.1502	NS
4	Salinity	−0.1718	NS	−0.1645	NS
5	OM	−0.1718	NS	−0.1645	NS
3. Soil Texture					
6	Sand	−0.1661	NS	0.2367	NS
7	Silt	0.0350	NS	−0.1366	NS
8	Clay	0.1889	NS	−0.1478	NS

4. Discussion

Multivariate techniques (i.e., classification and ordination) were conducted on tree and understory vegetation of 40 forest stands in the Shangla area. Ordination and cluster analyses have important advantages. They can yield corresponding results and provide deeper explanations of the ecological data, as reported by [35]. Frades and Matthiesen [36] reported that the cluster analysis approach is a quantitative and objective detailed categorization technique. Environmental factors play an essential role in understanding the vegetation pattern. The elevation is an environmental factor that could be considered a critical factor in understanding vegetation patterns distribution, as suggested by Lovett et al. [37] and Gajoti et al. [38]. Zhang et al. [39] reported that the vegetation distribution pattern is a basic tool for evaluating and managing forests. Other researchers investigated quantitative separation to illustrate the influence of environmental factors on the layers of plant communities and their distribution patterns [39]. Indeed, natural communities are distributed continuously and usually contain different plant communities at several succession stages related to environmental factors.

4.1. Classification

According to Ward's cluster analysis, the classified groups obtained from tree vegetation and understory vegetation data were associated with several factors, including topographic, edaphic, and soil physical parameters. In general, the results of classification and ordination, alongside the environmental variables, have improved the understanding of the vegetation communities and their locations in the study area. In the current investigation, the sequence of the natural vegetation of tree and understory communities could be explained in terms of several environmental characteristics. The vegetation groups found in the cluster analysis were almost dominated by a single species, and sometimes by two species. Group I (a) was composed of 21 stands dominated by a single species, *Pinus wallichiana,* showing a 100% average importance value, and was the largest group in this study. This group had an average elevation of 1953.1 ± 70 m and the highest slope angle (39.05 ± 1.4). A *Pinus wallichiana* pure community was reported by Ahmed, Tareen, and Tareen [13] from different climatic zones of Himalayan forests in Pakistan. They described a pure *Pinus wallichiana* stand at an approximately 2770 m elevation from Naltar-Gilgit and a 3100 m elevation from Takht-e-Suleiman. Saima et al. [40] found that the *Pinus wallichiana* community was associated with two species, *Taxus wallichiana* and *Abies pindrow,* in the Ayubia National Park.

Ahmad et al. [41] reported that *Pinus wallichiana* was common in 22 forests in Pakistan at an elevation range from 1950 to 2700 m with a slope of 23° to 25°. Khan and Hussain [34] and Wahab, Moinuddin, Nasrullah, and Sarangzai [20] examined the communities of the same species in the Dir District and Chitral at approximately 1875 m and 2559 m elevations, respectively. Siddiqui, Ahmed, Shaukat, and Khan [17] reported five different *Pinus wallichiana*-dominated stands in the moist temperate areas of the Himalayan mountains range in Pakistan, and assumed that this species preferred to grow on relatively medium elevations of 2368 m and low slope angles of 29°. Akbar [42] described *Pinus wallichiana* tree forests at a 3169 m elevation with a low slope angle of 28°. Ilyas, Shinwari, and Qureshi [16] described *Pinus wallichiana* tree forests from temperate mountain forests in Qalagi Hills Swat. Group I (b) was composed of *Abies pindrow* as a codominant tree growing in three stands with an elevation of 2203.3 ± 29 m and at an average slope. Group I (c) contained four stands, with *Abies pindrow* and *Quercus baloot* growing communities, with an average elevation and slope angle. In all these subgroups, *Pinus wallichiana* was the main species in these groups. *Pinus wallichiana* shows large ecological growing areas in different climatic zones. Wahab et al. [43] found *Pinus wallichiana* and *Cedrus deodara* species naturally growing close to the Afghanistan and Pakistan border. Group II contained five stands occupied mainly by *Abies pindrow* and followed by *Picea smithiana* rec, with the highest average elevation of 2691.2 ± 47 m with a 34 ± 7.4° slope angle. Ahmed, Tareen, and Tareen [13] described *Abies pindrow* stands at a 3450 m elevation with a 45° slope angle in the Rama District Astore. Wahab, Moinuddin, Nasrullah, and Sarangzai [20] observed *Abies pindrow* stands at a 2670 m elevation in the Dir District. Siddiqui, Ahmed, Shaukat, and Khan [17] reported *Abies pindrow* in 21 communities in the moist temperate areas of the region of Lalazar, Naran. They assumed that *Abies pindrow* preferred growing at high elevations (2617 m) and at high slope angles (36°). The natural distribution of *Abies pindrow* trees shows that this species can grow and form communities in different climatic regions.

Group III was described in five stands containing *Cedrus deodara, Pinus wallichiana, Picea smithiana, Abies pindrow,* and broad-leaved species of *Quercus baloot* with an average elevation and slope angle. Similar communities were described by Wahab, Ahmed, and Khan [43], who reported *Cedrus deodara* and *Pinus wallichiana* communities at the Pakistan and Afghanistan border. Ahmad, Abdul, and Akbar [41] found *Cedrus deodara* forests from approximately 1650 m to 2770 m and at slope angles from 12° to 50°. Siddiqui, Ahmed, Shaukat, and Khan [17] reported *Cedrus deodara* and *Pinus wallichiana* natural forests in the moist temperate areas of Pakistan. Siddiqui, Ahmed, Shaukat, and Khan [17] studied the *Cedrus deodara* communities growing naturally in the Gol National Park, Chitral District. Siddiqui, Ahmed, Shaukat, and Khan [17] suggested that *Cedrus deodara* is the predominant species in the forest of the moist temperate areas of the Himalayan mountains. Group IV contained *Pinus roxburghii* pure stands, which was the smallest group in two different sites, with a low average elevation of 1374.5 ± 76.5 and a 35 ± 5° slope, and represented the dry condition of the area. Malik and Malik [44] reported that this species was common in the Azad Jammu and Kashmir areas. Ahmed et al. (2006) and Siddiqui, Ahmed, Shaukat, and Khan [17] observed *Pinus roxburghii* natural communities in the subtropical areas of Hindukush and the Himalayan mountains of Pakistan. Wahab, Moinuddin, Nasrullah, and Sarangzai [20] also reported the *Pinus roxburghii* natural community from the Dir District.

The understory vegetation cluster analysis revealed four significant groups. Diverse shapes were described in the vegetation of these forests, but some of them were common. The dominant species of the understory vegetation in the current study were *Asplenium ceterach, Fragaria orientalis, Cenchrus penusaliformis, Fragaria nubicola, Rubus fruticosus, Digeteria sanguinalis,* and *Solanum nigrum*. Group I was composed of two subgroups: group I (a) and group I (b). This group was predominated by *Digeteria sanguinalis, Verbascum Thapsus, Fragaria nubicola,* and *Pinus wallichiana* seedlings in group I (a), while *Polygonatum multiflorium* was the main plant species associated with *Fragaria nubicola*, commonly growing in both subgroups. Group II was composed mainly of *Tagatis minuta*, with a 55% average frequency, followed by the *Phragmatis karka*, with a 50% frequency. This group was found at the highest mean elevation of 2645.75 m. Group III was the smallest and contained *Fragaria nubicola* with a 45% average frequency. This group has a low mean elevation of 1509.83 m and a low slope. Group IV contained mainly *Anaphalis scopulosa*, with a 47.5% average frequency, a mean elevation of 1792.14 m, and the highest slope angle of 40.71°. Environmental factors, including the edaphic and photographic variables of the understory species, were common among the groups.

4.2. Ordination

Ordination techniques are commonly applied to study the relationship between the vegetation pattern composition and the gradients of the underlying environment [45,46]. Ordination might be one of the easiest methods to determine species commonly growing in areas and to associate them with other species. Furthermore, it describes how the species composition is changed in these natural communities with fluctuations of the elevation range. Two basic approaches were complementary to each other, which were classification and ordination, and could be applied to the natural communities [35]. The PCA ordination approach was used to investigate the compositional variations in the gradients of the environment in the tree vegetation ecological data from other taxonomic studies [47–50]. McCune and Mefford [51] considered PCA to be an essential and effective approach for the ordination and evaluation of the data of homogenous communities. This is mainly an Eigen analysis, in which the sum of the Eigenvalues is primarily equal to the sum of the variance of all the variables in a data set. PCA provides reasonable and true indicators of the relationship among vegetation species in natural communities. Nonmetric multidimensional scaling (NMS) was used to study the different environmental variables that correlate with the species composition of understory vegetation. The NMS is considered a true nonparametric ordination approach to identify the best-reduced space portrayal of environmental relationships. These methods are commonly used to identify possible similarities in a data set, classify the information, and order them. Kenkel [1] considered NMS a highly effective approach for evaluating data sets showing a low diversity.

The current investigations showed four separated groups of tree vegetation on ordination axes dominated by conifer species. These results might show acceptable correspondence between the ordination and cluster analysis results of the tree and understory vegetation studied. Group I was composed of three subgroups: Group I (a, b, and c). The groups were composed mainly of *Pinus wallichiana,* and an association with other cospecies was found on all three axes axis, 1–2, 1–3, and 2–3. These three subgroups were naturally growing on the average mean elevations of 1953, 2203, and 2171, which may provide sufficient evidence that this is the suitable elevation range and slope angle for them. Group II showed a high mean elevation of 2691 m, which is suitable for growing *Abies pindrow* and steep slopes. Group III, mainly composed of *Cedrus deodara,* was associated with *Pinus wallichiana, Picea smithiana, Abies pindrow,* and an angiospermic species, *Quercus baloot,* located on an average mean elevation (2188 m). Siddiqui, Ahmed, Shaukat, and Khan [17] also recorded *Cedrus deodara* on a moderate slope on all exposures equally and suggested that exposure is not a controlling factor for the existence of *Cedrus deodara.* Group IV was the smallest group, composed of two monospecific stands with a very low mean elevation of 1374 m. Similarly, four main groups of tree vegetation were derived for understory vegetation based on the nonmetric multidimensional scaling (NMS) ordination technique. These four groups were superimposed on two axes, plotted between axes 1 and 2, showing a continuous pattern among these axes.

In the present study, environmental variables with PCA ordination showed that elevation is significantly ($p < 0.05$) associated with axis 1 and highly significant ($p < 0.001$) with axis 3 of tree vegetation, and a highly significant ($p < 0.001$) correlation showed in the understory vegetation with the NMS ordination. Siddiqui, Ahmed, Shaukat, and Khan [17] also calculated a highly significant correlation on axis 1 with the DCA ordination of the tree and axis 3 (in the understory vegetation). Khan [21] reported a significant correlation on axis 1 with the DCA ordination of tree vegetation and axis 1 and 3 of the understory vegetation from the Chitral District. Wahab, Moinuddin, Nasrullah, and Sarangzai [20] reported a significant correlation of elevation in all three axes of tree vegetation and axis 1 of the understory vegetation from the Dir District when applying the NMS ordination. Akbar [42] reported a significant correlation of elevation for the understory vegetation in axis 1 with the DCA ordination. Their results agreed with these findings. Holdridge [52] and García [53] showed that elevation has a great extent of control in climatic conditions, particularly in temperature and precipitation. They also observed that an increase in altitude causes a decrease in temperature, water-holding capacity, soil fertility, and plant cover. Ver Hoef et al. [54] revealed a correlation with vegetation on ordination axes and suggested that the elevation and study area are the first two factors regulating the vegetation composition. In other environmental variables, the maximum water-holding capacity and soil moisture showed significant correlation on axis 3 in the tree vegetation data; similarly, in the soil texture, silt had a significant correlation in axes 1 and 2.

5. Conclusions

As a part of the moist temperate forests of Pakistan, the tree association groups were similar to the other findings in Pakistan. On the other hand, there is a considerable difference among the groups of the understory vegetation. Despite some disturbances, the classification and ordination showed a similar association of the tree species, which are the controlling agent of the environment of the Shangla forests. Moreover, in these natural mature forests, elevation and soil moisture play an essential role in determining the association. These forests provide a better environment, carbon storage, biodiversity, and other resources to the local population. These forests should be maintained and used sustainably. In the future, seedling growth, regeneration patterns, and health status should be carried out to protect and manage these forests.

Author Contributions: Conceptualization and methodology, J.I., N.S., M.A., W.Z. and A.K.; software, validation, and formal analysis, D.O.E.-A., H.S., H.O.E. and S.P.; investigation, resources, and data curation, J.I., N.S., M.A., W.Z. and A.K.; writing—original draft preparation, writing—review and editing, visualization, supervision, project administration, and funding acquisition, S J.I., N.S., M.A., W.Z., A.K., D.O.E.-A., H.S., H.O.E. and S.P.; visualization, methodology, and writing—review and editing, W.Z. and A.A. All authors have read and agreed to the published version of the manuscript.

Funding: This research was funded by King Saud University (RSP-2021/118).

Institutional Review Board Statement: Not applicable.

Informed Consent Statement: Not applicable.

Data Availability Statement: Not applicable.

Acknowledgments: The authors extend their deep appreciation to the Researchers Supporting Project number (RSP-2021/118), King Saud University, Riyadh, Saudi Arabia. The authors acknowledge Precision Agriculture Laboratory, Department of Pomology, Faculty of Agriculture (El-Shatby), Alexandria University.

Conflicts of Interest: The authors declare no conflict of interest.

References

1. Kenkel, N.C. The personal computer as a tool for data analysis in population and community ecology. *Abstr. Bot.* **1986**, *10*, 79–86.
2. Austin, M.P.; Belbin, L.; Meyers, J.A.; Doherty, M.D.; Luoto, M. Evaluation of statistical models used for predicting plant species distributions: Role of artificial data and theory. *Ecol. Model.* **2006**, *199*, 197–216. [CrossRef]
3. Shaukat, S.S.; Qadir, S.A. Multivariate analysis of the vegetation of calcarious hills around Karachi. *Vegetatio* **1971**, *23*, 235–253. [CrossRef]
4. Ahmed, M. Multivariate analysis of the vegetation around Skardu. *Agri. Pak.* **1976**, *26*, 177–187.
5. Moinuddin, A. Vegetation of some foothills of Himalayan range in Pakistan. *Pak. J. Bot.* **1986**, *18*, 261–269.
6. Enright, N.J.; Miller, B.P. Livestock grazing impacts on desert vegetation, Khirthar National Park, Pakistan. *Rangel. Ecol. Manag.* **2007**, *60*, 680–684. [CrossRef]
7. Manghwar, H.; Hussain, A. Mechanism of tobacco osmotin gene in plant responses to biotic and abiotic stress tolerance: A brief history. *Biocell* **2022**, *46*, 623. [CrossRef]
8. Manghwar, H.; Hussain, A.; Ali, Q.; Liu, F. Brassinosteroids (BRs) role in plant development and coping with different stresses. *Int. J. Mol. Sci.* **2022**, *23*, 1012. [CrossRef]
9. Manghwar, H.; Hussain, A.; Ali, Q.; Saleem, M.H.; Abualreesh, M.H.; Alatawi, A.; Ali, S.; Munis, M.F.H. Disease severity, resistance analysis, and expression profiling of pathogenesis-related protein genes after the inoculation of Fusarium equiseti in wheat. *Agronomy* **2021**, *11*, 2124. [CrossRef]
10. Manghwar, H.; Hussain, A.; Ullah, A.; Gul, S.; Shaban, M.; Khan, A.H.; Ali, M.; Sani, S.G.A.S.; Chaudhary, H.J.; Munis, M.F.H. Expression analysis of defense related genes in wheat and maize against Bipolaris sorokiniana. *Physiol. Mol. Plant Pathol.* **2018**, *103*, 36–46. [CrossRef]
11. Naz, H.S.; Gul, S.; Chaudhary, H.J.; Munis, M.F.H. First report of Rhizopus oryzae causing fruit rot of *Citrus medica* L. in Pakistan. *J. Plant Pathol.* **2015**, *97*, 215.
12. Shaukat, S.S.; Uddin, M. An application of canonical correlation and principal component analysis to the study of desert environment. *Abstr. Bot.* **1989**, *13*, 17–45.
13. Ahmed, M.; Tareen, R.; Tareen, K. Vegetation ordination of Chilton in Balochistan. *Mod. Trends Plant Sci. Res. Pak.* **1987**, 39–43.
14. Hussain, F.; Ahmed, M.; Shaheen, G.; Durrant, M.J. Phytosociology of the vanishing tropical deciduous forest in district Swabi, Pakistan. II. Ordination. *Pak. J. Bot.* **1994**, *26*, 149–160.
15. Ahmad, S.S.; Fazal, S.; Valeem, E.E.; Khan, Z.I.; Sarwar, G.; Iqbal, Z. Evaluation of ecological aspects of roadside vegetation around Havalian city using multivariate techniques. *Pak. J. Bot* **2009**, *41*, 53–60.
16. Ilyas, M.; Shinwari, Z.K.; Qureshi, R. Vegetation composition and threats to the montane temperate forest ecosystem of Qalagai hills, Swat, Khyber Pakhtunkhwa, Pakistan. *Pak. J. Bot.* **2012**, *44*, 113–122.
17. Siddiqui, M.F.; Ahmed, M.; Shaukat, S.S.; Khan, N. Advanced multivariate techniques to investigate vegetation-environmental complex of pine forests of moist temperate areas of Pakistan. *Pak. J. Bot* **2010**, *42*, 267–293.

18. Siddiqui, M.F.; Shaukat, S.S.; Ahmed, M.; Khan, N.; Khan, I.A. Vegetation-Environment relationship of conifer dominating forests of moist temperate belt of Himalayan and Hindukush regions of Pakistan. *Pak. J. Bot* **2013**, *45*, 577–592.
19. Rashid, A.; Swati, M.F.; Sher, H.; Al-Yemeni, M.N. Phytoecological evaluation with detail floristic appraisal of the vegetation arround Malam Jabba, Swat, Pakistan. *Asian Pac. J. Trop. Biomed.* **2011**, *1*, 461–467. [CrossRef]
20. Wahab, M.; Moinuddin, A.; Nasrullah, K.; Sarangzai, A.M. A phytosociological study of pine forests from district Dir, Pakistan. *Int. J. Biol. Biotechnol.* **2011**, *7*, 219–226.
21. Khan, N. Vegetation ecology and Dendrochronology of Chitral. Ph.D. Thesis, Federal Urdu University of Arts Sciences & Technology, Islamabad, Pakistan, 2011.
22. Huo, H.; Feng, Q.; Su, Y.H. The influences of canopy species and topographic variables on understory species diversity and composition in coniferous forests. *Sci. World J.* **2014**, *2014*, 252489. [CrossRef] [PubMed]
23. Ares, A.; Neill, A.R.; Puettmann, K.J. Understory abundance, species diversity and functional attribute response to thinning in coniferous stands. *For. Ecol. Manag.* **2010**, *260*, 1104–1113. [CrossRef]
24. Spanner, M.A.; Pierce, L.L.; Peterson, D.L.; Running, S.W. Remote sensing of temperate coniferous forest leaf area index The influence of canopy closure, understory vegetation and background reflectance. *Remote Sens.* **1990**, *11*, 95–111. [CrossRef]
25. Cottam, G.; Curtis, J.T. The use of distance measures in phytosociological sampling. *Ecology* **1956**, *37*, 451–460. [CrossRef]
26. Dahdouh-Guebas, F.; Koedam, N. Empirical estimate of the reliability of the use of the Point-Centred Quarter Method (PCQM): Solutions to ambiguous field situations and description of the PCQM+ protocol. *For. Ecol. Manag.* **2006**, *228*, 1–18. [CrossRef]
27. de Lucena, R.F.P.; de Medeiros, P.M.; de Lima Araújo, E.; Alves, A.G.C.; de Albuquerque, U.P. The ecological apparency hypothesis and the importance of useful plants in rural communities from Northeastern Brazil: An assessment based on use value. *J. Environ. Manag.* **2012**, *96*, 106–115. [CrossRef]
28. Eaton, S.V. Physical Properties of Soil. *Bot. Gaz.* **1931**, 380.
29. Dean, W.E. Determination of carbonate and organic matter in calcareous sediments and sedimentary rocks by loss on ignition; comparison with other methods. *J. Sediment. Res.* **1974**, *44*, 242–248.
30. Jackson, M.L. *Soil Chemical Analysis*; Constable and Co. Ltd.: London, UK, 1962.
31. McCune, B.; Grace, J.B.; Urban, D.L. *Analysis of Ecological Communities*; MjM software design: Gleneden Beach, OR, USA, 2002; Volume 28.
32. Minchin, P.R. An evaluation of the relative robustness of techniques for ecological ordination. In *Proceedings of the Theory and Models in Vegetation Science: Proceedings of Symposium, Uppsala, Sweden, July 8–13, 1985*; Prentice, I.C., van der Maarel, E., Eds.; Springer Netherlands: Dordrecht, Holand, 1987; pp. 89–107.
33. Tansley, A.G.; Chipp, T.F. *Aims and Methods in the Study of Vegetation*; White friars Press: London, UK, 1926.
34. Khan, M.; Hussain, F. Classification and ordination of vegetation in Tehsil Takht-e-Nasrati, District Karak, Khyber Pakhtunkhawa, Pakistan. *J. Ecol. Nat. Environ.* **2013**, *5*, 30–39. [CrossRef]
35. Greig-Smith, P. *Quantitative Plant Ecology*, 3rd ed.; Black-Well Scientific Publications: London, UK, 1983.
36. Frades, I.; Matthiesen, R. Overview on techniques in cluster analysis. *Bioinform. Methods Clin. Res.* **2010**, *593*, 81–107.
37. Lovett, J.C.; Clarke, G.P.; Moore, R.; Morrey, G.H. Elevational distribution of restricted range forest tree taxa in eastern Tanzania. *Biodivers. Conserv.* **2001**, *10*, 541–550. [CrossRef]
38. Gajoti, T.E.; Haciyev, V.; Javanshir, A.; Nosrati, H.; Haghighi, A.R.; Eimanifar, A.; Stewart, N., Jr. Vegetation analysis of Sutan-Chay Basin in Arasbaran. *Am. J. Agric. Biol. Sci.* **2010**, *5*, 357–362. [CrossRef]
39. Zhang, Z.L.; He, Y.Q.; Pang, H.X. Variation of glaciers in response to ENSO in the Mount Yulong. *J. Glaciol Geogryology* **2004**, *26*, 294–297.
40. Saima, S.; Dasti, A.A.; Hussain, F.; Wazir, S.M.; Malik, S.A. Floristic compositions along an 18-km long transect in Ayubia National Park district Abbottabad, Pakistan. *Pak. J. Bot.* **2009**, *41*, 2115–2127.
41. Ahmad, S.S.; Abdul, W.; Akbar, K.F. Multivariate classification and data analysis of vegetation along motorway (M-2), Pakistan. *Pak. J. Bot.* **2010**, *42*, 1173–1185.
42. Akbar, M. Forest Vegetation and Dendrochronology of Gilgit, Astore and Skardu Districts of Northern Areas (Gilgit-Baltistan), Pakistan. Ph.D. Thesis, Federal Urdu University of Arts, Science & Technology, Islamabad, Pakistan, 2013.
43. Wahab, M.; Ahmed, M.; Khan, N. Phytosociology and dynamics of some pine forests of Afghanistan. *Pak. J. Bot.* **2008**, *40*, 1071–1079.
44. Malik, N.Z.; Malik, Z.H. Present status of subtropical chir pine vegetation of Kotli Hills, Azad Jammu and Kashmir. *J. Res. Sci.* **2004**, *15*, 85–90.
45. Ali, S.M.; Malik, R.N. Spatial patterns of vegetation with underlying soil properties prevailing along drain side areas in Islamabad City. *Pak. J. Bot.* **2010**, *42*, 2397–2410.
46. Bai, X.; Sadia, S.; Yu, J. Community composition and structure along the environmental gradients of *Larix gmelinii* forest in northeast China. *Pak. J. Bot.* **2021**, *53*, 1845–1850. [CrossRef]
47. Ullah, F.; Gao, Y.; Sari, İ.; Jiao, R.; Saqib, S.; Gao, X. Macro-Morphological and Ecological Variation in Rosa sericea Complex. *Agronomy* **2022**, *12*, 1078. [CrossRef]
48. Tabaripour, R.; Keshavarzi, M.; Ullah, F. Micromorphological characters variation of lemma and palea in subtribe of Loliinae (Poaceae). *Microsc. Res. Tech.* **2022**, *85*, 117–134. [CrossRef] [PubMed]

49. Ullah, F.; Ayaz, A.; Saqib, S.; Parmar, G.; Bahadur, S.; Zaman, W. Taxonomic implication of leaf epidermal anatomy of selected taxa of Scrophulariaceae from Pakistan. *Microsc. Res. Tech.* **2021**, *84*, 521–530. [CrossRef] [PubMed]
50. Ullah, F.; Gao, Y.; Jiao, R.-F.; Gao, X.-F. Comparative taxonomic variation in fruits and seeds' surface morphology among populations of alpine Rosa sericea complex (Rosaceae). *Microsc. Res. Tech.* **2021**, *84*, 2337–2350. [CrossRef]
51. McCune, B.; Mefford, M.J. *PC-ORD Multivariate Analysis of Ecological Data, Version 5.31;* MjM Solfware Gleneden Beach: Gleneden Beach, OR, USA, 2006.
52. Holdridge, L.R. *Ecología Basada En Zonas De Vida;* Instituto Interamericano de Cooperacion´ para la Agricultura: San Jose, Costa Rica, 1978.
53. García, E. Climates of the State of Veracruz according to the climatic classification of Koppen modified by the author. *An. Del Inst. De Biologia. Ser. Bot. Mex.* **1970**, *41*, 3–41.
54. Ver Hoef, J.M.; Neiland, B.J.; Glenn-Lewin, D.C. Vegetation gradient analysis of two sites in southeast Alaska. *Northwest Sci.* **1988**, *62*, 171–180.

agronomy

Article

A Dated Phylogeny of the Pantropical Genus *Dalbergia* L.f. (Leguminosae: Papilionoideae) and Its Implications for Historical Biogeography

Fabien Robert Rahaingoson [1,2], Oyetola Oyebanji [1,2], Gregory W. Stull [1], Rong Zhang [1,*] and Ting-Shuang Yi [1,*]

[1] Germplasm Bank of Wild Species, Kunming Institute of Botany, Chinese Academy of Sciences, Kunming 650201, China; fabienrobertrahaingo@mail.kib.ac.cn (F.R.R.); oyetola@mail.kib.ac.cn (O.O.); gwstull@gmail.com (G.W.S.)

[2] University of Chinese Academy of Sciences, Beijing 100049, China

* Correspondence: zhangronga@mail.kib.ac.cn (R.Z.); tingshuangyi@mail.kib.ac.cn (T.-S.Y.)

Abstract: The genus *Dalbergia* has a pantropical distribution and comprises approximately 250 species. Previous phylogenetic studies on the genus revealed that *Dalbergia* is monophyletic and is sister to *Machaerium* and *Aeschynomene* sect. *Ochopodium*. However, due to limited samples or DNA regions in these studies, relationships among the major clades are still unresolved, and divergence dates and biogeographical history of the genus have not been addressed. In this study, phylogenetic analyses of *Dalbergia* were conducted using broad taxon sampling and a combined dataset of two plastid DNA markers (*mat*K and *rbc*L) and one nuclear marker (ITS). We evaluated the infrageneric classification of the genus based on the reconstructed tree, and investigated biogeographical history of this genus through molecular dating and ancestral area reconstruction analyses. The monophyly of *Dalbergia* was strongly supported and the genus was resolved into five major clades with high support, several of which correspond to the previous recognized sections. We inferred that *Dalbergia* originated in South America during the Early Miocene (*c.* 22.9 Ma) and achieved its current pantropical distribution through multiple recent transoceanic long-distance dispersals (LDD). We highlighted the important historical events which may explain the pantropical distribution pattern of *Dalbergia*.

Keywords: *Dalbergia*; Early Miocene; ITS; long-distance dispersal; *mat*K; monophyletic; *rbc*L

Citation: Rahaingoson, F.R.; Oyebanji, O.; Stull, G.W.; Zhang, R.; Yi, T.-S. A Dated Phylogeny of the Pantropical Genus *Dalbergia* L.f. (Leguminosae: Papilionoideae) and Its Implications for Historical Biogeography. *Agronomy* **2022**, *12*, 1612. https://doi.org/10.3390/agronomy12071612

Academic Editors: Alessio Papini, Mushtaq Ahmad, Fazal Ullah and Wajid Zaman

Received: 16 June 2022
Accepted: 1 July 2022
Published: 4 July 2022

1. Introduction

The pantropical genus *Dalbergia* L.f. includes *c.* 250 species with centers of diversity in Central and South America, Africa, Madagascar, and Asia [1]. *Dalbergia* species grow in diverse habitats including tropical rain forests, dry forests, savannas, costal dunes, and rocky outcrops [2–4]. The *Dalbergia* species display a high diversity of life forms, including trees, shrubs, and woody lianas. The genus is economically important for its high-quality timber known as rosewood (e.g., Brazilian species *D. nigra* (Vell.) Allemão ex Benth., Chinese species *D. odorifera* T.C.Chen, Madagasar species *D. baronii* Baker, and Thailand species *D. cochinchinensis* Pierre), and blackwood (e.g., *D. melanoxylon* Guill. & Perr.), which are used for construction works, fine furniture and musical instruments [5]. Many *Dalbergia* species have also been used in traditional medicine and phytochemical studies [6]. Overexploitation, illegal logging, and habitat fragmentation have severely decreased the population sizes of many *Dalbergia* species.

Studies on the intrageneric classification of *Dalbergia* are limited. Only Bentham (1860) [7] carried out large-scale taxonomic study of the genus, and divided the 64 *Dalbergia* species known at that time into six series. Later, this work was followed by subsequent regional monographs based exclusively on morphological characteristics. In the Neotropics, the Brazilian species of *Dalbergia* were studied by Carvalho (1997) [2]. He divided the 41 Brazilian *Dalbergia* species into five sections based on inflorescence and fruit types.

Prain (1904) [8] classified the 86 South East Asian *Dalbergia* species into two subgenera, five sections and 24 series. Thothathri (1987) [9] categorized the 46 *Dalbergia* species present in the Indian subcontinent into four sections and seven series, based on androecium and fruit types.

Only a few molecular phylogenetic studies have been carried out to address the phylogenetic position of *Dalbergia* and its intrageneric relationships. Phylogenetic analysis of the tribe Dalbergieae, based on molecular and morphological data, placed *Dalbergia* within the Dalbergioid clade [10]. Using plastid *trn*L and *mat*K, and nuclear ribosomal ITS sequences data, *Dalbergia* were observed to be monophyletic and sister to a clade comprising the genus *Machaerium* Pers. and *Aeschynomene* sect. *Ochopodium* [11,12]. The above-mentioned studies have highlighted the relationships between *Dalbergia* and its relatives and included a limited sampling of *Dalbergia* species, leaving the interspecific relationships within the genus largely unknown. The first comprehensive phylogenetic study of the genus *Dalbergia* was conducted by Vatanparast et al. (2013) [13] using ITS sequences and 64 species representing most of the recognized sections: i.e., sects. *Dalbergia* L.f., *Triptolemea* (Mart. ex Benth.) Benth, *Selenolobium* Benth, *Ecastaphyllum* (P. Browne) Ducke (Carvalho 1997), and *Dalbergaria* Prain. Their study confirmed the monophyly of a clade comprising *Dalbergia*, *Machaerium* and *Aeschynomene* sect. *Ochopodium*, but the relationships among them were unresolved. In addition, *Dalbergia* was resolved into five major clades, but many relationships among them and among species were poorly resolved. The previously recognized sects. *Triptolemea*, *Ecastaphyllum* and *Dalbergaria* were shown to be monophyletic, whereas sects. *Dalbergia* and *Selenolobium* were non-monophyletic. Therefore, more samples and molecular markers are needed to resolve the phylogenetic relationships among *Dalbergia*.

To date, there has been no formal biogeographical study of *Dalbergia* based on its phylogenetic scheme. Vatanparast et al. (2013) [13] hypothesized that *Dalbergia* originated in the New World and suggested that multiple migrations through LDD across oceans might account for its pantropical distribution. However, they did not conduct formal molecular dating or biogeographical analyses, so the origin date, original center, and the evolutionary history leading to the current disjunct distribution pattern of *Dalbergia* are still largely unknown. Therefore, formal dating and biogeological analyses should be conducted to infer the origin and account for the disjunct distribution pattern of the genus.

In this study, we used two plastid markers (*mat*K and *rbc*L) and one nuclear marker (ITS) representing 93 species of *Dalbergia* across its major lineages and entire geographical range, to reconstruct the phylogenetic relationships, estimate divergence times using multiple fossil calibrations and explore the original center and subsequent dispersal history of the genus *Dalbergia*.

2. Materials and Methods

2.1. *Taxon Sampling and Sequence Alignment*

Sequences of two chloroplast markers (*mat*K and *rbc*L) and one nuclear marker (ITS) representing 93 *Dalbergia* species and six outgroup species were downloaded from GenBank. Our samples represent the major clades and cover the entire geographic range distribution of *Dalbergia*. The outgroups include four species from the genus *Aeschynomene* L., one from *Machaerium* Pers., and one from *Pterocarpus* Jacq. If there was more than one sequence for the same marker of the species, we kept the longest. Details of samples used in this study are given in Table S1.

The *mat*K, *rbc*L and ITS loci were individually aligned using MAFFT v.7.4.0 [14] as implemented in Geneious v.8.1.9 [15] under default parameters. Alignments of the three loci were further individually inspected and manually adjusted in Geneious.

2.2. *Phylogenetic Reconstruction of Dalbergia*

In order to evaluate the conflicts among datasets, Incongruence Length Difference (ILD) testing [16] was performed in PAUP * v.4.0b.10 [17] between plastid (*rbc*L *and mat*K) and nuclear regions (ITS), using a heuristic search with 1000 replicates, random taxa-addition, tree bisection and reconnection (TBR) branch-swapping saving 15 trees per replicate.

Prior to conducting the phylogenetic analyses, nucleotide substitution models were selected for each DNA region using the Akaike Information Criterion (AIC) implemented in jModelTest v.2.1.6 [18] (Table S2).

Phylogenetic analyses of the concatenated dataset were performed using Maximum Likelihood (ML) and Bayesian inference (BI). The ML analysis was performed using the IQ-Tree Web server [19]. We specified the best-fit model (GTR + I + G) selected by jModelTest and used the ultrafast bootstrap algorithm (UFBoot2) with 1000 replicates to assess branch support (-bb 1000), combined with a search of the best-scoring ML tree under default parameters. BI analyses were conducted using MrBayes v.3.2.6 [20]. We linked and specified GTR + I + G as the best-fit model according to the optimal scheme selected by jModelTest. Two independent Bayesian runs with four chains of Markov Chain Monte Carlo (MCMC) were run for 25 million generations, sampling every 10,000 generations. Chain convergence was checked in Tracer v.1.6 [21] by examining log likelihood plots and ensuring that Effective Sample Size (ESS) values were well above 200. After discarding 25% of the trees as burn-in, a majority rule consensus tree was constructed using TreeAnnotator v.2.3.2 [22]. Outputs of all phylogenetic analysis were read using FigTree v.1.4.2 [23] and nodes with ultrafast bootstrap (BS) > 95% [24] and posterior probability (PP) $\geq$ 0.90 [20] were considered well supported. We also conducted phylogenetic analyses of ML and BI for each DNA region using the best-fit model selected by jModelTest (Table S2). Other parameters were set similarly.

2.3. Divergence Time Estimation

Divergence time estimations were generated on the BI tree of the concatenated dataset using BEAST v.1.8.4 [22], which was modeled under a Yule process using a random starting tree and an uncorrelated relaxed clock. Within BEAST, a Birth-Death model was employed for tree priors, the GTR + G + I evolution model with four gamma categories was applied based on AIC results from jModelTest, and other parameters were set as default values. Four calibration points were used, with the root node calibrated at the maximum age of 96.33 Ma (Legumes stem) based on the meta-calibration study of flowering plants by Magallón et al. (2015) [25], as shown in Table 1. The prior distributions of the root node and the other calibration points were set as log normal with a standard deviation of 1.0.

Table 1. Fossils used as calibration points to generate a time-calibrated phylogenetic tree of *Dalbergia*. All ages in millions of years (Ma) and set to a minimum, except for Label A which was set to a maximum.

Label	Node Constrained (MRCA)	Species	Morphology	Age	References
A	Legume stem	*Polygala californica–Cercis occidentalis*		96.33	Magallon et al. (2015)
B	*Styphnolobium* stem	*Styphnolobium japonicum–Pickeringia montana*	Leaf and fruit	40	Lavin et al. (2005)
C	*Tipuana* stem	*Tipuana tipu–Pterocarpus indicus*	Fruit	10	Lavin et al. (2005)
D	*Dalbergia* stem	*Dalbergia hupeana–Machaerium lunatum*	Leaf	40	Lavin et al. (2005)

The BEAST file was generated in BEAUti v.1.8.4 [22]. Then, we conducted two runs of four Markov chains for 40 million generations with sampling every 1000 generations. The output files were examined in Tracer to evaluate convergence of the runs and the ESS ($\geq$200) for all parameters. The runs were combined using LogCombiner v.1.8.2 [22]. Following the removal of the first 20% samples as burn-in, the sampled posterior trees were summarized using TreeAnnotator v.1.8.4 [22] to generate a maximum clade credibility (MCC) tree and calculate the mean ages, 95% highest posterior density intervals (95% HPD), and PP. The chronogram was visualized and annotated using FigTree.

2.4. Ancestral Area Estimations

Ancestral area estimations were conducted with the maximum likelihood framework using the package BioGeoBEARS v.1.1.2 [26] in RASP v.4.2 [27]. The analysis was carried out on the chronogram tree from the BEAST analysis without the outgroups. Three models were tested, including Dispersal–Extinction–Cladogenesis (DEC; [28]), Dispersal Vicariance Analysis like (DIVALIKE; [29]), and Bayesian Analysis like (BAYAREALIKE; [30]). Given the ongoing debate around the use of founder-event speciation + j, we did not implement this parameter in our reconstructions [31]. Using the model selection function, the best-fit model was selected by comparing the AIC criterion among all models (Table S3). The maximum areas parameter was set according to the maximum number of areas occupied by any extant species in the dataset. Consequently, a maximum of three areas was chosen.

Five biogeographic areas were defined, based on the respective distribution of the species: A—Australasia; B—Africa; C—Madagascar; D—Central America (including Florida and Caribbean); E—South America. We compiled the information about species distribution from the literature [2,32–35], herbarium specimens and online databases (www.gbif.org, accessed on 2 May 2021; www.plantsoftheworldonline.org, accessed on 3 March 2021).

3. Results

3.1. Phylogenetic Analyses

The ILD test showed no significant incongruence between nuclear and plastid datasets ($p = 0.37$). Alignment length, numbers of variable sites, and parsimony informative sites for each marker and the concatenated dataset are provided in Table 2.

Table 2. Features of the DNA data sets used in this study (bp = base pairs).

DNA Region	Alignment Length (bp)	Number of Variable Sites		Number of Potentially Informative Sites	
*rbc*L	695	17	(2.44%)	24	(3.45%)
*mat*K	1070	74	(6.91%)	119	(11.12%)
ITS	1041	105	(10.08%)	338	(32.46%)
Concatenated dataset	2806	196	(6.98%)	481	(17.14%)

Topology of the BI tree was largely congruent with the ML tree described (Figures 1 and S1). Thus, we mapped support values of the BI tree onto the ML tree (Figure 1). The genus *Dalbergia* was strongly supported as monophyletic (BS = 100, PP = 1.00) and five major clades were recovered, labelled as clades I–V (Figure 1). All clades were strongly supported by all our analyses, except clade IV which was moderately supported in the ML analysis with BS = 89 (Figure 1). Clade I comprising two South American species (*D. miscolobium* Benth. and *D. spruceana* Benth.) was resolved as sister to all other *Dalbergia* species with strong support (BS = 98, PP = 1.00). Clade II comprising eight South American species was resolved as the second divergent clade with strong support (BS = 99, PP = 1.00). Clades I and II include species from sects. *Dalbergia* and *Selenolobium*, and all species of these two clades are mainly from the Neotropical regions. Clade III includes three well-supported subclades (III-a, III-b and III-c). Subclade III-a, comprising three Afro-American species (seven accessions) with the African *D. adamii* Berhaut being sister to the Neotropical sect. *Ecastaphyllum*, was resolved as sister to a clade comprising subclade III-b and subclade III-c. All species in subclades III-b and III-c are from the Australasian region (Figure 1). Clade IV is composed of two highly supported subclades (IV-a and IV-b). Within subclade IV-a, an African–Malagasy clade (M1) was recovered as sister to an entire Australasian clade. Within the subclade IV-b, an African–Malagasy clade (M2) was supported as sister to sect. *Dalbergaria* from the Australasian region. Similarly, Clade V was strongly supported (BS = 98, PP = 1.00) and split into four subclades (V-a, V-b, V-c, and V-d). Within subclade V-a, *D. nigra* (Vell.) Allemao ex Benth. from South America were resolved as sister to a clade comprising two species from Australasia (*D. sandakanensis* Surnamo & H.Ohashi and *D. bintuluensis* Surnamo & H.Ohashi) and two from Africa (*D. armata* E.Mey. and *D. hostilis* Benth.). Subclade V-b comprised two species from Australasia and two from Africa, and was supported as sister to the remaining species in Clade V with low support (BS = 67, PP = 0.57). All species in Subclade V-c are from Australasia, but monophyly of Subclade V-c was weakly supported. Finally, within subclade V-d, *D. bracteolata* Baker from Africa and Madagascar was supported as sister to the remaining species of this subclade, only in the ML tree (BS = 95). In addition, subclade V-d includes the monophyletic sect. *Triptolemea* from South America, the African-Malagasy clade (M3) and some dispersive lineages from Central America, Africa and Australasia.

3.2. Divergence Times and Biogeographical Analyses

The stem age of *Dalbergia* was inferred to be *c.* 40.7 Ma (95% HPD: 41.2–40.2 Ma) and the crown age to be *c.* 22.9 Ma (95% HPD: 25.9–19.9 Ma; nodes 1 and 2 respectively, Figure 2). Diversification of the main clades occurred from the Early to Late Miocene (nodes 3, 4, 5, 7 and 9, Figure 2)—i.e., from *c.* 19.0 Ma (95% HPD: 21.7–16.3 Ma) to *c.* 6.7 Ma (95% HPD: 9.5–4.5 Ma). Clades I and II (species from sects *Dalbergia* and *Selenolobium*) diverged during the Late Miocene *c.* 6.7 Ma (95% HPD: 9.5–4.5 Ma; node 3, Figure 2) and the Middle Miocene *c.* 15.9 Ma (95% HPD: 19.4–12.6 Ma; node 4; Figure 2), respectively. The divergence of sect. *Ecastaphyllum* was dated in the Pleistocene around 2.2 Ma (95% HPD: 3.4–1.2 Ma; node 6, Figure 2). Sect. *Dalbergaria* diverged in the Middle Miocene around 14.2 Ma (95% HPD: 16.6–12.1 Ma; node 8, Figure 2) and sect. *Triptolemea* was dated to the Pliocene *c.* 3.3 Ma (95% HPD: 4.7–2.1 Ma; node 10, Figure 2).

Ancestral range estimation recovered the DEC model as the best-fit model (lnL = −150.4 and AICwt = 1; Table S3). This result showed that long distance dispersal may have played an important role in the biogeographical history of *Dalbergia*, as shown by the values obtained for the two parameters of the analysis (d = 0.0063 and e = 0.0028, Table S3).

The ancestral area reconstruction showed that South America was the ancestral area of *Dalbergia* (node 1, Figure 3; Table 3) with high probability (p = 100%). Subsequently, it expanded into other regions through dispersal (Figures 3 and 4). South America was also inferred as the ancestral range for both Clades I and II (node 2 and 3, Figure 3; Table 3) with p = 100% for the two nodes. The DEC model inferred an Australasia–South America origin for Clade III (node 4, Figure 3; Table 3). Clades IV and V were estimated to be from Australasia (nodes 9 and 13, Figure 3; Table 3) with moderate probability p = 75% and p = 63%, respectively.

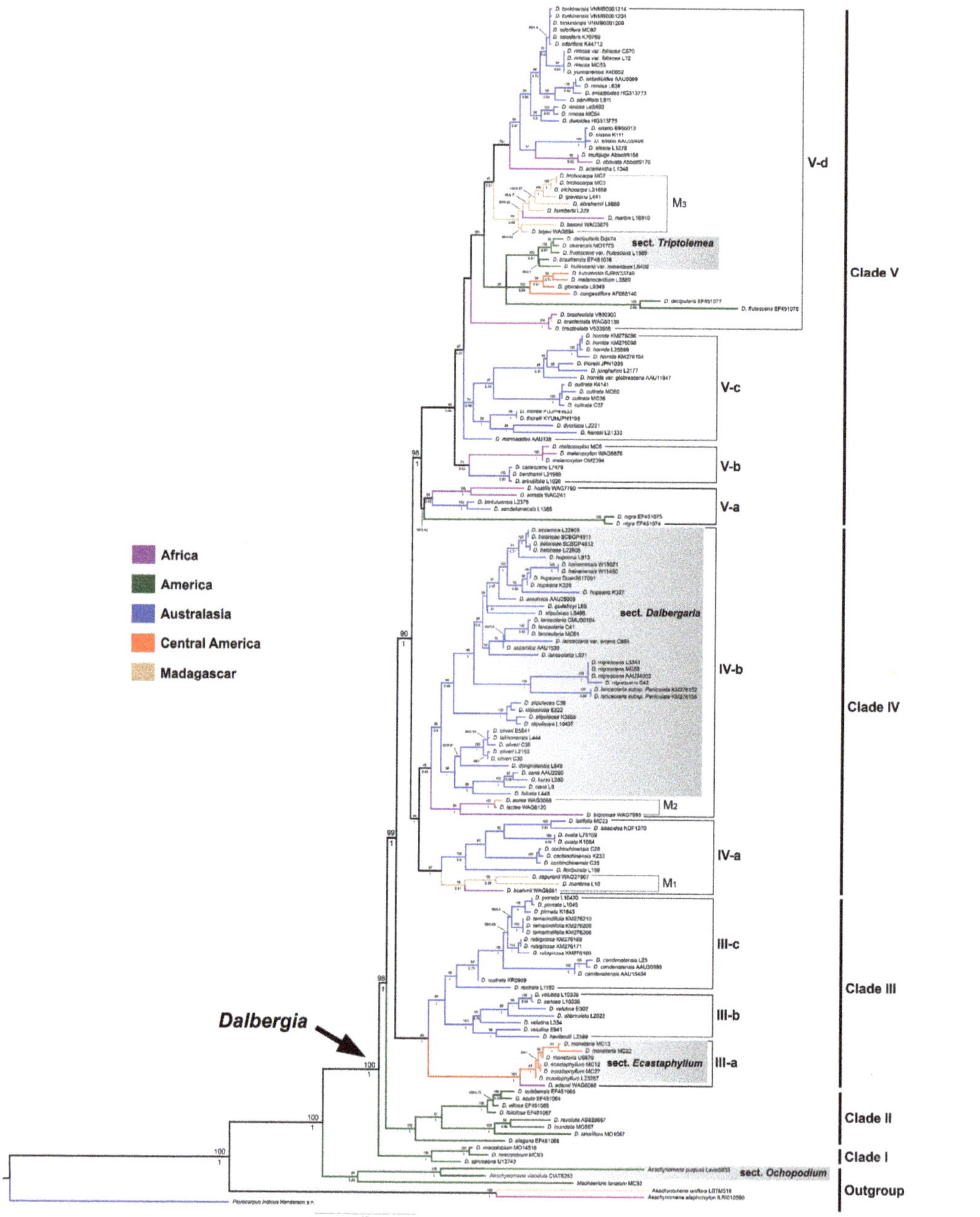

Figure 1. Majority rule consensus tree inferred from the ML analysis based on the concatenated data matrix (*mat*K, *rbc*L and ITS) showing the relationships among 93 species of *Dalbergia* and 6 outgroups. M1, M2, and M3 represents the Madagascar clades. Numbers along branches are ML bootstrap values (up) and BI posterior probabilities (down), respectively. A dash means the topology is not supported by the BI tree. Values <50% and <0.5 are not shown.

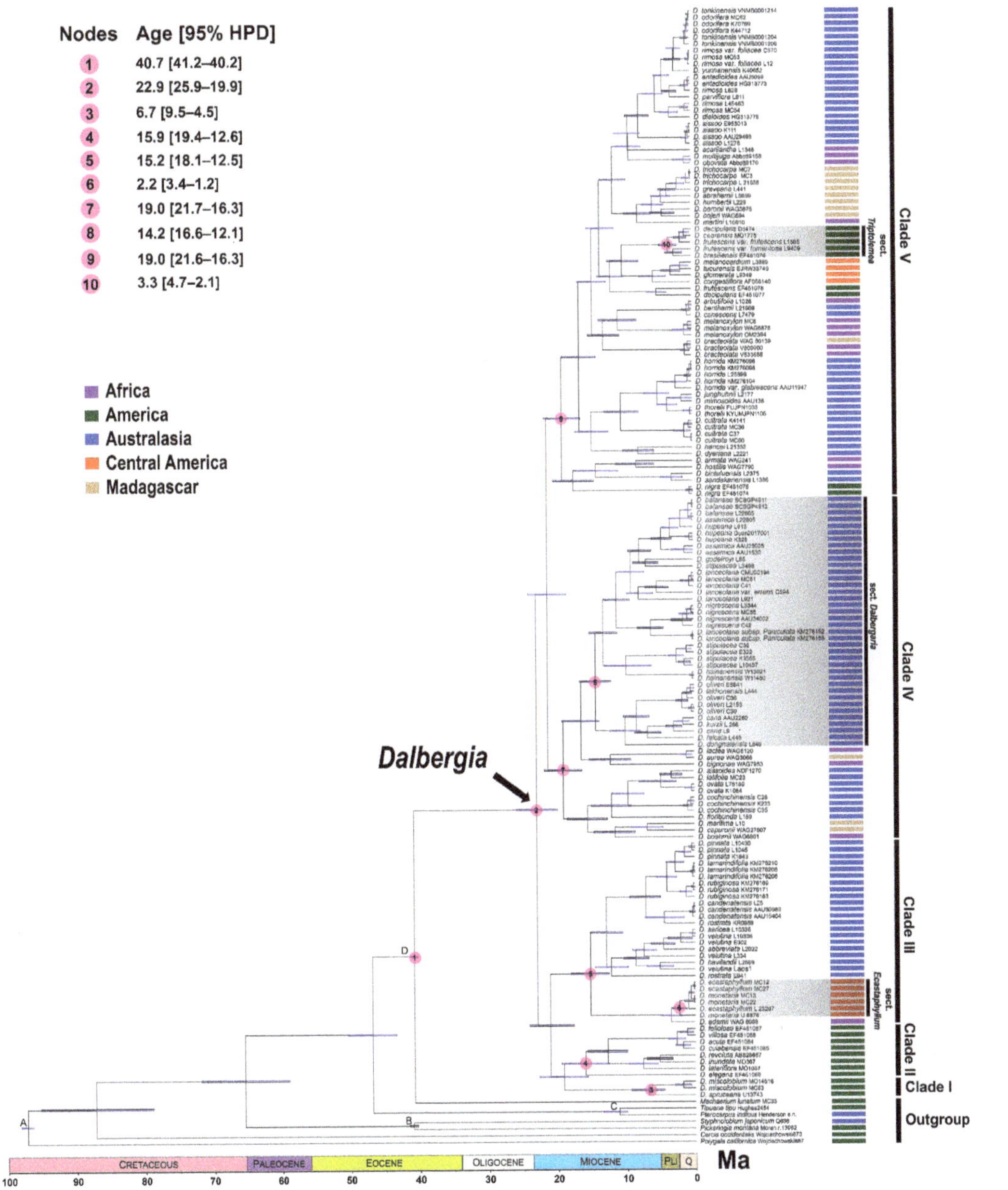

Figure 2. Chronogram of *Dalbergia* inferred from the BI tree using BEAST. The shaded blue horizontal bars show 95% HPD for the divergence times. Labels A–D correspond to the calibration points used (see Table 1 for details). Nodes of interests are marked as 1–10.

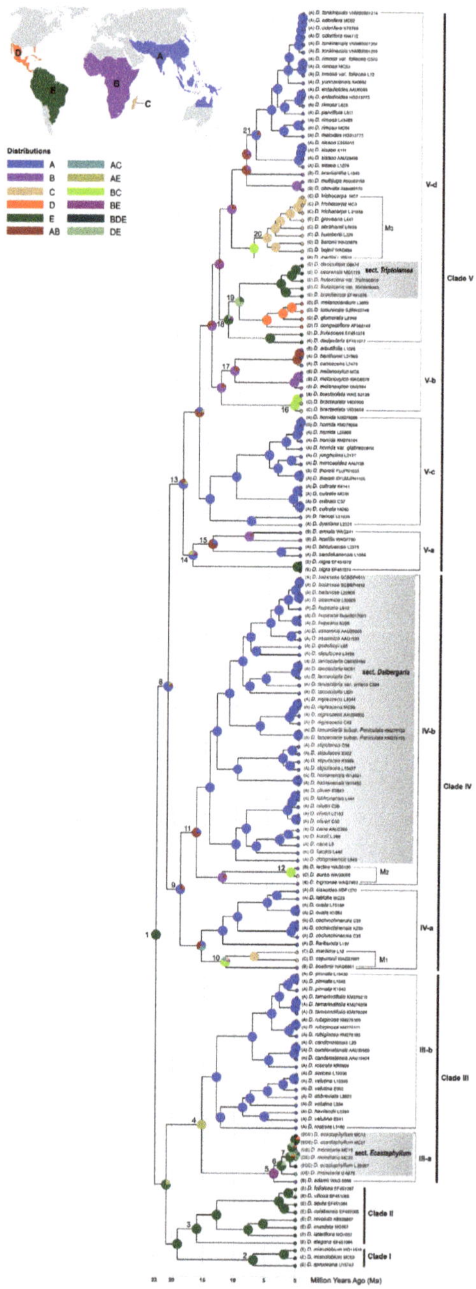

Figure 3. Ancestral area reconstructions under the dispersal–extinction cladogenesis (DEC) model in the BioGeoBEARS package as implemented in RASP; the relative probabilities of alternative ancestral areas are shown by pie charts at each node. Nodes of interests are marked 1–21. Area abbreviations are as follows: A—Australasia; B—Africa; C—Madagascar; D—Central America (including Florida and the Caribbean); E—South America. M1, M2, and M3 represent the Madagascar clades.

Table 3. Results of divergence time and ancestral area estimation for some major nodes in *Dalbergia*. The 95% highest posterior density (HPD) values of the divergence time estimations are provided in square brackets. The estimated ancestral areas of each node as inferred by BioGeoBEARS using the DEC model are shown, with rounded probabilities given as percentages. When only one area is shown, the probability is 100%. Areas are abbreviated as follows: A—Australasia; B—Africa; C—Madagascar; D—Central America (including Florida and the Caribbean); E—South America. Node numbers refer to Figure 3.

Node	Estimated Divergence Time in Ma with [95% HPD]	Estimated Ancestral Area (DEC)
1	22.9 [25.9–19.9]	E
2	6.7 [9.5–4.5]	E
3	15.9 [19.4–12.6]	E
4	15.2 [18.1–12.5]	AE
5	3.4 [5.1–2.0]	B
6	2.2 [3.4–1.2]	E 78; DE 22
7	0.9 [1.7–0.3]	E 68; DE 21; D 11
8	21.1 [24.0–18.5]	A 77; AE 12; AB 11
9	18.9 [21.7–16.3]	A 75; AB 25
10	11.5 [14.6–8.6]	BC 55; AC 20; C 17; B 8
11	16.3 [19.2–13.8]	AB 79; A 21.62
12	1.0 [2.1–0.3]	BC
13	18.9 [21.6–16.3]	A 63; AB 20; AE 17
14	17.3 [20.3–14.4]	A 52; AE 26; B 11; AB 11
15	14.1 [17.3–10.9]	AB 76; A 24
16	1.0 [1.7–0.4]	BC
17	10.7 [13.6–8.0]	B 75 AB 25
18	11.7 [13.8–9.0]	E 72; DE 15; BE 14
19	10.0 [12.3–7.7]	DE 74; E 26
20	5.6 [7.3–4]	C
21	7.5 [9.4–5.9]	A 84; AB 16

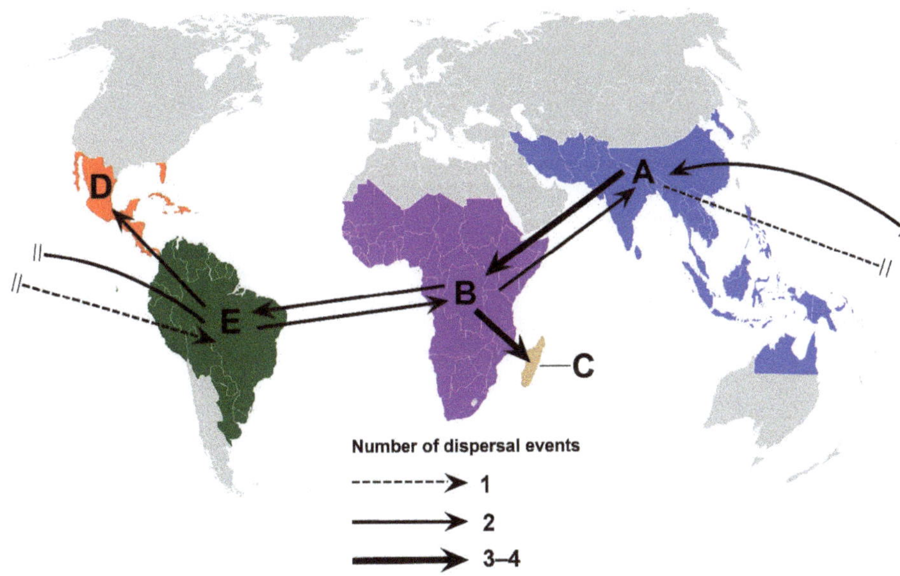

Figure 4. Major dispersal/migration routes suggested by DEC analyses of *Dalbergia*. Arrows indicate the dispersal direction; line thickness is proportional to the number of events. Labels A, B, C, D and E represent Australasia, Africa, Madagascar, Central America (including Florida and the Caribbean) and South America, respectively.

4. Discussion

4.1. Phylogenetic Relationships and Taxonomic Implications

Our study confirmed the monophyly of *Dalbergia* and resolved the genus to be sister to a clade comprising *Aeschynomene* sect. *Ochopodium* and *Machaerium*, as suggested in previous studies which focused on the relationships between *Dalbergia* and relatives [11,12]. Our analysis resolved the genera into five major clades, which is largely congruent with previously published phylogeny using nuclear DNA region ITS of 64 *Dalbergia* species [13]. However, our results strongly supported each clade and their relationships, except for clade IV in the ML tree (BS = 89, Figure 1), whereas these supports were weak in previous studies [13]. In addition, resolution and support for the vast majority of the nodes in our study have been improved compared to previous studies, e.g., Vatanparast et al. (2013) [13] and Ribeiro et al. (2007) [12]. This is attributed to our increased selection of samples and markers compared to previous studies, which were based on limited samples [11,12] or only one marker [13].

Within the five previously defined sections, four are from the Neotropics and only one from Australasia. Among those of the Neotropics, sect. *Ecastaphyllum* characterized by racemose or paniculate inflorescences and orbicular to suborbicular or reniform fruits, and sect. *Triptolemea* with cymose inflorescences and thin samaroid pods, were seen in our analyses to be monophyletic, as previously suggested in studies using both morphological and molecular data [12,13,36]. However, sect. *Dalbergia,* characterized by paniculate inflorescences and samaroid pods, and sect. *Selonolobium,* with racemose or paniculate inflorescences and crescent or kidney-shaped fruits, are non-monophyletic. Species of these two sections resolved into two diverged clades (Clade I and Clade II; Figures 1 and S1), each clade comprising species from each of these two sections. One exception is *D. nigra* from sect. *Dalbergia,* which has distinct phylogenetic position in Clade V according to our analyses. This species also has very distinct traits including a calyx with a glabrous tube and pilose teeth, obovate standard petals, and dark brown glossy fruits lacking prominent venation, as reported by Carvalho (1989) [36]. Among the species from Australasia, our results strongly supported sect. *Dalbergaria* to be monophyletic, whereas this support value was low in a previous study [13]. This section has Southeast Asian distribution and is characterized by reflexed standard petals and stamens that are usually in two bundles of five. In our study, many species were not included within any of the five above-mentioned sections. Thus, a through revision of the intrageneric *Dalbergia* classification integrating morphological traits and denser sampling phylogeny is urgently needed.

4.2. Origin and Biogeographical Diversification of Dalbergia

South America was inferred in our study to be the origin center of *Dalbergia* (Figure 3; Table 3), which is slightly different from the hypothesis proposed by Vatanparast et al. (2013) [13], who suggested a Neotropical (Central and South America) origin for this genus. This is probably due to differences in sampling and the definition of biogeographic regions. Despite these differences, South America was identified as an important region for the origin of *Dalbergia*. Our estimated stem age of *Dalbergia* (*c.* 40.7 Ma) is largely consistent with the *c.* 40.4–43.0 Ma estimated by Lavin et al. (2005) [37] which was based on the whole legumes family. However, our estimated crown age (*c.* 22.9 Ma; Early Miocene; see node 2, Figure 2) for *Dalbergia* is much older than *c.* 14.7 Ma (Middle Miocene) inferred by Hung et al. (2020) [38], who only sampled six representative species from the most basal Clade I and other clades of *Dalbergia*. Additionally, the use of different fossil calibrations probably contributed to the difference in the estimation of crown age between the two studies. Multiple fossil species have been reported; *Dalbergia phleboptera* is the earliest detected *Dalbergia* fossil species dated to 23.0–27.8 Ma (Late Oligocene) from France [39], then 15.97–23.03 Ma (Early Miocene) fossil species *D. nostratum* from Slovakia [40], and the later Miocene fossil (5.33–11.61 Ma) from China [41]. However, these fossils could not be confidently placed within the genus because of their limited morphological traits, thus they were not used time dating in our study.

Vicariance has been widely applied to explain the intercontinental distributions of biological taxa across broad oceans [42], but a series of molecular phylogenetic and biogeographical studies suggested that LDD should account for most of these disjunctions, mainly because the estimated divergence times are much younger than the continental breakup times inferred from geological data [43]. Thus, the finding that *Dalbergia* arose during the Early Miocene allows us to dismiss the hypotheses based on vicariance, and to take into account LDD as the most possible explanation for the pantropical pattern disjunct distribution of this genus. In addition, some previous legumes studies showed that LDD played an important role in the biogeography of several legume taxa, including *Apios* [44], *Canavalia* [45], *Zornia* [46], the tribe Fabeae [47] and legumes in general [48,49].

Our analysis suggested at least seven migrations between the Neotropics (South America) and the Paleotropics. Four of these were inferred to be from the Neotropics to the Paleotropics and took place during the Early Miocene to the Pleistocene (*c.* 21.6–0.9 Ma; nodes 4, 5, 7 and 8, Figures 3 and 4; Table 3), while three migrations from the Paleotropics to the Neotropics were inferred and were dated from the Early Miocene to the Pleistocene (*c.* 17.3–2.2 Ma; nodes 6, 14 and 18, Figure 3; Table 3). The estimated ages of these migrations are too young to support any hypothesis involving continental drift and early Tertiary biotic interchange. This suggests that the most probable scenario which to explain these tropical disjunctions involved transoceanic LDD by ocean currents. Generally, *Dalbergia* species present samaroid pods and buoyant seeds which are adapted to water dispersal [2,50].

Among the Neotropics, at least two dispersal events between South America and Central America have been inferred. One migration from South America to Central America was dated in the late Miocene at *c.* 10.0 Ma (nodes 19, Figures 3 and 4; Table 3). Although the emergence time of the Isthmus of Panama remains contentious, the authors assumed that the latest date for closure of this landmass which connects South and Central America was around 3.5 Ma [51–53]. Therefore, this first migration was probably through transoceanic LDD via ocean currents, birds, or wind. This is consistent with a wave of plant dispersal between South and Central America predating the closure of the Isthmus [51,54]. In addition, several examples of putatively pre-Isthmus dispersal from South America to Central and North America have been documented [55–58]. The second migration took place in the Early Pleistocene at *c.* 2.2 Ma (nodes 6 and 7, Figure 3; Table 3), after the closure of the Isthmus of Panama, and led to the distribution of sect. *Ecastaphyllum* across South and Central America. Thus, this dispersion probably followed the Isthmus which established an important route of migration between South, Central and North America, giving rise to the Great American Biotic Interchange [43,59].

At least two dispersal events from Africa to Australasia and four reverse dispersals from Australasia to Africa in the Early to Late Miocene were inferred in our analyses (nodes 9, 11, 13, 14, 15 and 21, Figures 3 and 4; Table 3). Boreotropical migration [60], rafting of the Indian subcontinent [61], transoceanic dispersal [62,63], or Miocene overland migration across the Arabian Peninsula [64–66] have all been invoked to explain intercontinental dispersal between Africa and Australasia. The young age of the migrations between Africa and Australasia in our analyses favors Miocene overland migration or transoceanic LDD. During the Early to the Middle Miocene, a land connection was formed between Africa and Southwest Asia which corresponded with a global warming phase that peaked from 17 to 15 Ma [67–69]. This probably played an important role in these migrations, and several studies have shown similar patterns of dispersal [43,66,70]. However, many *Dalbergia* species are adapted to water dispersal [2] and have winged fruits [71]. Therefore, transoceanic LDD between Africa and Australasia cannot be ruled out.

The Malagasy *Dalbergia* were resolved in three clades (M1, M2 and M3), nested with other species from Africa (Figures 1 and 3). This occurrence in three distinct clades and the inference of an African species as sister to each Malagasy clade suggests at least three independent migrations from Africa to Madagascar during the Late Miocene to the

Pleistocene (*c.* 11.5–1.0 Ma; nodes 10, 12 and 20, Figures 3 and 4; Table 3). Given these dates, it is evident that these splits occurred after Madagascar separated from Africa [72]. Thus, both migrations from continental Africa to Madagascar were probably achieved through LDD across the Mozambique channel, which supports the predominant biogeographical pattern found by Yoder and Nowak (2006) [73]. In addition, LDD from continental Africa to Madagascar has been suggested for many taxa of animals and plants [63,74–77].

5. Conclusions

Although our study corroborates previous findings (e.g., the monophyly of *Dalbergia* and the presence of five major clades), it sheds new light on the relationships among the major clades of the genus. These were resolved in our phylogenetic analyses with strong support, and an increase of support for the vast majority of the nodes (around 85% of the nodes). In addition, we have provided the first detailed divergence times and biogeographical history study of the genus. We inferred a Middle Eocene and Early Miocene origin for the stem and crown of *Dalbergia*, respectively. The genus originated in South America and achieved its present-day pantropical distribution largely through recent transoceanic long-distance dispersal. However, taxon sampling included in this study included 39% of the total species diversity of the genus. Therefore, we recommend future studies to harness more loci and increased taxonomic sampling for deeper resolution of the phylogeny of the genus *Dalbergia*.

Supplementary Materials: The following supporting information can be downloaded at: https://www.mdpi.com/article/10.3390/agronomy12071612/s1, Figure S1: Bayesian 50% consensus tree of *Dalbergia* resulting from the combined nuclear (ITS) and plastid (*mat*K and *rbc*L) datasets. Bayesian posterior probability (only values >0.50) are presented above the branches; Table S1: Species list, voucher information and GenBank data accession number of the taxa used for the analysis; Table S2: Nucleotide substitution models selected using jModelTest under the Aikake Information Criterion (AIC) for each DNA region; Table S3: Likelihood (LnL) and Akaike Information Criterion (AIC) scores from each of the models tested in BioGeoBEARS implemented in RASP v.4.2 for the ancestral area estimations analyses. The best model is highlighted in bold.

Author Contributions: F.R.R. and T.-S.Y. designed the research; F.R.R., O.O. and R.Z. performed the research and analyzed the data; F.R.R., G.W.S. and T.-S.Y. wrote the first version of the manuscript which was reviewed and agreed by all co-authors. All authors have read and agreed to the published version of the manuscript.

Funding: This research was supported by the National Natural Science Foundation of China, key international (regional) cooperative research project (No. 31720103903), the Strategic Priority Research Program of Chinese Academy of Sciences (XDB31000000), the Science and Technology Basic Resources Investigation Program of China (2019FY100900), the Large-scale Scientific Facilities of the Chinese Academy of Sciences (No. 2017-LSF-GBOWS-02), and the National Natural Science Foundation of China (Project No. 31270274), the Yunling International High-end Experts Program of Yunnan Province, China (grant No. YNQR-GDWG-2017-002 to P.S.S. and T.-S.Y. and YNQR-GDWG-2018-012 to D.E.S. and T.-S.Y.), China Postdoctoral Science Foundation (2020M683391) and Postdoctoral Directional Training Foundation of Yunnan Province.

Informed Consent Statement: Not applicable.

Data Availability Statement: Not applicable.

Acknowledgments: We are grateful to the Germplasm Bank of Wild Species in Kunming Institute of Botany (KIB), including the iFlora high performance computing center, for facilitating this study. We thank Maria Vorontsova from Royal Botanical Gardens Kew, Richmond, UK for the review of the previous version of the manuscript.

Conflicts of Interest: The authors declare no conflict of interest.

References

1. Klitgaard, B.; Lavin, M. Tribe Dalbergieae sens. lat. In *Legumes of the World*; Royal Botanic Gardens Kew: Richmond, UK, 2005; pp. 307–335.
2. de Carvalho, A. A Synopsis of the Genus *Dalbergia* (Fabaceae: Dalbergieae) in Brazil. *Brittonia* **1997**, *49*, 87–109. [CrossRef]
3. Du Puy, D.J.; Labat, J.; Rabevohitra, R.; Villiers, J.; Bosser, J.; Moat, J. *The Leguminosae of Madagascar*; Royal Botanic Gardens Kew: Richmond, UK, 2002; p. 737.
4. Mabberley, D. *Mabberley's Plant-Book: A Portable Dictionary of Plants, Their Classifications, and Uses*; Cambridge University Press: Cambridge, UK, 2008; p. 1021.
5. Lewis, G.P.; Schrire, B.; Mackinder, B.; Lock, M. *Legumes of the World*; Royal Botanic Gardens Kew: Richmond, UK, 2005; p. 577.
6. Saha, S.; Shilpi, J.A.; Mondal, H.; Hossain, F.; Anisuzzman, M.; Hasan, M.M.; Cordell, G.A. Ethnomedicinal, Phytochemical, and Pharmacological Profile of the Genus *Dalbergia* L.f. (Fabaceae). *Phytopharmacology* **2013**, *2*, 291–346.
7. Bentham, G. Synopsis of Dalbergieae, a Tribe of Leguminosae. *Bot. J. Linn. Soc.* **1860**, *4*, 1–128. [CrossRef]
8. Prain, D. The Species of *Dalbergia* of Southeastern Asia. *Ann. Roy. Bot. Gard.* **1904**, *10*, 1–114.
9. Thothathri, K. Taxonomic Revision of the Tribe Dalbergieae. In *The Indian Subcontinent*; Botanical Survey of India: Culcutta, India, 1987; pp. 231–239.
10. Lavin, M.; Pennington, R.T.; Klitgaard, B.B.; Sprent, J.I.; de Lima, H.C.; Gasson, P.E. The Dalbergioid Legumes (Fabaceae): Delimitation of a Pantropical Monophyletic Clade. *Am. J. Bot.* **2001**, *88*, 503–533. [CrossRef]
11. Cardoso, D.B.; Ramos, G.; Barbosa São-Mateus, W.M.; Paganucci de Queiroz, L. *Aeschynomene chicocesariana*, a Striking New Unifoliolate Legume Species from the Brazilian Chapada Diamantina and its Phylogenetic Placement in the Dalbergioid Clade. *Syst. Bot.* **2019**, *44*, 810–817. [CrossRef]
12. Ribeiro, R.A.; Lavin, M.; Lemos-Filho, J.P.; Santos, F.R.d.; Lovato, M.B. The Genus *Machaerium* (Leguminosae) is More Closely Related to *Aeschynomene* sect. Ochopodium than to Dalbergia: Inferences from Combined Sequence Data. *Syst. Bot.* **2007**, *32*, 762–771.
13. Vatanparast, M.; Klitgård, B.B.; Adema, F.A.; Pennington, R.T.; Yahara, T.; Kajita, T. First Molecular Phylogeny of the Pantropical Genus *Dalbergia*: Implications for Infrageneric Circumscription and Biogeography. *S. Afr. J. Bot.* **2013**, *89*, 143–149. [CrossRef]
14. Katoh, K.; Standley, D.M. MAFFT Multiple Sequence Alignment Software Version 7: Improvements in Performance and Usability. *Mol. Biol. Evol.* **2013**, *30*, 772–780. [CrossRef]
15. Kearse, M.; Moir, R.; Wilson, A.; Stones-Havas, S.; Cheung, M.; Sturrock, S.; Buxton, S.; Cooper, A.; Markowitz, S.; Duran, C. Geneious Basic: An Integrated and Extendable Desktop Software Platform for the Organization and Analysis of Sequence Data. *Bioinformatics* **2012**, *28*, 1647–1649. [CrossRef]
16. Farris, J.S.; Kallersjo, M.; Kluge, A.G.; Bult, C. Testing Significance of Incongruence. *Cladistics* **1994**, *10*, 315–319. [CrossRef]
17. Swofford, D.L. *PAUP*. Phylogenetic Analysis Using Parsimony (*and Other Methods), Version 4.0b10*; Sinauer: Sunderland, MA, USA, 2002.
18. Darriba, D.; Taboada, G.L.; Doallo, R.; Posada, D. jModelTest 2: More Models, New Heuristics and Parallel Computing. *Nat. Methods* **2012**, *9*, 772. [CrossRef] [PubMed]
19. Trifinopoulos, J.; Nguyen, L.-T.; von Haeseler, A.; Minh, B.Q. W-IQ-TREE: A Fast Online Phylogenetic Tool for Maximum Likelihood Analysis. *Nucleic Acids Res.* **2016**, *44*, W232–W235. [CrossRef] [PubMed]
20. Ronquist, F.; Teslenko, M.; Van Der Mark, P.; Ayres, D.L.; Darling, A.; Höhna, S.; Larget, B.; Liu, L.; Suchard, M.A.; Huelsenbeck, J.P. MrBayes 3.2: Efficient Bayesian Phylogenetic Inference and Model Choice Across a Large Model Space. *Syst. Biol.* **2012**, *61*, 539–542. [CrossRef]
21. Rambaut, A.; Drummond, A.J. Tracer Version 1.6. 2013. Available online: http://tree.bio.ed.ac.uk/software (accessed on 11 December 2013).
22. Drummond, A.J.; Suchard, M.A.; Xie, D.; Rambaut, A. Bayesian Phylogenetics with BEAUti and the BEAST 1.7. *Mol. Biol. Evol.* **2012**, *29*, 1969–1973. [CrossRef]
23. Rambaut, A. FigTree Version 1.4.2. 2014. Available online: http://tree.bio.ed.ac.uk/software/Figtree (accessed on 9 July 2014).
24. Kress, W.J.; Prince, L.M.; Williams, K.J. The Phylogeny and a New Classification of the Gingers (Zingiberaceae): Evidence from Molecular Data. *Am. J. Bot.* **2002**, *89*, 1682–1696. [CrossRef]
25. Magallón, S.; Gómez-Acevedo, S.; Sánchez-Reyes, L.L.; Hernández-Hernández, T. A Metacalibrated Time-Tree Documents the Early Rise of Flowering Plant Phylogenetic Diversity. *New Phytol.* **2015**, *207*, 437–453. [CrossRef]
26. Matzke, N.J. BioGeoBEARS: BioGeography with Bayesian (and Likelihood) Evolutionary Analysis in R Scripts. R Package, Version 0.2.1. 2013. Available online: http://phylo.wikidot.com/biogeobears (accessed on 2 January 2013).
27. Yu, Y.; Blair, C.; He, X. RASP 4: Ancestral State Reconstruction Tool for Multiple Genes and Characters. *Mol. Biol. Evol.* **2020**, *37*, 604–606. [CrossRef]
28. Ree, R.H.; Smith, S.A. Maximum Likelihood Inference of Geographic Range Evolution by Dispersal, Local Extinction, and Cladogenesis. *Syst. Biol.* **2008**, *57*, 4–14. [CrossRef]
29. Ronquist, F. Dispersal-Vicariance Analysis: A New Approach to the Quantification of Historical Biogeography. *Syst. Biol.* **1997**, *46*, 195–203. [CrossRef]
30. Landis, M.J.; Matzke, N.J.; Moore, B.R.; Huelsenbeck, J.P. Bayesian Analysis of Biogeography When the Number of Areas is Large. *Syst. Biol.* **2013**, *62*, 789–804. [CrossRef] [PubMed]

31. Ree, R.H.; Sanmartín, I. Conceptual and Statistical Problems with the DEC+J Model of Founder-Event Speciation and its Comparison with DEC via Model Selection. *J. Biogeogr.* **2018**, *45*, 741–749. [CrossRef]
32. Murthy, K.; Rani, S.S.; Pullaiah, T. Genus *Dalbergia* L.f. (Leguminosae: Faboideae) in Eastern Ghats. *J. Econ. Taxon. Bot.* **2000**, *24*, 133–139.
33. Niyomdham, C. An Account of *Dalbergia* (Leguminosae-Papilionoideae) in Thailand. *Thai For. Bull.* **2002**, *30*, 124–166.
34. Sunarno, B.; Ohashi, H. *Dalbergia* (Leguminosae) of Sulawesi, Indonesia. *J. Jpn. Bot.* **1996**, *71*, 241–248.
35. Sunarno, B.; Ohashi, H. *Dalbergia* (Leguminosae) of Borneo. *J. Jpn. Bot.* **1997**, *72*, 198–220.
36. Carvalho, d.A.e.M. Systematic Studies of the Genus *Dalbergia* L.f. in Brazil. Ph.D. Thesis, University of Reading, Reading, UK, 1989.
37. Lavin, M.; Herendeen, P.S.; Wojciechowski, M.F. Evolutionary Rates Analysis of Leguminosae Implicates a Rapid Diversification of Lineages During the Tertiary. *Syst. Biol.* **2005**, *54*, 575–594. [CrossRef] [PubMed]
38. Hung, T.H.; So, T.; Sreng, S.; Thammavong, B.; Boounithiphonh, C.; Boshier, D.H.; MacKay, J.J. Reference Transcriptomes and Comparative Analyses of Six Species in the Threatened Rosewood Genus *Dalbergia*. *Sci. Rep.* **2020**, *10*, 17749. [CrossRef]
39. Saporta, d.G. *Dalbergia phleboptera* Saporta. Muséum National d'Histoire Naturelle (MNHN), Paris, France. 2015. Available online: http://coldb.mnhn.fr/catalognumber/mnhn/f/14084 (accessed on 14 December 2015).
40. Kučerová, J. Miocénna Flóra z Lokalít Kalonda a Mučín. *Acta Geol. Slovaca* **2009**, *1*, 65–70.
41. Guo, S.; Zhou, Z. The Megafossil Legumes from China. In *Advances in Legume Systematics Part 4*; Royal Botanic Gardens Kew: Richmond, UK, 1994; pp. 207–223.
42. Lomolino, M.V. Four Darwinian Themes on the Origin, Evolution and Preservation of Island Life. *J. Biogeogr.* **2010**, *37*, 985–994. [CrossRef]
43. Jin, J.J.; Yang, M.Q.; Fritsch, P.W.; van Velzen, R.; Li, D.Z.; Yi, T.S. Born Migrators: Historical Biogeography of the Cosmopolitan Family Cannabaceae. *J. Syst. Evol.* **2020**, *58*, 461–473. [CrossRef]
44. Li, J.; Jiang, J.; Stel, H.V.; Homkes, A.; Corajod, J.; Brown, K.; Chen, Z. Phylogenetics and Biogeography of *Apios* (Fabaceae) Inferred from Sequences of Nuclear and Plastid Genes. *Int. J. Plant Sci.* **2014**, *175*, 764–780. [CrossRef]
45. Vatanparast, M.; Takayama, K.; Sousa, M.S.; Tateishi, Y.; Kajita, T. Origin of Hawaiian Endemic Species of *Canavalia* (Fabaceae) from Sea-Dispersed Species Revealed by Chloroplast and Nuclear DNA Sequences. *J. Jpn. Bot.* **2011**, *86*, 15–25.
46. Fortuna-Perez, A.P.; da Silva, M.J.; de Queiroz, L.P.; Lewis, G.P.; Simões, A.O.; de Azevedo Tozzi, A.M.G.; Sarkinen, T.; de Souza, A.P. Phylogeny and Biogeography of the Genus *Zornia* (Leguminosae: Papilionoideae: Dalbergieae). *Taxon* **2013**, *62*, 723–732. [CrossRef]
47. Schaefer, H.; Hechenleitner, P.; Santos-Guerra, A.; de Sequeira, M.M.; Pennington, R.T.; Kenicer, G.; Carine, M.A. Systematics, Biogeography, and Character Evolution of the Legume Tribe Fabeae with Special Focus on the Middle-Atlantic Island Lineages. *BMC Evol. Biol.* **2012**, *12*, 250. [CrossRef]
48. Lavin, M.; Schrire, B.P.; Lewis, G.; Pennington, R.T.; Delgado–Salinas, A.; Thulin, M.; Hughes, C.E.; Matos, A.B.; Wojciechowski, M.F. Metacommunity Process Rather than Continental Tectonic History Better Explains Geographically Structured Phylogenies in Legumes. *Philos. Trans. R. Soc. London Ser. B Biol. Sci.* **2004**, *359*, 1509–1522. [CrossRef]
49. Pennington, R.T.; Richardson, J.E.; Lavin, M. Insights into the Historical Construction of Species-Rich Biomes from Dated Plant Phylogenies, Neutral Ecological Theory and Phylogenetic Community Structure. *New Phytol.* **2006**, *172*, 605–616. [CrossRef]
50. Croat, T.B. *Flora of Barro Colorado Island, Vol. VIII*; Stanford University Press: Stanford, CA, USA, 1978; p. 943.
51. Bacon, C.D.; Silvestro, D.; Jaramillo, C.; Smith, B.T.; Chakrabarty, P.; Antonelli, A. Biological Evidence Supports an Early and Complex Emergence of the Isthmus of Panama. *Proc. Natl. Acad. Sci. USA* **2015**, *112*, 6110–6115. [CrossRef]
52. Hoorn, C.; Flantua, S. An Early Start for the Panama Land Bridge. *Science* **2015**, *348*, 186–187. [CrossRef]
53. Montes, C.; Cardona, A.; Jaramillo, C.; Pardo, A.; Silva, J.; Valencia, V.; Ayala, C.; Pérez-Angel, L.; Rodriguez-Parra, L.; Ramirez, V. Middle Miocene Closure of the Central American Seaway. *Science* **2015**, *348*, 226–229. [CrossRef]
54. Cody, S.; Richardson, J.E.; Rull, V.; Ellis, C.; Pennington, R.T. The Great American Biotic Interchange Revisited. *Ecography* **2010**, *33*, 326–332. [CrossRef]
55. Dupin, J.; Matzke, N.J.; Särkinen, T.; Knapp, S.; Olmstead, R.G.; Bohs, L.; Smith, S.D. Bayesian Estimation of the Global Biogeographical History of the Solanaceae. *J. Biogeogr.* **2017**, *44*, 887–899. [CrossRef]
56. Grose, S.O.; Olmstead, R. Evolution of a Charismatic Neotropical Clade: Molecular Phylogeny of *Tabebuia* s. l., Crescentieae, and Allied Genera (Bignoniaceae). *Syst. Bot.* **2007**, *32*, 650–659. [CrossRef]
57. Olmstead, R.G. Phylogeny and Biogeography in Solanaceae, Verbenaceae and Bignoniaceae: A Comparison of Continental and Intercontinental Diversification Patterns. *Bot. J. Linn. Soc.* **2013**, *171*, 80–102. [CrossRef]
58. Davis, C.C.; Bell, C.D.; Fritsch, P.W.; Mathews, S. Phylogeny of *Acridocarpus-Brachylophon* (Malpighiaceae): Implications for Tertiary Tropical Floras and Afroasian Biogeography. *Evolution* **2002**, *56*, 2395–2405. [CrossRef]
59. Hoorn, C.; Wesselingh, F.; Ter Steege, H.; Bermudez, M.; Mora, A.; Sevink, J.; Sanmartín, I.; Sanchez-Meseguer, A.; Anderson, C.; Figueiredo, J. Amazonia Through Time: Andean Uplift, Climate Change, Landscape Evolution, and Biodiversity. *Science* **2010**, *330*, 927–931. [CrossRef]
60. Davis, C.C.; Bell, C.D.; Mathews, S.; Donoghue, M.J. Laurasian Migration Explains Gondwanan Disjunctions: Evidence from Malpighiaceae. *Proc. Natl. Acad. Sci. USA* **2002**, *99*, 6833–6837. [CrossRef]
61. Conti, E.; Eriksson, T.; Schönenberger, J.; Sytsma, K.J.; Baum, D.A. Early Tertiary Out-of-India Dispersal of Crypteroniaceae: Evidence from Phylogeny and Molecular Dating. *Evolution* **2002**, *56*, 1931–1942. [CrossRef]

62. Samonds, K.E.; Godfrey, L.R.; Ali, J.R.; Goodman, S.M.; Vences, M.; Sutherland, M.R.; Irwin, M.T.; Krause, D.W. Spatial and Temporal Arrival Patterns of Madagascar's Vertebrate Fauna Explained by Distance, Ocean Currents, and Ancestor Type. *Proc. Natl. Acad. Sci. USA* **2012**, *109*, 5352–5357. [CrossRef]
63. Yuan, Y.-M.; Wohlhauser, S.; Möller, M.; Klackenberg, J.; Callmander, M.W.; Küpfer, P. Phylogeny and Biogeography of *Exacum* (Gentianaceae): A Disjunctive Distribution in the Indian Ocean Basin Resulted from Long Distance Dispersal and Extensive Radiation. *Syst. Biol.* **2005**, *54*, 21–34. [CrossRef]
64. Kosuch, J.; Vences, M.; Dubois, A.; Ohler, A.; Böhme, W. Out of Asia: Mitochondrial DNA Evidence for an Oriental Origin of Tiger Frogs, Genus *Hoplobatrachus*. *Mol. Phylogenet. Evol.* **2001**, *21*, 398–407. [CrossRef] [PubMed]
65. Yu, X.-Q.; Maki, M.; Drew, B.T.; Paton, A.J.; Li, H.-W.; Zhao, J.-L.; Conran, J.G.; Li, J. Phylogeny and historical biogeography of Isodon (Lamiaceae): Rapid Radiation in South-West China and Miocene Overland Dispersal into Africa. *Mol. Phylogenet. Evol.* **2014**, *77*, 183–194. [CrossRef] [PubMed]
66. Zhou, L.; Su, Y.C.; Thomas, D.C.; Saunders, R.M. "Out-of-Africa" Dispersal of Tropical Floras During the Miocene Climatic Optimum: Evidence from *Uvaria* (Annonaceae). *J. Biogeogr.* **2012**, *39*, 322–335. [CrossRef]
67. Popov, S.; Rögl, F.; Rozanov, A.; Steininger, F.F.; Shcherba, I.; Kovac, M. Lithological-Paleogeographic Maps of Paratethys (Maps 1–10). *Cour. Forschungsinst. Senckenberg* **2004**, *250*, 46.
68. Rögl, F. Palaeogeographic Considerations for Mediterranean and Paratethys Seaways (Oligocene to Miocene). *Ann. Naturhist. Mus. Wien A* **1997**, *99*, 279–310.
69. Zachos, J.; Pagani, M.; Sloan, L.; Thomas, E.; Billups, K. Trends, Rhythms, and Aberrations in Global Climate 65 Ma to Present. *Science* **2001**, *292*, 686–693. [CrossRef] [PubMed]
70. Huang, X.; Deng, T.; Moore, M.J.; Wang, H.; Li, Z.; Lin, N.; Yusupov, Z.; Tojibaev, K.S.; Wang, Y.; Sun, H. Tropical Asian Origin, Boreotropical Migration and Long-Distance Dispersal in Nettles (Urticeae, Urticaceae). *Mol. Phylogenet. Evol.* **2019**, *137*, 190–199. [CrossRef]
71. Van der Pijl, L. *Principles of Dispersal in Higher Plants*; Springer: New York, NY, USA, 1982; p. 153.
72. McLoughlin, S. The Breakup History of Gondwana and its Impact on Pre-Cenozoic Floristic Provincialism. *Aust. J. Bot.* **2001**, *49*, 271–300. [CrossRef]
73. Yoder, A.D.; Nowak, M.D. Has Vicariance or Dispersal Been the Predominant Biogeographic Force in Madagascar? Only Time Will Tell. *Annu. Rev. Ecol. Evol. Syst.* **2006**, *37*, 405–431. [CrossRef]
74. Baum, D.A.; Small, R.L.; Wendel, J.F. Biogeography and Floral Evolution of Baobabs Adansonia, Bombacaceae as Inferred from Multiple Data Sets. *Syst. Biol.* **1998**, *47*, 181–207. [CrossRef]
75. Clayton, J.W.; Soltis, P.S.; Soltis, D.E. Recent Long-Distance Dispersal Overshadows Ancient Biogeographical Patterns in a Pantropical Angiosperm Family (Simaroubaceae, Sapindales). *Syst. Biol.* **2009**, *58*, 395–410. [CrossRef] [PubMed]
76. Meimberg, H.; Wistuba, A.; Dittrich, P.; Heubl, G. Molecular Phylogeny of Nepenthaceae Based on Cladistic Analysis of Plastid trnK Intron Sequence Data. *Plant Biol.* **2001**, *3*, 164–175. [CrossRef]
77. Renner, S.S. Multiple Miocene Melastomataceae Dispersal between Madagascar, Africa and India. *Philos. Trans. R. Soc. London Ser. B Biol. Sci.* **2004**, *359*, 1485–1494. [CrossRef] [PubMed]

agronomy

Article

Pollen Morphological Peculiarities of Selected Mimosoideae Taxa of Hainan Island and Their Taxonomic Relevance

Saraj Bahadur [1], Sehrish Taj [2], Wenxing Long [1,*] and Uzma Hanif [3]

1 Wuzhishan National Long-Term Forest Ecosystem Monitoring Research Station, College of Forestry, Hainan University, Haikou 570228, China; sirajbahadur14@gmail.com
2 State key Laboratory of Marine Resource Utilization in South China Sea, Haikou 570228, China; 184105@hainanu.edu.cn
3 Department of Botany, Government College University, Lahore 54000, Pakistan; uzmahanif@gcu.edu.pk
* Correspondence: oklong@hainanu.edu.cn

Abstract: Mimosoideae is one of the taxonomically complex subfamilies of Fabaceae. Several studies have reported the pollen morphology of Mimosoideae taxa and their taxonomic relevance, but no such study was found specifically for Hainan Island in southern China. Therefore, the present study was designed to investigate the selected Mimosoideae taxa and explore the new palynological traits to support and strengthen the systematics of Mimosoideae using multiple microscopic techniques. The polar axis, equatorial diameter of the pollen grains, colpus length and width were measured. The smallest pollen grain size was found in *Mimosa pudica* (7.8 × 7.75 µm), while the largest pollen size was found in *Albizia lebback* (87.54 × 77.97 µm). Similarly, significant variation was found in the exine and colpus surface patterns. The subfamily Mimosoideae is considered eurypalynous because of the variation in pollen traits. In addition, variation was also found in the quantitative traits. Comparatively, the pollen features were found to be helpful at the genus and species levels, as well as in the correct identification and discrimination of the taxa. Hence, this study gives a detailed account of the pollen morphologies of certain selected taxa of Mimosoideae collected from different geographical regions on Hainan Island. The pollen morphological traits were proven to have significant taxonomic potential and can be used as additional tools for the correct identification and discrimination of Mimosoideae taxa. These results will provide the basis for further systematic studies.

Keywords: Mimosoideae taxa; pollen morphology; taxonomic relevance; light microscopy; scanning electron microscopy

Citation: Bahadur, S.; Taj, S.; Long, W.; Hanif, U. Pollen Morphological Peculiarities of Selected Mimosoideae Taxa of Hainan Island and Their Taxonomic Relevance. *Agronomy* **2022**, *12*, 1122. https://doi.org/10.3390/agronomy12051122

Academic Editors: Alessio Papini, Mushtaq Ahmad, Fazal Ullah and Wajid Zaman

Received: 17 March 2022
Accepted: 24 April 2022
Published: 6 May 2022

Publisher's Note: MDPI stays neutral with regard to jurisdictional claims in published maps and institutional affiliations.

1. Introduction

Fabaceae is one of the most diverse families, being found in most ecosystems around the globe. It has been divided into three subfamilies—Papilionoideae, Caesalpinioideae and Mimosoideae [1]. In terms of modern Fabaceae classifications, the Legume Phylogeny Working Group (LPWG 2017) [2] recognized six subfamilies, whereby the traditional subfamily Mimosoideae was presented as a clade nested in the Caesalpinioideae family with a new level of discrimination. Hence, our study is based on the classifications made by Lewis et al. [1]. Furthermore, the recent classifications, which only propose changes at the subfamily level and have not yet solved the taxonomic problem of the tribes in the Mimosoid clad, do not affect the interpretation of our results. Mimosoideae is a subfamily of Fabaceae, although is sometimes treated as a separate family [3]. Fabaceae contains about 18,000 species belonging to 697 genera [4]. Fabaceae is the second largest family among the angiosperms after Asteraceae [5] and contains three subfamilies, i.e., Mimosoideae, Papilionoideae and Caesalpinioideae [6]. Mimosoideae comprises 3100 species and 58 genera around the world [7]. About 79 genera are distributed worldwide, including in tropical, subtropical, arid and semiarid regions, as well as in rocky soil [8]. The species has numerous filaments and stamens and a freely downwards-extending base [9]. The group was given the rank of a family by Hutchinson [10], which is rarely used by taxonomists. The Mimosoideae family is broadly dispersed due to its high reproductive rates, making the plants accomplish in producing a lot of viable seeds [11].

The Mimosoideae taxa are used for multiple purposes. For example, in the dry season, they can be widely used as animal foods when there are no pastures, and some taxa are used by local communities for medicinal purposes, firewood and fencing stacks [12]. Before European colonization, the root and bark of Mimosa species are used by the indigenous peoples of Northern Brazil to prepare a hallucinogenic drink [13]. The legumes are used as crops, forages and green manures. These Leguminoceae plants are also used to synthesize a large range of natural products, including dyes, flavorings and poisons [14]. The medicinal uses of the Fabaceae family in Bangladesh have been studied previously [15]. The usefulness of edible and medicinally important Fabaceae taxa was reported in Argentina [16]. This family is of great importance in urban and indigenous communities all over the world. *Acacia mangium* and *A. auriculiformis* are multipurpose species and are widely used for fuelwood, timber, tanning, ornamental horticulture and agroforestry [17,18]. Similarly, *Mimosa pudica* is used to treat various ailments, i.e., inflammation, vaginal and uterine complaints, asthma and fatigue [19,20]. Albizia lebbeck has been used to treat diarrhea, anxiety, allergies, skin diseases and asthma [21].

Light and scanning electron microscopy plays a crucial role in the determination of the fine details of plant materials. For example, the pollen traits are important in the classification of various plants groups, and taxonomists mostly depend on these traits to define species boundaries for various taxonomic ranks [22–33]. This microscopy has a wide range of applications in other fields of diverse biological sciences [34,35]. The pollen morphological features are significant in the evolutionary and phylogenetic studies of various plants groups [36].

The investigation of Mimosoid pollen primarily focuses on the characteristics observed via light microscopy [37,38]. There have been no taxonomic micro-morphological studies of Mimosoid pollen, although the limited available literature suggests that exine patterns may have more systematic potential than those of the pollen units or apertures [39]. Most of the Mimosaceae taxa are taxonomically complex because of the close macro-morphological features [40]. Hence, palynological studies may help to solve the taxonomic problems, correctly identify species and defining species boundaries at various taxonomic levels. Palynological studies of the Mimosaceae taxa have been sporadically conducted in different geographical regions around the world [37,41] Some peculiar pollen characteristics, as well as variations in the pollen traits, have long been known. For example, El Ghazali et al. [42] described the intraspecific variation in pollen morphology and identified various pollen types in *Mimosa* taxa. Similarly, geographical variations in pollen morphology of the Mimosaceae taxa have been studied previously in Sudan [43], with reports that exine sculpturing is almost constant. Pollen diversity and its relevance in the systematics of the Mimosaceae taxa of Pakistan have been analyzed to solve taxonomic problems [41,42,44,45]. Variations in the pollen morphology of *Leucaena* and closely related species were illustrated and described by Hughes [46]. Caccavari and Dome [9] analyzed the pollen morphologies of 77 American *Acacia* taxa and suggested that pollen features can be used as distinguishing factors in the generic limitation process. Hughes [46] analyzed variations in pollen morphology of the *Leucaena* species, which helped show relationships and define generic boundaries within Mimosoideae taxa. The pollen morphologies of 21 *Calliandra* taxa were investigated by De Assis-Ribeiro dos Santos and De Oliveira-Romão [47], who concluded that some of the polyad characters were useful in corroborating previous infrageneric classifications of the genus. In China, a study performed on the pollen morphologies of Mimosoideae taxa and varieties using a scanning electron microscope, which highlighted the similarities among the genera, mainly in terms of the exine, aperture and size [48]. The pollen traits, such as the size and shape of the polyads, exine pattern and presence of apertures, may also be of taxonomic relevance and could be explored further [49]. However, no pollen morphological studies specifically on Hainan Island have been documented yet. Hence, to fill this gap, further palynological studies are needed.

It is hoped that this palynological study will not only add to the systematic information about certain selected Mimosaceae taxa, but will also provide the basis for further phylogenetic studies. The present study aims (1) to provide descriptions of the pollen types of selected taxa of Mimosoideae and outline their taxonomic relevance and (2) to strengthen and support the taxonomy of Mimosoideae and define species boundaries using both light microscopy and scanning electron microscopy.

2. Material and Methods

2.1. Collection and Identification

For the collection of Mimosoideae taxa, various field trips were arranged in the year 2021 to various geographical regions on Hainan Island (Table 1). The island is characterized by a tropical monsoon climate with a wet season ranging from May to October and a dry season ranging from November to April. The annual average precipitation ranges from 2350 to 2651 mm [50–52]. The species were collected and identified with the help of the available literature Flora of China efloras.org World Flora Online.

Table 1. List of Mimosoideae taxa collected from different geographical regions on Hainan Island.

Species	Locality	Altitude	Sea Level
Acacia auriculiformis A.Cunn. ex Benth.	Sanya	N: 18.26415 E: 109. 52084	119 m
Acacia confusa Merr.	**Wanning**, Botanical Garden, Haikou	N: 18.69396 E:110.23170	118 m
Acacia mangium Willd.	Sanya, **Chang Jiang**	N: 19.33905 E: 108.210	33 m
Albizia julibrissin Durazz.	Wanning	N: 18.69396 E:110.23170	118 m
Albizia lebbeck (L.)	Haikou	N: 20.062363 E: 110.3186713	3 m
Calliandra haematocephala Hassk.	Sanya	N: 18.26415 E: 109.52084	119 m
Entada phaseoloides (Linn.) Merr.	Wanning, Botanical Garden	N: 18.69396 E:110.23170	118 m
Leucaena leucocephala (Lam.)	**Haikou**, Sanya	N: 20.062363 E: 110.3186713	3 m
Mimosa bimucronata (DC.) Kuntze	Haikou, Wanning	N: 18.69396 E:110.23170	118 m
Mimosa diplotricha C. Wright	Beihualing, Haikou	N:19.00262 E: 109.81523	471 m
Mimosa pudica Linn.	Haikou, Wanning, Sanya, **Chang Jiang**	N: 19.33905 E: 108.210	33 m

2.2. Light Microscopy

For the light microscopic study, the filaments were first separated from the flowers and crushed on a glass slide by adding 2–3 drops of acetic acid. The debris was removed through a needle and then a cover slip was placed on it. The prepared slide was then observed under a light microscope equipped with a digital camera. The pollen micrograph was taken and its various taxonomic features were observed. For pollen descriptions, we mostly followed the terminology used by Punt et al. [53].

2.3. Scanning Electron Microscopy

The anthers were separated from the mature flower as performed by Ali et al. [23]. The anthers were then crushed on the glass slide by adding a few drops of acetic acid and transferred into Eppendorf tubes. With the help of a micropipette, pollen samples from the tubes were taken and placed on a metallic stub attached with double-sided sticky tape. The prepared samples were sputtered with platinum for twenty-five minutes in a Leica Mikrosystem made in Austria with a high-vacuum coater (ACE600) and observed under SEM (Thermo Scientific, Model: verios g 4 uc) Manufacturer Seimer Technology was installed in the analytical and testing center of Hainan University, Haikou, China.

2.4. Quantitative Analysis

For quantitative analysis, we measured about ten pollen grains in each sample. The polar axis and equatorial diameter of the polyads and colpus length and width were measured using Image J software.

3. Results and Discussion

The Mimosaceae taxa were analyzed using both light and scanning electron microscopic techniques. LM and SEM micrographs are illustrated in Figures 1–5. Details of pollen traits of each species are given below.

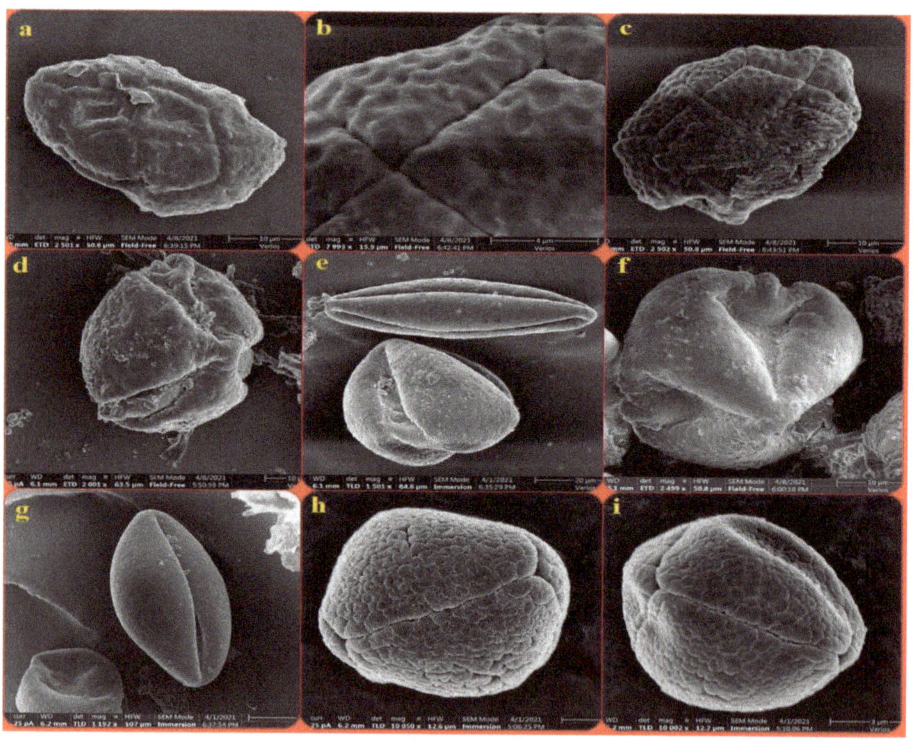

Figure 1. *Acacia confusa* (**a–c**), showing (**a**) general view of the pollen and (**b**) close view of the exine sculpturing. *Entada phaseoloides* (**d–g**), showing (**d**) oblique polar view, (**e**) equatorial view, (**f**) sunken colpus and (**g**) zonocolpus. *Mimosa pudica* Wanning (**h**,**i**), showing oblique equatorial view. Scale bars: a = 10 μm; b = 4 μm; c = 10 μm; d = 10 μm; e = 10 μm; f,g = 10 μm; h,i = 3 μm.

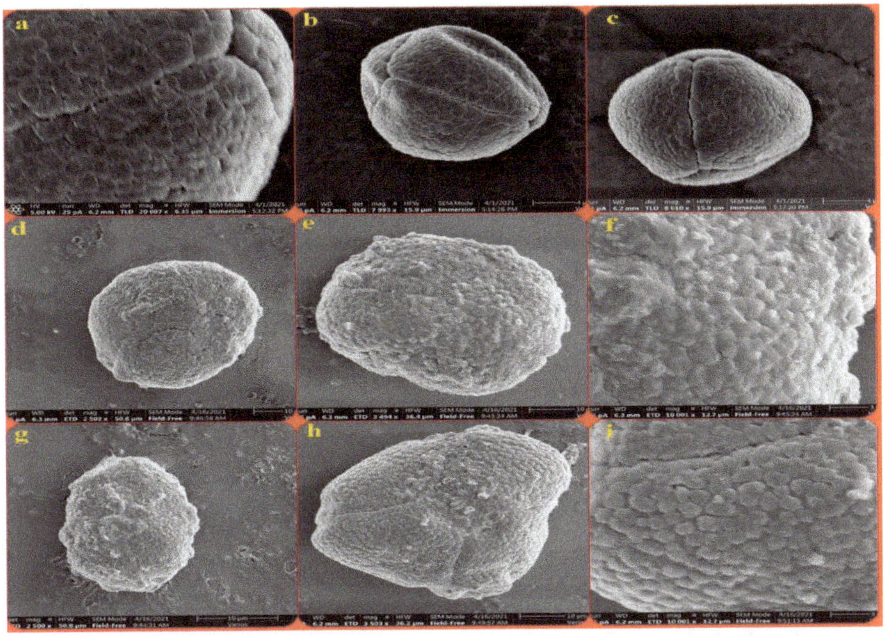

Figure 2. SEM micrographs showing the pollen grains of *Mimosa pudica* Wanning, showing (**a**) close view of exine sculpturing, (**b**) the polar area and (**c**) equatorial view. *Mimosa diplotricha* Beihualing (**d–g**), showing (**d**) equatorial view, (**e**) polar view, (**f**) close view of the exine and (**g**) prolate-shaped pollen. *Mimosa diplotricha* Haikou (**h**,**i**), showing (**h**) polar view and (**i**) close view of the exine sculpturing. Scale bars: a = 1 μm; b,c = 4 μm; c = 10 μm; d,e = 10 μm; f = 3 μm; g,h = 10 μm; i = 3 μm.

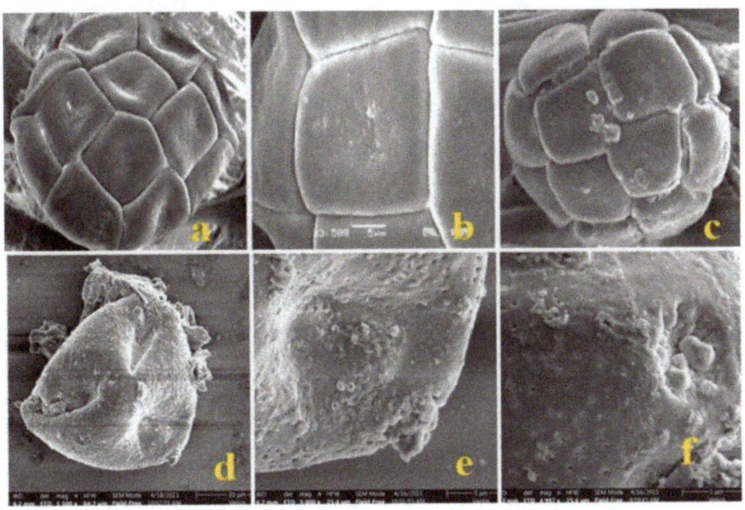

Figure 3. The SEM micrographs of *Albizia lebback* pollen grain (**a–c**), showing (**a**) quadrangular-shaped pollen in polar view outline, (**b**) a close view of an individual grain of the polyad and (**c**) inaperturate pollen. *Leucaena leucocephala* (**d–f**), showing a (**d**) polar view, (**e**) close view of the colpus surface membrane and (**f**) the apocolpium region. Scales: (**a**) 10 μm; (**b**) 5 μm; (**c**) 10 μm; (**d**) 30 μm; (**e–f**) 5 μm.

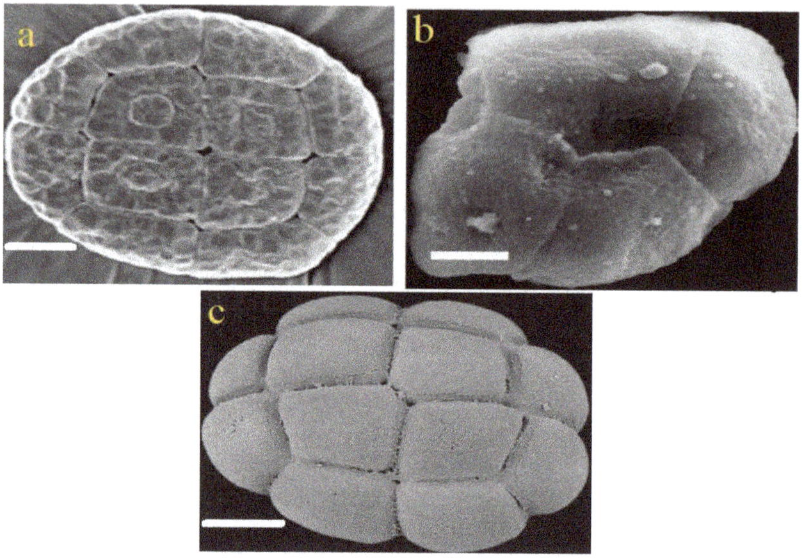

Figure 4. SEM micrographs of the pollen: *Acacia mangium* (**a**) showing octad polyads; *Mimosa bimucronata* (**b**) showing bitetrad polyads; *Albizia julibrissin* (**c**) showing 12-grain polyads. Scale bars: a = 10 μm; b = 2 μm; c = 10 μm.

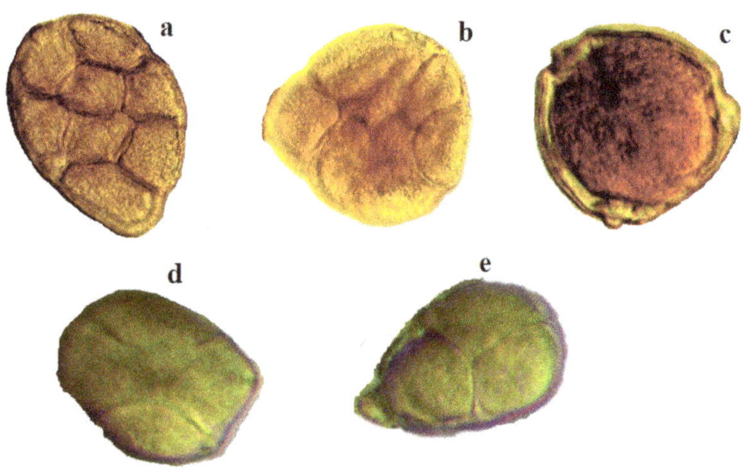

Figure 5. Light micrographs of pollen. *Calliandra haematocephala* (**a**,**b**), showing (**a**) equatorial view and (**b**) polar view. *Entada phaseoloides,* (**c**) showing polar view. *Mimosa diplotricha.* (**d**) showing tetrad-type pollen. *Mimosa pudica* (**e**), showing triyad-type pollen. Scale bar = 50 μm.

3.1. *Acacia auricifoliformis*

The pollen of *A. auricifoliformis* was a polyad comprising 16 grains. Eight grains are arranged in the center, which is surrounded by eight periphery cells. Each grain was to be found inaperturate, while the exine sculpturing was found to be psilate. The average diameter of the polyads was noted as 32 μm. Similarly, some other species of Acacia have been described previously and the polyad type was also reported to be pollen (16–32-celled), with or without distinct apertures [54–56]. The general description of polyads and the interrelationships of the grains in *Acacia* have also been discussed previously [57].

3.2. *Acacia confusa*

The pollen was shed in polyads (12 grains), the outline in the polar view was rhomboidal, the symmetry was bilateral and the polarity was heteropolar. The aperture condition was noted as inaperturate. Exine sculpturing was found as mega-reticulate and rarely scabrate. The reticula were thick and flattened, having a rarely scabrate structure, while the lumina was filled with a somewhat micro-scabrate structure (Figure 1a–c). The pollen size was noted as 23.19 × 13.88 μm, while the individual grain diameter was 5.745 μm.

3.3. *Acacia mangium*

The pollen of *A. mangium* was polyad of 12 grains (Figure 4a). The four grains were arranged in the center, surrounded by eight periphery grains. The exine sculpturing was found to be megareticulate and heterobrochate. An operculum was observed in the central grains of the polyads. The average diameter of the polyad was noted as 35.61 μm (Figure 1a). To the best of our knowledge, no descriptive palynological studies have been reported yet. However, Wang et al. [58] documented only the isolation and characterization of flower-specific transcripts in *Acacia mangium*.

3.4. *Albizia julibrissin*

The pollen type was polyad, comprising 12 grains. The outline in polar view was elliptical. The central 4 grains were arranged in tetragonal form. The exine sculpturing was psilate (Figure 3a–c). The pollen size was noted as 71.81 × 81.01 μm, while the individual grain diameter was 25.13 μm (Figure 4c). The pollen of *A. julibrissin* was closely related to that of *A. lebback* and *A. procera*. For example, Khan et al. [41] documented the pollen diversity and its implications in Mimosaceous taxa using light and scanning electron microscopic techniques, and reported psilate–poveolate and sparsely psilate–scabrate exine sculpturing in *A. procera* pollen. Hence, these traits were found to be useful in delimiting and defining species boundaries within the genus *Albizia*.

3.5. *Albizia lebback*

The pollen type was polyad (12-grains), while the outline in polar view was spheroidal. The central 4 grains were arranged in tetragonal form. The exine sculpturing was psilate to rugulate (Figure 3a–c). The pollen size was 87.54 × 77.97 μm, while the individual grain diameter was 24.35 μm. However, in comparison with the previous results, the 16-celled polyad-type pollen grains reported by Agashe and Caulton [59] were not corroborated by our findings. Moreover, a minimum length of 28 μm and maximum length of 57.4 μm and a minimum width of 30.5 μm and maximum width of 43.05 μm were reported by Parveen and Qaiser [44] in the pollen grains of *A.lebback*. Similarly, the aperture and exine sculpturing of *Albizia* was found to be smooth or almost smooth and different from *Calliandra, Cylindrokelupha* and *Zygia,* due to the presence of tubercles [48].

3.6. *Calliandra haematocephala*

Pollen was shed as octad polyads (8 grains), while the outline in the polar view was a pyramid. The two central large-sized grains are in the center and the other grains lie in the periphery. As found in the previous study by Guinet [39], individual pollen grains from the same polyads showed a marked heteromorphic structure, with peripheral grains differing from the central grains in shape, size and less obviously ornamentation. The exine sculpturing was found to be verrucate, aerolate and somewhat reticulate towards the periphery. The pollen tip acute and the base was constricted. The surface of pollen was found to be undulate. According to Van Campo and Guinet [60], the pollen dispersal units of *Calliandra* are bi-tetrads and polyads. Similarly, Chen [61] mentioned that the bi-tetrad pollen of the *Calliandra* is family is rare in Angiosperms. Some other species of this genus also have polyad-type pollen [47]. The pollen grain octads, shed as polyads, were flattened and oval-shaped, while the longest axes ranged from 164 to 173 μm and the shortest axes ranged from 82 to 93 μm. Similarly, the tectate and perforate exine was found in the Philippines in *Calliandra haematocephala*. However, to clarify the differences in exine sculpturing of the appendix in detail, transmission electron microscopy will be needed.

3.7. *Entada phaseoloides*

The pollen type was monad, the outline in polar view was ovate and was elliptical in equatorial view, the symmetry was bilateral and the polarity was heteropolar. The aperture condition was noted as trizonocolpate. The orientation of the colpus was sunken and wide, having an obtuse tip, while the sculpturing was scabrate to verrucate and sometimes psilate. The exine sculpturing was perforate and rarely scabrate, with ubisch bodies found on its surface. The polar area was found to be very small. The pollen size was 34.61 × 26.76 μm. The colpus length was noted as 21.31 μm and the width was 6.72 μm. The polar area was 5.45 μm and the meso-colpium diameter was 25.33. Furthermore, Rao and Lee [62] documented the pollen flora of Singapore and Malaya and observed a tricolporate and subprolate structure, a size of 30 × 35μm and a faintly reticulate exine pattern of the *Entada spiralis* pollen. Similarly, the shape of *Entada scandens* pollen grains and other pollen morphological features of *Entada phaseoloides* were reported in a previous study [54].

3.8. *Leucaena leucocephala*

The pollen type was monad, the outline in polar view was triangular, the symmetry was bilateral and the polarity was isopolar. The aperture condition was noted as tricolpate. The orientation of the colpus was sunken and wide, having an obtuse tip, while the sculpturing was scabrate to rugulate. The exine sculpturing was perforate and rarely scabrate. The polar area was found to be very small. The pollen size was 21.48 × 21.35 μm. The colpus length was 9.27 μm and the width was 3.4 μm. In comparison with a previous study [45], where the exine sculpturing was found to be subpsilate, this was not corroborated by our findings, while the polar axis was 42.42 μm and the equatorial diameter was 31.5 μm. The colpus length was 17.01 μm and the width was 8.61 μm. The mesocolpium area was 23.31 μm, while the apocolpium was 39.95 μm. The sexine was thicker than the nexine. Hence, such variation may be due to variation in the pollen acetolysis techniques and the different geographical regions, i.e., the tropical rain forest in Hainan Island and subtropical regions in Pakistan having different altitudes and different environmental filters.

3.9. *Mimosa bimucronata*

The pollen was small in size and shed in polyad bitetrads containing 8 grains, i.e., four pollen grains in two planes (4 + 4). The prolate was spheroidal and tetragonal, the polarity was heteropolar and the symmetry was bilateral. Aperture condition was found as tri- to tetracolporate. The exine sculpturing was found to be rugulate and rarely scabrate (Figure 5b). The pollen size was noted as 16.7 × 12.6 μm. Similarly, 14 species of the Mimosoideae family of the Atlantic Forest in Brazil were investigated, and it was reported that pollen grains of *Mimosa bimucronata* are bitetrad with exine sculpturing that is rugulate, which supports our findings. However, bitetrad-type pollen is also found in *Mimosa elliptica* and *Mimosa pellita*. Hence, further palynological studies are needed to explore new palynological traits, which will help to solve the taxonomic problem within the *Mimosa* genus.

3.10. *Mimosa diplotricha*

The first sample of this species was collected from a tropical rainforest in Beihualing. The pollen was shed as tetrad polyads, the shape was tetrahedral, the symmetry was bilateral and the polarity was heteropolar. The aperture condition was noted as inaperturate. The uniplanar tetrad with the proximal sides of two individual units is directly connected, while the other two units are separated. The pollen size was noted as 19.1 × 22.39 μm, while the individual grain diameter was 14.28 μm. The exine sculpturing was predominantly areolate and verrucate, rarely being scabrate and gemmate. Ubisch bodies were found as supra-tectal elements (Figure 2d–g). The second sample was collected from Haikou, and an additional perforate, scabrate and densely verrucate exine sculpturing was found, while the center of the polyad was observed as concave, which was found to be different from the former species. The pollen size was noted as 22.21 × 22.11 μm, while the individual grain diameter was 13.47 μm (Figure 2h,i). In conclusion, variation was found in these two populations of *Mimosa diplotricha*. This may have been due to the variations in altitude and environmental factors such as temperature and humidity. The previous study reported that the number of grains per polyad and their shape are useful taxonomic traits in the characterization of *Mimosa* taxa [63]. Hence, our study confirmed that the number of grains per polyad and their shape are valuable taxonomic characteristics for the characterization of Mimosaceae taxa. Additionally, the structure of the polyads may protect the pollen grains from dehydration in dry habitats.

3.11. *Mimosa pudica*

The first sample was collected from Wanning. The pollen was shed as tetrad polyads, the shape was tetrahedral, the symmetry was bilateral and the polarity was heteropolar. The aperture condition was noted as inaperturate. The uniplanar tetrad with the proximal sides of two individual units is directly connected, while the other two units are separated. The exine sculpturing was predominantly areolate and psilate, with various sized verrucae having an isodiametric shape (Figure 1h,i, Figure 2a–c). The pollen size was noted as 7.8 × 7.75 μm. The individual grain diameter was 3.85 μm and the diameter of each verruca was 0.64 μm. In comparison with previous studies, the exine sculpturing was characterized by tubercles that were inconsistent with our results, except for the tetrads type of pollen noted by Agashe and Caulton [59]. Inaperturate tetrahedral- and tetrad-type pollen with an average diameter of 9 μm in *M. pudica* samples in Singapore has also been reported previously [62].

In Fabaceae, the pollen evolution ranges from simple to tetrads and polyads. The simple grains were confirmed in our study, for example for the pollen from *Entada phaseoloides* and *Leucaena leucocephala*. The pollen morphology of the *Mimosa* genus has been extensively studied. For instance, pollen grains of about 255 species have been described, of which 247 were studied analyzed by light microscopy, 216 by scanning electron microscopy and 19 by transmission electron microscopy [37,41,63–68]. These studies documented detailed descriptions of the pollen morphologies of many *Mimosa* taxa, showing the variation in the different types of pollen within the genus, i.e., polyads, tetrads and octads in various shapes and sizes, with verrucate to microverrucate sculpturing, porate apertures in the subdistal position and the presence of an operculum and annulus [69]. However, in the present study, this was not noted in *Mimosa bimucronata*, where the bitetrad 8-grain polyads were arranged in two planes (4 + 4) with a rugulate exine structure.

Jamwal et al. [70] investigated the characteristic features of *Albizia* pollen, i.e., polyad (16-celled), prolate–spheroidal, tricolporate, isopolar and radially symmetric. Sizes ranged from 57 μm × 49 μm. The tectum was previously investigated and shown to be psilate to faveolate. Similarly, Aftab and Perveen [45] conducted a palynological study of cultivated trees in Karachi, Pakistan, and observed 14 celled polyads, which were bilaterally symmetrical, with pollen sizes pf 79.59 μm × 74.235 μm and a tectum subpsilate for *Albizia lebback*. In comparison with this, 12 grains per polyad were observed in the present study for *A. lebback*, meaning our findings did not corroborate theirs.

In *Acacia*, polymorphism has been found in the grains of each polyad (numbering 4, 8, 12, or 16 grains) in Australian species, while 16 to 32 grains were found in polyads in African species [71]. Five species of Pakistani *Acacia* were studied by Perveen and Qaiser [44], who concluded that morphological differences in pollen of the Mimosoideae species are significant at the tribe and genera levels. Caccavari and Dome [9] analyzed the pollen morphologies of 77 American *Acacia* species and suggested that pollen features can be used as a distinguishing factor for the generic limitation of *Acacia*. Similarly, Rajurkar et al. [72], reported 16 grains per polyad in *Acacia* species and variations in their morphological traits, i.e., size, shape and exine pattern, which were found to be helpful in genera- and species-level identification. However, in comparison with these findings, 16 grains per polyad were found in the present study for *Acacia auricifoliformis* and 12 grains per polyad were found in *A. confusa* and *A. mangium*. The latter species can be delimited from the former due to the presence of operculum on the central grain surface of the polyad. Hence, our study also confirmed that morphological traits of the pollen of *Acacia* species are useful at the generic and specific levels.

The pollen morphology of *Leuceana leucocephala* species was noted as a perforate and scabrate exine pattern, while pollen units of monad and tricolpate types were found in the present study. In comparison with a previously published study [46], the variation in pollen morphology of the *Leucaena* species was found to be helpful in showing the relationship and defining generic boundaries within Mimosoideae. For instance, they reported perforate and punctate exine sculpturing with pollen units of monad and tricolporate types in *Leucaena leucocephala*. However, psilate to scabrate exine sculpturing in Pakistani *L. leucocephala* was found by Khan et al. [41]. These findings mostly support our study results.

The pollen morphologies of 11 *Calliandra* taxa were investigated by De Assis-Ribeiro dos Santos and De Oliveira-Romão [47], who concluded that some of the polyad characteristics are useful in corroborating previous infrageneric classifications of the genus. Several palynological studies included *Calliandra* taxa. Sorsa [37], gathered data from nine species. Similarly, Guinet and Hernández [73] made a comparison between *Calliandra* and *Zapoteca*, the former of which has eight grains per polyad, while the latter has 16 grains per polyad. Our study corroborated these findings (8-grain polyads) in *Calliandra haematocephala*. Other studies [74–76] recorded the pollen morphology of one or species of the genus.

4. Conclusions

Taxa from the subfamily Mimosoideae collected from different geographical regions on Hainan Island were analyzed using both light microscopy and scanning electron microscopy. Both LM and SEM morphological descriptions of the pollen can be helpful for plant taxonomists to correctly identify and discriminate the Mimosoideae taxa at generic and specific levels. Significant variation was found in both the qualitative and quantitative traits of the pollen, confirming the Mimosoideae species as eurypalynous. The pollen traits were significantly proven to have taxonomic potential that will support and strengthen the systematics of this subfamily. This study will also provide a basis for further phylogenetic and molecular studies to strengthen the systematics of the Mimosoideae subfamily. Heteromorphism was found in the shape of the monads in the pollen association (polyads, bitetrad, octads). Additionally, exines exhibited heteromorphism in the colpus surface membrane ornamentation of the studied taxa. Interspecies variation was present in the exine ultrastructure of the pollen grains of the studied Mimosoideae taxa. Furthermore, palynological, anatomical, molecular and phylogenetic studies are needed to support and strengthen the systematics of the Mimosoideae subfamily.

Author Contributions: All authors contributed to the study conception and design. Funding acquisition, provision of resources, study design, supervision of the project and critical review of the manuscript were performed by W.L. Material preparation, data collection and analyses were performed by S.B. Material preparation and analysis and interpretation of the data were performed by S.T. The first draft of the manuscript was written by S.B.,S.T. and U.H. wrote and revised the manuscript. All authors have read and agreed to the published version of the manuscript.

Funding: This research project was supported by the National Natural Science Foundation of China (31870508, 31660163, 32171772).

Institutional Review Board Statement: Not applicable.

Informed Consent Statement: Not applicable.

Data Availability Statement: The data presented in this study are available within the article or on request from Saraj Bahadur.

Acknowledgments: This research article is part of the Ph.D. thesis of the first author. We are thankful to Hainan University for providing the SEM facility and the College of Forestry for providing the LM facility. We are also thankful to our labmates from the College of Forestry for their useful suggestions and help, especially in providing materials.

Conflicts of Interest: The authors declare no conflict of interest.

References

1. Lewis, G.; Schrire, B.; Mackinder, B.; Lock, M. *Legumes of the World*; The Bath Press (CPI Group): The Royal Botanic Gardens, Kew, UK, 2005; p. 577.
2. Legume Phylogeny Working Group. A new subfamily classification of the Leguminosae based on a taxonomically comprehensive phylogeny. *Taxon* **2017**, *66*, 44–47. [CrossRef]
3. Shinwari, Z.K.; Jamil, K.; Zahra, N.B. Molecular systematics of selected genera of subfamily Mimosoideae-Fabaceae. *Pak. J. Bot.* **2014**, *46*, 591–598.
4. Polhill, R.M. Evolution and systematics of the *Leguminosae*. *Advences Legume Syst.* **1981**, *1*, 1–26.
5. Vural, C.; Ekici, M.; Akan, H.; Aytaç, Z. Seed morphology and its systematic implications for genus Astragalus L. sections Onobrychoidei DC., Uliginosi Gray and Ornithopodium Bunge (*Fabaceae*). *Plant Syst. Evol.* **2008**, *274*, 255–263. [CrossRef]
6. Fawzi, N.M. Macro-and micromorphological seed characteristics of some selected species of Caesalpinioideae-Leguminosae. *Res. J. Bot.* **2011**, *6*, 68–77. [CrossRef]
7. Mabberley, D.J. *The Plant-Book: A Portable Dictionary of the Vascular Plants*; Cambridge University Press: Cambridge, UK, 1997.
8. Murphy, D.J.; Brown, G.K.; Miller, J.T.; Ladiges, P.Y. Molecular phylogeny of *Acacia* Mill. (Mimosoideae: Leguminosae): Evidence for major clades and informal classification. *Taxon* **2010**, *59*, 7–19. [CrossRef]
9. Caccavari, M.; Dome, E. An Account of Morphology and Structural Characterization of American Mimosoideae Pollen. Part I: Tribe Acaceae. *Palynology* **2000**, *24*, 231–248. [CrossRef]
10. Hutchinson, J. *The Genera of Flowering Plants. Dicotylédones*; Oxford University Press: London, UK, 1964; Volume 1.
11. Panicker, K.T.C.; Sreedevi, P. Studies in the pollen morphology of Mimosaceae monad, tetrad and octad taxa. *J. Palynol.* **2004**, *40*, 9–21.
12. Costa, J.A.S.; Nunes, T.S.; Ferreira, A.P.L.; Stradmann, M.T.S.; Queiroz, L.P. *Leguminosasforrageiras da Caatinga: Espéciesimportantespara as Comunidadesrurais do Sertão da Bahia*; UniversidadeEstadual de Feira de Santana, SASOP: Feira de Santana, Brazil, 2002; pp. 1–8.
13. Souza, R.S.O. Jurema-preta (*Mimosa tenuiflora* (Willd.)Poir.): A review of its traditional use, phytochymistry and pharmacology. *Braz. Arch. Biol. Technol.* **2008**, *51*, 937–947. [CrossRef]
14. Ahmad, F.; Anwar, F.; Hira, S. Review on medicinal importance of Fabaceae family. *Pharmacologyonline* **2016**, *3*, 151–157.
15. Rahman, A.H.M.M.; Parvin, M.I.A. Study of medicinal uses on Fabaceae family at Rajshahi, Bangladesh. *Res. Plant Sci.* **2014**, *2*, 6–8.
16. Molares, S.; Ladio, A. The usefulness of edible and medicinal Fabaceae in Argentine and Chilean Patagonia: Environmental availability and other sources of supply. *Evid. Based Complement Alternat. Med.* **2012**, *2012*, 901918. [CrossRef]
17. Turnbull, J.W. (Ed.) Australian vegetation. In *Multipurpose Australian Trees and Shrubs: Lesser-Known Species for Fuelwood and Agroforestry*; Australian Centre for International Agricultural Research: Canberra, Australia, 1986; Volume 14, pp. 29–44.
18. Sornsathapornkul, P.; Owens, J.N. Pollination Biology in a Tropical Acacia Hybrid (*A. mangium* Willd. × *A. auriculiformis* A. Cunn. ex Benth.). *Ann Bot.* **1998**, *81*, 631–645. [CrossRef]
19. Joseph, B.; George, J.; Mohan, J. Pharmacology and traditional uses of *Mimosa pudica*. *Int. J. Pharm. Sci. Drug Res.* **2013**, *5*, 41–44.
20. Mambou, A.H.M.Y.; Bila, R.B.; Wanyu, B.Y.; Nkwemeh, C.N.; Jugha, V.T.; Toukam, L.L.; Bopda, O.S.M.; Taiwe, G.S.; Bum, E.N. Antifatigue properties of an aqueous extract of *Mimosa pudica* Linn. (Fabaceae) in mice subjected to weight loaded force swimming test. *GSC Biol. Pharma. Sci.* **2022**, *18*, 224–233. [CrossRef]
21. Arasu, M.V.; Al-Dhabi, N.A.; Choi, K.C.; Bensy, A.D.V.; Rajaselvam, J. Bioactive potential of *Albizia lebbeck* extract against phytopathogens and protective properties on tomato plant against speck disease in greenhouse. *Phy. Mol. Plant Patho.* **2022**, *117*, 101750. [CrossRef]
22. Bahadur, S.; Ahmad, M.; Mir, S.; Zafar, M.; Sultana, S.; Ashfaq, S.; Arfan, M. Identification of monocot flora using pollen features through scanning electron microscopy. *Microsc. Res. Tech.* **2018**, *81*, 599–613. [CrossRef]
23. Ali, M.; Liu, Y.J.; Xia, Q.P.; Bahadur, S.; Hussain, A.; Shao, J.W.; Shuaib, M. Pollen micromorphology of eastern Chinese Polygonatum and its role in taxonomy by using scanning electron microscopy. *Microsc. Res. Tech.* **2021**, *84*, 1451–1461. [CrossRef]
24. Ashfaq, S.; Zafar, M.; Ahmad, M.; Sultana, S.; Bahadur, S.; Khan, A.; Shah, A. Microscopic investigations of palynological features of convolvulaceous species from arid zone of Pakistan. *Microsc. Res. Tech.* **2018**, *81*, 228–239. [CrossRef]
25. Ashfaq, S.; Ahmad, M.; Zafar, M.; Sultana, S.; Bahadur, S.; Ahmed, S.N.; Gul, S.; Nazish, M. Pollen morphology of family Solanaceae and its taxonomic significance. *An. Acad. Bras. Cienc.* **2020**, *92*, e20181221. [CrossRef]
26. Bahadur, S.; Taj, S.; Ahmad, M.; Zafar, M.; Gul, S.; Shuaib, M.; Butt, M.A.; Hanif, U.; Nizamani, M.M.; Hussain, F.; et al. Authentication of the therapeutic Lamiaceae taxa by using pollen traits observed under scanning electron microscopy. *Microsc. Res. Tech.* **2022**, *84*, 1–19. [CrossRef]

27. Bahadur, S.; Taj, S.; Long, W.; Ahmad, M. Pollen morphology and its implication in the taxonomy of some selected taxa of the bi and tri-ovulate Euphorbiaceae of the Hainan Island by using multiple microscopic techniques. *Microsc. Res. Tech.* **2022**, *26*, 1–16. [CrossRef]
28. Butt, M.A.; Zafar, M.; Ahmad, M.; Kayani, S.; Bahadur, S.; Ullah, F.; Khatoon, S. The use of taxonomic studies to the identification of wetlands weeds. *Adv. Weed Sci.* **2021**, *39*, e222645. [CrossRef]
29. Hameed, A.; Zafar, M.; Ahmad, M.; Sultana, S.; Bahadur, S.; Anjum, F.; Shuaib, M.; Taj, S.; Irm, M.; Altaf, M.A. Chemo-taxonomic and biological potential of highly therapeutic plant Pedicularis groenlandica Retz. using multiple microscopic techniques. *Microsc. Res. Tech.* **2021**, *84*, 2890–2905. [CrossRef]
30. Javed, S.; Javaid, A.; Hanif, U.; Bahadur, S.; Sultana, S.; Shuaib, M.; Ali, S. Effect of necrotrophic fungus and PGPR on the comparative histochemistry of Vigna radiata by using multiple microscopic techniques. *Microsc. Res. Tech.* **2021**, *84*, 2737–2748. [CrossRef]
31. Nabila; Ahmad, M.; Zafar, M.; Bahadur, S.; Sultana, S.; Taj, S.; Celep, F.; Majeed, S. Palynomorphological diversity among the Asteraceous honeybee flora: An aid to the correct taxonomic identification using multiple microscopic techniques. *Microsc. Res. Tech.* **2022**, *85*, 570–590. [CrossRef]
32. Ullah, F.; Ahmad, M.; Zafar, M.; Parveen, B.; Ashfaq, S.; Bahadur, S.; Safdar, Q.; Bin Safdar, L.; Alam, F.; Luqman, M. Pollen morphology and its taxonomic potential in some selected taxa of Caesalpiniaceae observed under light microscopy and scanning electron microscopy. *Microsc. Res. Tech.* **2021**, *85*, 1410–1420. [CrossRef]
33. Zaman, W.; Ahmad, M.; Zafar, M.; Amina, H.; Lubna; Ullah, F.; Bahadur, S.; Ayaz, A.; Saqib, S.; Begum, N.; et al. The quest for some novel antifertility herbals used as male contraceptives in district Shangla, Pakistan. *Acta Ecol. Sin.* **2020**, *40*, 102–112. [CrossRef]
34. Rasool, S.; Faheem, M.; Hanif, U.; Bahadur, S.; Taj, S.; Liaqat, F.; Pereira, L.; Liaqat, I.; Shaheen, S.; Shuaib, M.; et al. Toxicological effects of the chemical and green ZnO NPs on Cyprinus carpio L. observed under light and scanning electron microscopy. *Microsc. Res. Tech.* **2021**, *85*, 848–860. [CrossRef]
35. Rubab, S.; Rizwani, G.H.; Bahadur, S.; Shah, M.; Alsamadany, H.; Alzahrani, Y.; Shuaib, M.; Hershan, A.; Hobani, Y.H.; Shah, A.A. Determination of the GC–MS analysis of seed oil and assessment of pharmacokinetics of leaf extract of *Camellia sinensis* L. *J. King Saud Univ. Sci.* **2020**, *32*, 3138–3144. [CrossRef]
36. Park, J.M.; Song, U.S. Pollen morphology of the genus *Rhododendron* (Ericaceae) in Korea. *J. Korean Soc. Forest Sci.* **2010**, *99*, 663–672.
37. Sorsa, P. Pollen morphological studies on the *Mimosaceae. Ann. Bot. Fenn.* **1969**, *6*, 1–34.
38. Guinet, P.H. Les Mimosacees, Etude de palynologiefundamentale, correlations, evolution. *Inst. Fr. Pondichery Trav. Sci. Tech.* **1969**, *9*, 1–293.
39. Guinet, P. Mimosoideae: The characters of their pollen grains. In *Advances in Legume Systematics*; Polhill, R.R., Raven, P.H., Eds.; The Bath Press (CPI Group): The Royal Botanic Gardens, Kew, UK, 1981; Volume 2, pp. 835–855.
40. Khan, S.U.; Zafar, M.; Ahmad, M.; Anjum, F.; Sultana, S.; Kilic, O.; Ozdemir, F.A.; Nazir, A.; Yaseen, G.; Aabidin, S.Z.U. Pollen micromorphological analysis of tribe *Acacieae* (*Mimosaceae*) with LM and SEM techniques. *Microsc. Res. Tech.* **2019**, *82*, 1610–1620. [CrossRef]
41. Khan, S.U.; Zafar, M.; Ullah, R.; Shahat, A.A.; Ahmad, M.; Sultana, S.; Malik, K. Pollen diversity and its implications to the systematics of mimosaceous species by LM and SEM. *Microsc. Res. Tech.* **2021**, *84*, 42–55. [CrossRef]
42. El Ghazali, G.E.; Satti, A.M.; Tsuji, S.I. Intra-specific pollen polymorphism in *Mimosa pigra* (Mimosaceae). *Grana* **1997**, *36*, 279–283. [CrossRef]
43. Kordofani, M.; Ingrouille, M. Geographical variation in the pollen of *Acacia* (Mimosaceae) in Sudan. *Grana* **1992**, *31*, 113–118. [CrossRef]
44. Perveen, A.; Qaiser, M. Pollen Flora of Pakistan-XI. *Leguminosae* (Subfamily: Mimosoideae). *Turk J Botany.* **1998**, *22*, 151–156.
45. Aftab, R.; Perveen, A. A palynological study of some cultivated trees from Karachi. *Pak. J. Bot.* **2006**, *38*, 15–28.
46. Hughes, C.E. Variation in anther and pollen morphology in *Leucaena* Benth. (Leguminosae-Mimosoideae). *Bot. J. Linn. Soc.* **1997**, *123*, 177–196. [CrossRef]
47. De Assis Ribeiro dos Santos, F.; de Oliveira Romão, C. Pollen morphology of some species of *Calliandra* Benth. (Leguminosae-Mimosoideae) from Bahia, Brazil. *Grana* **2008**, *47*, 101–116. [CrossRef]
48. Zhi-Min, L.I. Studies on the Pollen Morphology of Some Genera in Mimosaceae from China. *Plant Divers.* **1994**, *16*, 1–3.
49. Simon, M.F.; Grether, R.; Queiroz, L.P.; Särkinen, T.E.; Dutra VF Hughes, C.E. The evolutionary history of *Mimosa* (Leguminosae): Toward a phylogeny of the sensitive plants. *Am. J. Bot.* **2011**, *98*, 1201–1221. [CrossRef]
50. Long, W.-X.; Ding, Y.; Zang, R.-G.; Yang, M.; Chen, S.-W. Environmental characteristics of tropical cloud forests in the rainy season in Bawangling National Nature Reserve on Hainan Island, South China. *Chin. J. Plant Ecol.* **2011**, *35*, 137. [CrossRef]
51. Long, W.; Yang, X.; Li, D. Patterns of species diversity and soil nutrients along a chronosequence of vegetation recovery in Hainan Island, South China. *Ecol. Res.* **2012**, *27*, 561–568. [CrossRef]
52. Wang, X.X.; Long, W.X.; Yang, X.B.; Xiong, M.H.; Kang, Y.; Huang, J.; Wang, X.; Hong, X.J.; Zhou, Z.L.; Lu, Y.Q.; et al. Patterns of plant diversity within and among three tropical cloud forest communities in Hainan Island. *Chin. J. Plant Ecol.* **2016**, *40*, 469.
53. Punt, W.; Hoen, P.P.; Blackmore, S.; Nilsson, S.; Le Thomas, A. Glossary of pollen and spore terminology. *Rev. Palaeobot. Palynol.* **2007**, *143*, 1–81. [CrossRef]

54. Erdtman, G. *Pollen Morph Ology and Plant Taxonomy. Angiosperms*; Chronica Botanica Co.: Waltham, MA, USA, 1952; Volume 539, p. 261.
55. Hyde, H.A.; Adams, K.F. *An Atlas of Airborn Pollen Grains*; Macmillan & Co.: London, UK, 1958; 112p.
56. Brown, G.K.; Murphy, D.J.; Miller, J.T.; Ladiges, P.Y. Acacia ss and its relationship among tropical legumes, tribe Ingeae (Leguminosae: Mimosoideae). *Syst Bot.* **2008**, *33*, 739–751. [CrossRef]
57. Wodehouse, R.P. Preparation of pollen for microscopic examination. *Bull. Torrey Bot. Club.* **1933**, *60*, 417–421. [CrossRef]
58. Wang, X.J.; Cao, X.L.; Hong, Y. Isolation and characterization of flower-specific transcripts in *Acacia mangium*. *Tree Physiol.* **2005**, *25*, 167–178. [CrossRef]
59. Agashe, S.N.; Caulton, E. *Pollen and Spores: Applications with Special Emphasis on Aerobiology and Allergy*, 1st ed.; CRC Press: Boca Raton, FL, USA, 2009; p. 412. [CrossRef]
60. Van Campo, M.; Guinet, P. Les pollen compose's. L'exemplede^sMimosace´es. *Pollen Spores* **1961**, *8*, 201–218.
61. Chen, Y.-Y. Studies on the development of polyad grains of *Calliandra haematocephala* Hassk. with electron microscopic technique. *Taiwania* **1973**, *17*, 18–28.
62. Rao, A.N.; Lee, Y.K. Studies on Singapore pollen. *Pac. Sci.* **1970**, *24*, 255–268.
63. Santos-Silva, J.; Simon, M.; Tozzi, A.M.G.D.A. Pollen diversity and its phylogenetic implications in Mimosa ser. Leiocarpae Benth. (Leguminosae, Mimosoideae). *Grana* **2013**, *52*, 15–25. [CrossRef]
64. Caccavari, M.A. Granos de polen de las leguminosas de la Argentina IV. Genero Mimosa. *Bol. Soc. Argent. Bot.* **1985**, *24*, 151–167.
65. Caccavari, M.A. Estudio de los caracteres del polen en las MimosaLepidotae. *Pollen Spores* **1986**, *28*, 29–42.
66. Caccavari, M.A. 1987. Study of pollen grains in Mimosa Glanduliferae. In Proceedings of the VII Argentine Symposium on Paleobotany and Palynology, Buenos Aires, Argentina, 30 August 2007; pp. 141–145.
67. Caccavari, M.A. Ultraestructura del polen de Mimosa (Mimosoideae-Leguminosae). *Pollen Spores* **1988**, *30*, 275–296.
68. Caccavari, M.A. Pollen morphology and structure of tropical and subtropical American genera of the Piptadenia-group (Leguminosae: Mimosoideae). *Grana* **2022**, *41*, 130–141. [CrossRef]
69. Medina-Acosta, M.; Grether, R.; Martínez-Bernal, A.; Ramírez-Arriaga, E. Comparative study of pollen morphology and exine ultrastructure in tetrads, octads and polyads of the genus *Mimosa* (Leguminosae). *Palynology* **2019**, *43*, 188–212. [CrossRef]
70. Jamwal, R. Palynological investigations on some selected bee forage plants of family fabaceae using light and scanning electron microscopy from himachal pradesh, India. *Plants Arch.* **2021**, *21*, 1047–1053. [CrossRef]
71. Kenrick, J.; Knox, R.B. Function of the polyad in reproduction of Acacia. *Ann. Bot.* **1982**, *50*, 721–727. [CrossRef]
72. Rajurkar, A.V.; Tidke, J.A.; Jadhav, S.S. Palynomorphological Studies on Family Mimosaceae. *Int. J. Pharma. Bio. Sci.* **2013**, *4*, 273–279.
73. Guinet, P.; Hernández, H.M. Pollen characters in the genera Zapoteca and Calliandra (Leguminosae, Mimosoideae), their systematic and phylogenetic relevance. *Pollen Spores* **1989**, *31*, 5–22.
74. Thulin, M.; Guinet PHunde, A. *Calliandra* (Leguminosae) in continental Africa. *Nord. J. Bot.* **1981**, *1*, 27–34. [CrossRef]
75. Niezgoda, C.J.; Feuer, S.M.; Nevling, L.I. Pollen ultrastructure of the tribe *Ingeae* (Mimosoideae-Leguminosae). *Am. J. Bot.* **1983**, *70*, 650–667. [CrossRef]
76. Hernández, H.M.; Sousa, S.M. Two new species of *Calliandra* (Leguminosae: Mimosoideae) from Southern Mexico. *Syst. Bot.* **1988**, *13*, 519–524. [CrossRef]

agronomy

MDPI

Article

Macro-Morphological and Ecological Variation in *Rosa sericea* Complex

Fazal Ullah [1,2], Yundong Gao [1,*], İdris Sari [3], Rui-Fang Jiao [1,4], Saddam Saqib [2,5] and Xin-Fen Gao [1,*]

1 CAS Key Laboratory of Mountain Ecological Restoration and Bioresource Utilization and Ecological Restoration, Biodiversity Conservation Key Laboratory of Sichuan Province, Chengdu Institute of Biology, Chinese Academy of Sciences, Chengdu 610041, China; fazalbotanist@gmail.com (F.U.); 2019322040062@stu.scu.edu.cn (R.J.)
2 University of Chinese Academy of Sciences, Beijing 100049, China; saddamsaqib.qau@gmail.com
3 Department of Biology, Faculty of Science and Art, Erzincan Binali Yıldırım University, Erzincan 24100, Turkey; botanical24@gmail.com
4 Key Laboratory of Bio-Resources and Eco-Environment of the Ministry of Education, College of Life Sciences, Sichuan University, Chengdu 610064, China
5 State Key Laboratory of Systematic and Evolutionary Botany, Institute of Botany, Chinese Academy of Sciences, Beijing 100093, China
* Correspondence: gaoyd@cib.ac.cn (Y.G.); xfgao@cib.ac.cn (X.G.)

Abstract: Taxon delimitation and identification are important in many areas of biology, especially in taxonomy and conservation. Taxonomic treatment is required to establish and justify recommendations in conservation biology for the group being considered. Imperfect and controversial taxonomy can undermine conservation assessment. We studied 71 populations; 665 individuals corresponding to the morphology of the *Rosa sericea* complex (including six taxa, one of which has extremely narrow distributions) were collected from sympatric, parapatric, and allopatric populations distributed in China. This study aims to investigate whether the complex species are macromorphologically different species and evaluate the rare taxa of the complex for conservation priority. The morphological characters and principal component analysis (PCA) of the *R. sericea* complex showed that the complex species have overlapping characters but can distinguish morphologically. The species of *R. sericea* complex systematics status based on previous DNA sequencing is controversial. The ecological habitat's current morphological characters only delimit the *R. morrisonensis* (in Taiwan). To evade mistakes in species conservation, we recommend that taxonomical knowledge be needed to ensure success in protecting target species. Thus, the complementarity of systemic and conservation assurance makes conservation actions more necessary for the complex's rare taxa. The ecological niche modelling (ENM) results showed that habitats of these conspecific taxa would be shrunken. With the presence of snapshots in time, the geography of taxa might decrease rapidly in representative entirely of the Geographic space (G-space) and Environmental space (E-space) that such taxa are bright to inhabit. So far, the significant inferences meant for the niche occupy the most incredible comparative research, taking the impermanent nature of taxa distributions and undertaking that such species are at a state of stability. If the artificially identified species (rarely distributed) are based on morphological identification, they must be conserved.

Keywords: species boundaries; species concept; taxonomic relationship; conservation; future climate change; ecology

Citation: Ullah, F.; Gao, Y.; Sari, İ.; Jiao, R.-F.; Saqib, S.; Gao, X.-F. Macro-Morphological and Ecological Variation in *Rosa sericea* Complex. *Agronomy* **2022**, *12*, 1078. https://doi.org/10.3390/agronomy12051078

Academic Editor: Marco Scortichini

Received: 25 January 2022
Accepted: 27 April 2022
Published: 29 April 2022

1. Introduction

The concept of species is crucial in evolutionary research and all biological thoughts. Species are the fundamental unit of life and biological diversity with a specific karyotype, morphology, DNA sequence, ecological niche, or behaviour [1]. Taxonomists made several efforts to describe species, beginning from simple morphology to genetics [2]. Earlier taxonomists such as Linnaeus defined species based on what they saw. Later on, this was formalized as a morphological or typological species concept. Reproductive isolation is considered the key standard for delimiting the species [3–5]. Yet, the limitation between closely related taxa becomes unclear with hybridization or continued gene flow [6]. Delimitation of species is important to distinguish between the species of the same nature, and diagnostic characters are helpful to separate all sorts of species boundaries in biological sciences. The idea of delimitation is known as the species concept [7,8]. Though, the species concept itself is under disputation. To understand the taxonomy of biodiversity, especially conservation biology, the delimitation of species is vital for the controversial taxa. Furthermore, species are important, especially in conserving the existing biodiversity in the face of the extinction disaster of the Anthropocene [9–11].

The complex group of plants exhibits diverse forms. Taxonomic boundaries between putative taxa are often concealed by a lack of identified fixed morphological differences, potential hybridization, and a lack of informative collections. The taxonomically complex taxa are sometimes difficult to differentiate and delimit due to their complex characters and distribution. The artificial taxonomy or the species delimit through confused characters sometimes leads the taxa in danger for biological conservation [12,13]. Misidentification could lead to underestimating or overestimating species richness, and these difficulties could entirely compromise the investigation. The poor taxonomy could risk understanding ecological patterns since they are based on richness and measurement of species turnover between sites [14,15]. Conservation actions are taken without accurate taxonomic identifications may impair the effective conservation of the target taxa. The goal of taxonomic studies should not be restricted to distinguishing and describing living things, and it seems impossible to talk about conservation without taxonomy [15]. For species delimitation, taxonomic variations are important [15].

Though systematic is usually regarded as the science of diversity, it played a limited role in developing current approaches to conservation biology. In contrast, ecological values and population genetic concepts are more widely incorporated into conservation theory and management strategies [16,17]. Different studies examined species delimitation based on morphological and geographical distribution [18–20]. However, these studies were limited to one or two factors, whereas species limitation is based on various drivers such as interaction with climate, evolutionary traits, morphological characteristics, and biogeographical characteristics. In other words, species delimitation and conservation can be addressed by focusing on different levels, from genes to populations, ecosystems, and species concepts [21,22]. Scientists favour biodiversity as more integrative based on genetic and genomic approaches [23]. Systematics offers a foundation of information for conservation biology equally valuable as that of population genetics [19,20]. Systematic is vital due to the linkage of a taxon to other relative species based on taxonomic and molecular characterizations [24]. Species delimitation without an accurate taxonomic identification is incomplete. However, funding agencies have neglected taxonomy nowadays because of being descriptive discipline [15]. Difficulties and hurdles in systematics, ecology, and taxonomic determination among species hamper research in areas of species definition [25]. For species delimitation, identification is important for the complex taxa [15]. Funding agencies need to provide financial assistance for taxonomy and ecology to conduct extensive field works to explore rare and endangered species, especially in megadiverse countries like China. The technique red lists would be viewed and used for poorly known species, depleted species, population decline, restricted ranges, and rarity based on all these criteria. Procedures associated with priority situations and the progress of national red lists are essential to justify some expectations in formulating the criteria [26]. Taxonomy, ecological distribution threats, and population knowledge are important for the red-list assessments and adequate capacity to process and analyze data. Both capacity and data are absent for numerous species-rich taxa, despite their great ecological importance [27,28].

The taxa of the *R. sericea* complex occur at a high latitudinal gradient from 1000 to more than 4000 m above sea level [6,29]. Based on the previous study, we hypothesized that the species of *R. sericea* complex showed few distinct macro-morphological characteristics. Our observation demonstrates that the studied species have clear morphological characters to differentiate the taxa in the complex. So, in this research, we applied population morphological and ecological data (precipitation, temperature, and nineteen bio factors) to observe the variation in the focal taxa and their distribution. This paper aims to distinguish between the species of the *R. sericea* complex using morphological and ecological characteristics and conservation status to attain the following objectives: (1) To delineate species boundaries in *R. sericea* complex by adopting more comprehensive approaches based on populations sampling; (2) To exclude vulnerably, threatened taxa utilizing the above process; and (3) To determine how accurate and efficient morphological variation is possible based on proper species delimitation.

2. Materials and Methods

2.1. Study Area, Populations Morphology, and Record of Samples

A total of 665 individuals of 71 populations of six taxa of *R. sericea* complex (*R. sericea* Lindley (RS) 36 populations, 311 individuals; *R. omeiensis* Rolfe (RO) 24 populations, 285 individuals; *R. sikangensis* T. T. Yu and T. C. Ku (RSK) 8 populations, 29 individuals; *R. mairei* H. Léveillé (RM), *R. morrisonensis* Hayata (RMO) with 10, 21 individual, respectively; *R. zhongdianensis* T. C. Ku (RZ) 1 population, 9 individuals) were collected from China (Table S1) and studied at the CDBI herbarium. RMO is the only taxon with one population collected from Taiwan (Figure 1). This work examined fifty morphological characteristics for each individual: twenty-six leaf characters, twenty-one flower and fruit characters, and three other traits. The mean values of the quantitative morphological characters were used for principal component analysis (PCA). Our studied populations showed that the species of this complex individuals sometimes share and grow in the same ecological habitat, except the species RMO (from Taiwan) and RM, and RZ from southwest China. During the collection, we press the samples in the newspapers in the field and bring them to CDBI for a detailed study. The population records and codes information are given in Table 1. The detailed morphological features were studied in the herbarium of the Chengdu Institute of Biology (CDBI). We used a binocular dissecting microscope to distinguish the study taxa based on detailed characters studied for a concise and clear character. The macro-morphological characters of each specimen studied herein are described to determine the species macromorphologically. We also selected some taxonomically important characters to distinguish the complex taxa species. We constructed dichotomous keys and taxonomic descriptions to identify these taxa easily (Table 2). Some characters were noted differently within the same population, while some features were observed overlap between the different species populations. Various qualitative and quantitative characters have been examined in detail.

2.2. PCA Analysis of Morphological and Environmental Factor

The PCA analysis was carried out for the morphological and environmental factors to understand the relationship between the studied taxa. The morphological characters were analyzed using qualitative, quantitative, and ecological characters to see the complex taxa's structure, relationship, and species boundaries. We did PCA analysis for macro morphological characteristics, qualitative and quantitatively, and the environmental factors (Figure S1).

2.3. Correlation between Quantitative Characters and Environmental Factors

The quantitative data and the 19 bio parameters (https://www.worldclim.org/, (accessed on 15 June 2021) were subjected to analysis of variance in R studio to evaluate the difference between quantitative characters and environmental factors relationship. We used the Pearson linear correlation bivariate between the quantitative and ecological characters of different morphological characters.

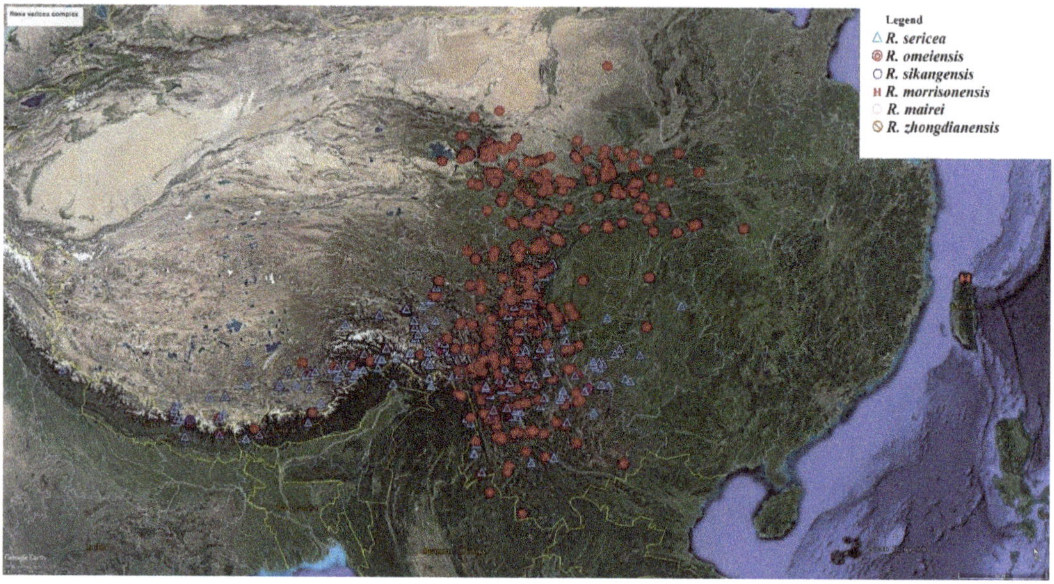

Figure 1. The distribution of different taxa of the *Rosa sericea* complex based on different herbaria samples. The map is based on different Herbaria records i.e., CDBI, PE, QTPMB, NAS, KUN, LZH, and KATH.

Table 1. Sample collection information for this research.

Species	Locality	Latitude	Longitude	Altitude	Population Code	Collection Year
RS	Tibet, Yadong	27.42	88.94	2786	GXF-16646	2018
RS	Tibet, Cuona	27.91	91.80	2935	GXF-16692	2018
RS	Tibet, Gyaca	29.13	92.69	3188	GXF-16734	2018
RS	Sichuan, Baoxing	30.77	102.72	2325	GXF-17017	2019
RO	Sichuan, Baoxing	30.83	102.72	3127	GXF-17018	2019
RO	Sichuan, Xianojin	30.89	102.65	3593	GXF-17022	2019
RO	Sichuan, Erlang	29.85	102.29	2885.69	GXF-17314	2020
RS	Sichuan, Erlang	29.85	102.26	2738	GXF-17316	2020
RS	Sichuan, Erlang	29.84	102.25	2497.53	GXF-17322	2020
RS	Sichuan, Erlang	29.8464	102.26	2409	GXF-17323	2020
RS	Sichuan, Erlang	29.84	102.25	2247.08	GXF-17324	2020
RO	Sichuan, Xiangcheng	29.12	99.99	3970.7	GXF-17373	2020
RS	Sichuan, Xiangcheng	29.14	99.97	3717.34	GXF-17375	2020
RO	Sichuan, Xiangcheng	29.13	99.97	3831.88	GXF-17376	2020
RO	Sichuan, Xiangcheng	29.14	99.96	3679.29	GXF-17377	2020
RS	Sichuan, Xiangcheng	29.15	99.93	3572	GXF-17380	2020
RS	Sichuan, Xiangcheng	29.15	99.93	3412.35	GXF-17383	2020
RS	Sichuan, Xiangcheng	29.14	99.91	3176.9	GXF-17384	2020
RS	Sichuan, Xiangcheng	29.11	99.91	2941.69	GXF-17385	2020
RS	Sichuan, Xiangcheng	28.97	99.84	2797.29	GXF-17386	2020
RS	Tibet, Yadong	27.45	88.92	2872.27	GXF-17510	2020
RS	Tibet, Yadong	27.51	88.95	3088.38	GXF-17512	2020
RS	Tibet, Yadong	27.55	89.00	3455.09	GXF-17513	2020
RS	Tibet, Yadong	27.51	88.95	3232.9	GXF-17514	2020
RS	Tibet, Jilong	28.78	85.30	3955.38	GXF-17538	2020

Table 1. *Cont.*

Species	Locality	Latitude	Longitude	Altitude	Population Code	Collection Year
RS	Tibet, Jilong	28.43	85.26	2919.29	GXF-17539	2020
RO	Tibet, Jilong	28.49	85.22	3167.12	GXF-17540	2020
RO	Tibet, Jilong	28.51	85.22	3335.99	GXF-17542	2020
RO	Tibet, Jilong	28.37	85.33	2749.6	GXF-17543	2020
RS	Tibet, Jilong	28.40	85.35	2751.14	GXF-17559	2020
RS	Tibet, Jilong	28.64	85.27	3733.56	GXF-17561	2020
RO	Tibet, Jilong	28.55	85.24	3580.76	GXF-17562	2020
RS	Tibet, Lhasa	29.74	91.15	3880.1	GXF-17571	2020
RS	Tibet, Ding Qing	31.21	95.79	3537.08	GXF-17575	2020
RS	Tibet, ChangDu	31.09	96.98	4094.1	GXF-17576	2020
RS	Tibet, ChangDu	31.10	97.00	3851.54	GXF-17578	2020
RS	Tibet, ChangDu	31.12	97.02	3707.31	GXF-17579	2020
RS	Tibet, ChangDu	31.16	97.02	3515.18	GXF-17580	2020
RS	Tibet, ChangDu	31.19	97.03	3334.63	GXF-17581	2020
RS	Tibet, ChangDu	30.6886	97.250197	4168	GXF-16771	2018
RS	Yunnan, Zhaotong	27.46603	104.14769	1750–1789	GXF-12800	2011
RS	Yunnan, Zhaotong	27.24207	104.1603	1450–1500	GXF-12804	2011
RS	Yunnan, Zhaotong	27.50387	105.12936	1380–1450	GXF-12818	2011
RS	Guizhou, Liupanshui	26.39813	104.45166	1970	GXF-12849	2011
RM	Yunnan, Qujing	26.4845	103.589875	2185	GXF-16785	2019
RO	Hubei, Xingshan	31.28560	110.18061	2150–2180	GXF-13074	2011
RO	Hubei, Yichang	31.27466	110.1341	2700	GXF-13088	2011
RO	Shaanxi, Ankang	32.1228	109.18652	2057	GXF-13160	2011
RO	Shaanxi, Xian	33.50604	108.48483	1800–1915	GXF-13168	2011
RO	Gansu, Pingliang	35.10787	106.21600	2110–2300	GXF-13220	2011
RO	Gansu, Baiyin	37.8084	103.44553	2618–2820	GXF-13238	2011
RO	Gansu, Lanzhou	35.47600	104.3323	2200–2600	GXF-13250	2011
RO	Gansu, Dingxi	35.57906	104.0568	2600	GXF-13259	2011
RO	Yunnan, Shangri-la	28.34458	99.50008	4250	GXF-13407	2011
RO	Yunnan, Dêqên	28.20123	99.5514	4180	GXF-13442	2011
RS	Tibet, Nagqu	31.11892	94.2563	3960	GXF-15366	2011
RO	Qinghai, Xining	36.59186	101.44413	2500	GXF-15470	2011
RO	Qinghai, Haidong	35.49680	102.41725	1950	GXF-15500	2011
RMO	Taiwan, Hualian	24.141085	121.283714	3180	GXF-15625	2012
RS	Guizhou, Bijie	27.65232	105.380261	1645	GXF-17006	2019
RO	Tibet, ChangDu	30.4127	97.1608	3600	GXF-12543	2010
RO	Tibet, Yadong	27.57816	89.02537	3800	GXF-16639	2018
RZ	Yunnan, Diqing	28.094916	99.481725	3007	GXF-17390	2020

2.4. Distribution and Ecological Characters

We pursued to relate and contrast the geographic ranges and environmental tolerance of putative taxa within the studied taxa. Our studied samples were only belonging to the geographical ranges of China. We obtained the environmental data while using our field collection data of given populations. The distribution data of 71 populations and the spatially unique data at 2.5 arc minutes (for future distribution) resolution ($5 * 5$ km^2 at the equator) of 63 recorded points. We used the ENMSDM package and China elevation data (30 arcs second for current), taken from the SRTM elevation data of 63 records. We took and used nineteen temperature, seasonality, and perception variables from WorldClim [30].

Table 2. Dichotomous taxonomic key of the Taxa of *Rosa sericea* complex based on macromorphological features.

1	+	number of leaflets minimum 5–11	2
	−	number of leaflets minimum 7	6
2	+	number of leaflets maximum 7–15, length of leaf minimum 12.5–50 mm, and maximum 20–80 mm, sepals ovate-lanceolate, abaxially sparsely pubescent or subglabrous, adaxially villous, margin entire, apex acuminate or acute	*R. sericea*
	−	number of leaflets maximum 9–17, length of leaf minimum 20–80 mm, and maximum 35–115 mm, sepals lanceolate, abaxially subglabrous, adaxially sparsely pubescent, margin entire, apex acuminate or long caudate	3
3	+	length of the first leaflet on apex minimum 5–20 mm, maximum 9–41.1, leaf margins single serrate	*R. omeiensis*
	−	length of the first leaflet on apex minimum 5.12–10.29 mm, maximum 8.11–19.78, leaf margins double serrate	4
4	+	leaf margin double tooth, thrones of petiole and rachis rare, sparse, medium, dense, length of first leaflets base minimum, 3.12–6.78 mm, and maximum, 5.14–11.23	*R. sikangensis*
	−	leaf margin single tooth, thrones of petiole and rachis absent, length of first leaflets base minimum 3–6 mm, and maximum, 5–9 mm	5
5	+	Hip colour red or bright to reddish, length of pedicel minimum 2–7 mm, and maximum 3–8 mm	*R. mairei*
	−	Hip colour orange to red-purple, length of pedicel minimum 4–7 mm, and maximum 5–10 mm	6
6	+	Minimum number of leaflets 7, maximum 9–11, length of leaflet ranges from 15–45 mm	*R. morrisonensis*
	−	Minimum number of leaflets 7–9, maximum 9–11, length of leaflets ranges from 21–47 mm	7
7	+	Sepal broadly ovate, apex shortly caudate, sepal size 6–10 mm, width 2–5 mm	*R. taronensis*
	−	Sepal lanceolate, apex acuminate, sepal size 9–17 mm, width 2.5–3.5 mm	*R. zhongdianensis*

2.5. Modelling the Species Complex Current and Future Distribution

2.5.1. Records of the Species Complex and Bioclimatic Variables

We collected 63 presence points (records) of the species complex during fieldworks from 2010 to 2020 (Table 1). After thinning records of the species using the enmSdm package [31] and China's elevation data (30 arc seconds) derived from Worldclim (https://www.worldclim.org, accessed on 15 June 2021), we obtained 62 records of the species complex. Nineteen bioclimatic variables are crucial in defining species' climatic niches downloaded from Worldclim dataset v2.1 (30 arc-seconds) [32] for the current period. Because the high correlation between variables is expected to affect model performance and increase uncertainty in model results. Spearman rank correlation (r_s) was calculated using R packages (scales and legendary) in R v4.04 to evaluate multicollinearity among the bioclimatic variables. In the analysis output, positive correlations ($r_s \geq 0.7$) between variable pairs were drawn in black and negative correlations ($r_s \leq -0.7$) were drawn in red.

2.5.2. MaxEnt Modelling of the Species Complex Distribution

We modelled the current and future distribution of the species complex using the MaxEnt method [33] implemented through Maxent v3.4.4 in the SDMtune package. SDMtune uses a particular object to compile the data for the analysis. This object, called SWD, bundles all the information related to each record, thereby reducing the risk of mistakes in further investigations [34]. After creating the SWD object, we trained a model using all 19 climatic variables. Then we considered the variable importance of the Jackknife test of the first model and correlations among bioclimatic variables to model species complex distribution with fewer variables. Because a study [35] revealed that seed dormancy breaking requirements and timing of seedling emergence of the species have a chilling need, Mean Temperature of Coldest Quarter (Bio11) was chosen as the predictor variable among highly correlated variables. In the end, paying regard to all of these predictions obtained, we decided to model the distribution of the species complex with Temperature Seasonality (Bio04), Bio11, and Precipitation of Coldest Quarter (Bio19) (Figure S2).

The results of SDMs under future climatic conditions are affected by a range of factors, including the choice of the statistical model, climate model range, and emission scenarios [36,37]. Although the Intergovernmental Panel on Climate Change (IPCC) considers all Global Climate Models (GCMs) equal, certain GCMs better represent some climate types and regions. For example, among the GCMs, BCCCSM (Beijing Climate Centre, Climate System Model) has higher reliability and has been better studied in China. The GCM used in the study was the BCC-CSM2-MR GCM [34] (2.5 arc minutes) obtained from (https://www.worldclim.org/data/cmip6/cmip6_clim2.5m.html, accessed on 15 June 2021). While the scenarios used were the Shared Socioeconomic Pathways (SSPs; ssp2 4.5 and ssp5 8.5) (CMIP6), which are now being used as important inputs for the latest climate models, feeding into the Intergovernmental Panel on Climate Change (IPCC) sixth assessment report is due to be published in 2021 (https://tntcat.iiasa.ac.at/SspDb/dsd?Action=htmlpage&page=welcome, accessed on 22 June 2021; https://www.ipcc.ch/assessment-report/ar6, accessed on 22 June 2021). The species complex's future distribution was predicted for 20-year periods (2021–2040, 2041–2060, 2061–2080, and 2081–2100).

2.5.3. Tuning Model Hyperparameters and Evaluation Model Performance

Tuning model hyperparameters is a long process, as it requires testing many combinations to identify the best-performing model [34]. To get the model with high predictive power, we tuned model parameters using 'gridSearch' function implemented in SDMtune. Since model evaluation measures the capacity of a given model to reflect "truth" and whether it can be applied under other conditions [38] to assess the performance of the models. We used receiver operating characteristic (ROC) curve analysis and the true skill statistic (TSS). Thresholds for interpreting the TSS values were defined as follows: Value ≥ 0.9 = best, $0.9 >$ Value ≥ 0.8 = very good, $0.8 >$ Value ≥ 0.7 = good and Value < 0.7 = weak [39].

In creating distribution maps of the species, five threshold values were used. Accordingly, '0–0.25' denotes unsuitable areas, '0.25–0.5' of very low suitability, '0.75–1' highly suitable. We also calculated the size of moderate and highly suitable (we call them suitable here) predicted by models to quickly assess future distribution areas of the species complex and to see if the species complex would have gain or loss in the size of suitable habitat areas in the future. To do that, first, we reclassified highly suitable sites as '1' and other areas as '0' using ArcMap v10.8. Second, we converted raster files into shape format. Lastly, we calculated the size of the regions from shape data. Distribution maps of the species complex and cartography works were created in QGIS v3.18.

To grasp more information about the Rsc climatic niche and distribution in China, we used the "rmaxent" and "ENMTools" that have some excellent features implemented in functions about the species niches (URL: https://github.com/johnbaums/rmaxent, accessed on 9 June 2021) [40]. In the R maxent, we modelled the species' climatic niche using all bioclimatic variables to reveal which variable/variables affect the species' complex distribution in China. In the ENMTools, we measured the spatial heterogeneity of the distribution of suitability scores from the model results we obtained. This feature returns Levins' two metrics of niche breadth (B1 and B2) [40].

3. Results

Different morphological characters have been studied in this article for the six taxa of the *R. sericea* complex (RO, RS, RMO, RM, RSK, and RZ) (Figure 2). In addition, we focused on the conservation of rare species of the complex.

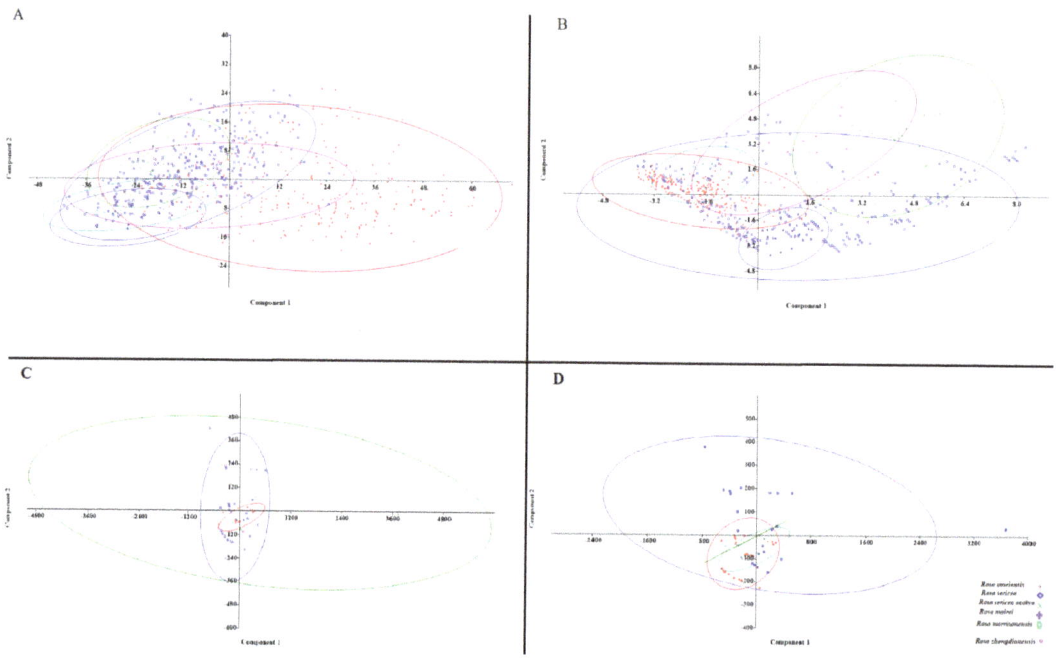

Figure 2. PCA analysis of quantitative morphological characteristics of *R. sericea* complex (**A**), PCA analysis of qualitative morphological features of *R. sericea* complex (**B**), PCA analysis of environmental characteristics population-based of *R. sericea* complex (**C**), PCA analysis of morphological characters and environmental factors of *R. sericea* complex (**D**).

3.1. Morphological Variation

Variation has been observed in different morphological characteristics of the studied taxa. The characters investigated in the focal taxa samples were somehow morphologically significant for delimiting the taxa into distinct taxon. We identified different morphological characters to differentiate the study taxa of the complex. A detailed morphological description of the study taxa based on qualitative and quantitative features was revised. In the present work, we studied 50 qualitative and quantitative characters of each individual, which showed some variation in the studied species. Moreover, these characters were analyzed with the environmental characters, and the correlation is given in (Figure 3). The vegetative morphology was mainly focused on the leaf morphological characters. One of the most important characteristics was the leaflet's morphology and leaflets in different individuals (species). The species boundaries in the complex taxa were evaluated based on morphological characters (Appendix A).

Each individual's qualitative and quantitative characteristics were examined in the CDBI herbarium and statistically analyzed. The description of these species was mainly based on the morphological characteristics of collected populations specimens from different geographical distribution ranges of the *R. sericea* complex. Population information is given in (Table 1), and morphological characters descriptions are given in (Appendix A). The studied species of this complex have tetramerous perianth. The height of these shrubby plants' species ranged from 1 to 4 m.

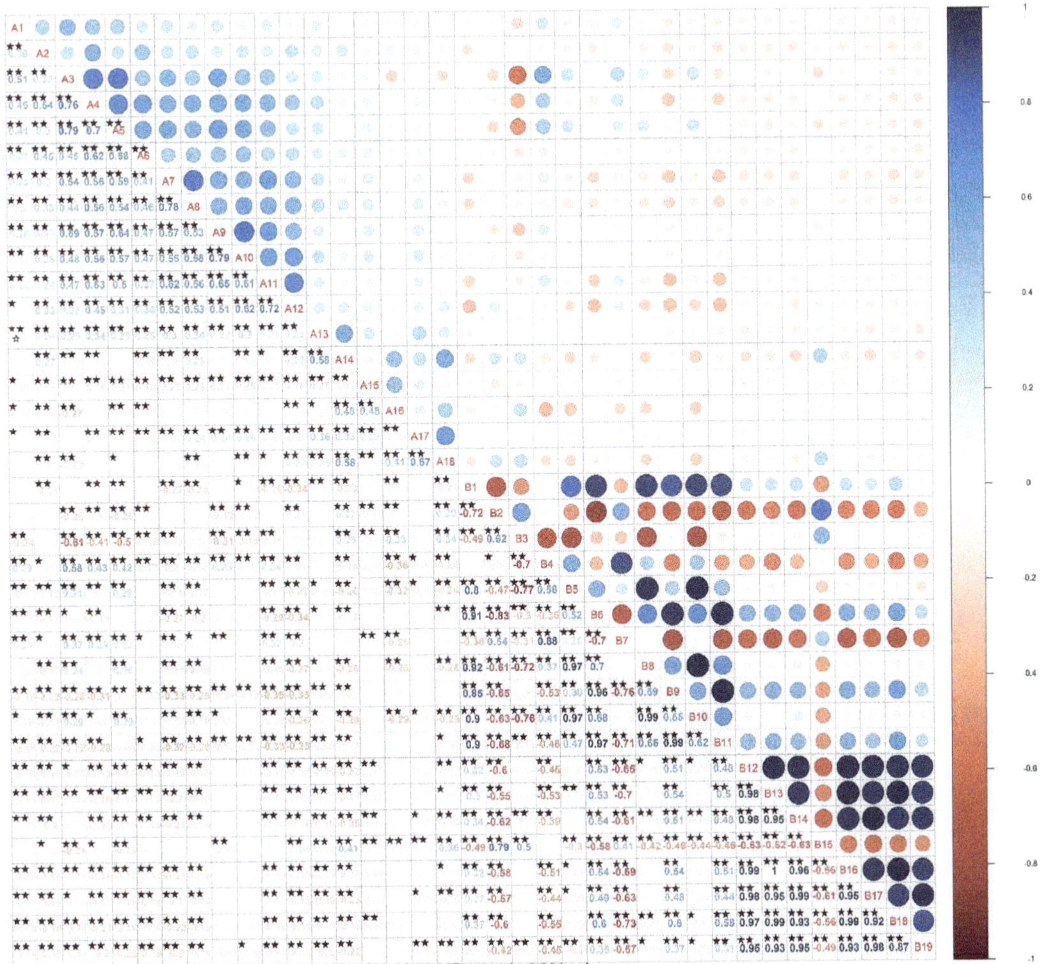

**Correlation is significant at the 0.01 level and * Correlation is significant at the 0.05 level.

Figure 3. Correlation between the 19 bio and quantitative characteristics (1 = Length of leaf minimum, 2 = Length of leaf maximum, 3 = length of the first leaflet on apex minimum, 4 = length of the first leaflet on apex maximum, 5 = width of the first leaflet on apex minimum, 6 = width of the first leaflet on apex maximum, 7 = length of the first leaflet on-base minimum, 8 = length of the first leaflet on-base maximum 9 = width of the first leaflet on-base minimum, 10 = width of the first leaflet on-base maximum, 11 = length of pedicel minimum, 12 = length of pedicel maximum, 13 = length of sepal minimum, 14 = length of sepal maximum, 15 = width of sepal minimum, 16 = width of sepal maximum, 17 = bio1, 18 = bio2, 19 = bio3, 20 = bio4, 21 = bio5, 22 = bio6, 23 = bio7, 24 = bio8, 25 = bio9, 26 = bio10, 27 = bio11, 28 = bio12, 29 = bio13, 30 = bio14, 31 = bio15, 32 = bio16, 33 = bio17, 34 = bio18, 35 = bio19) the bio 19 at 2.5 arc minute resolution and projected based on present-day climatic conditions the codes of bio 19 is according to https://www.worldclim.org/, accessed on 15 June 2021. The graphical representation showed a correlation between −1 and 1 from brown and red to bluish.

3.2. Ecological Niche Modelling

The complex taxa except for RS, and RO, while the taxa RSK, RM, RMO, and RZ have narrow ecological amplitude, and hence their distribution is rare. Based on our field observations and herbarium records, the population of the one taxon RZ (with nine individuals with one population) compiled as taxon of very narrow ecological distribution, followed by RM, RMO, and RSK. Some types specimens of this taxon were available in the herbarium. Phylogenetically these species are complex, and the evolutionary tree of these taxa quite complicated. This species are closely interrelated, and morphologically synapomorphies have been examined (Appendix A).

These taxa have phylogenetically complicated systematics. The species boundaries were difficult to differentiate [6,29], while the morphological characters showed that the species have distinct characteristics to define species boundaries. Due to the narrow ecological amplitude and rarity, these taxa should be conserved. The phylogenetic classification is essential if we are thinking of organizing the biodiversity in such a way as to establish conservation significance and improve informed conservation policies [41,42]. Additionally, revised classification is dreaded by conservation practitioners and other taxonomy users. The new classification of plants is based on molecular or phylogenetic analyses considered the basic basis of practical taxonomy [43,44].

From our fieldwork and analysis of morphological characters, and work previously done on the species of the complex [45], these taxa were clustered in the same clade [46]. These taxa's field observations and geographical distribution showed that RS and RO have a broad ecological amplitude. In contrast, RSK and RM have moderate distribution, while the taxonRZ have a narrow ecological amplitude and few populations in nature. The RMO has occurred in the isolated habitat in Taiwan. We have recorded one population with 21 individuals (Figure 1). Morphologically these species could be differentiated because they have shown morphological species concepts to distinguish them. We assessed the quantitative characteristics of these taxa showed variation in the studied taxa. Taxonomic distinctness based on morphological characters showed that and fulfilled the conservation criteria due to their narrow biogeographical distribution and rarity in nature. Populations extinction is one of the significant threats to plant diversity, foremost to range reductions, disintegration, and isolation, which reduces species abundance.

The current SDMs of the *R. sericea* complex predicted through the model provide very high success rates with training and test AUC values of 0.90 and 0.97, respectively. These findings indicate that the predictor variables used for SDMs were suitably selected, thus leading to very high prediction success. The AUC curves in developing *R. sericea* complex SDM under current environments are given (Figure 4). The TSS, AUC, and variable importance values for the complex conspecific under future climatic are given (Figure 4). The present finding obtained from the model, the sum of the three first potential distribution variables was 59.407%. The higher the percent contribution, the more important that a variable is for predicting the occurrence of the taxa. Bio 11 had the highest predictive contribution in the present study, 28.44%.

The current potential distribution of the taxa showed that the area with red indicates a highly suitable area for the complex species. In contrast, the blue indicates the site is not suitable for taxa (Figure 5). The BCC-CSM2-MR GCM and the SSPs (ssp2 4.5 and ssp5 8.5) (CMIP6) based models predicting future habitat suitability of the *R. sericea* complex taxa are given in (Figures 6 and 7). Concluding from these futures modelling, the results indicate that the species may have difficulties shortly. Some of the rare taxa of the complex will be extinct due to their narrow ecological amplitude.

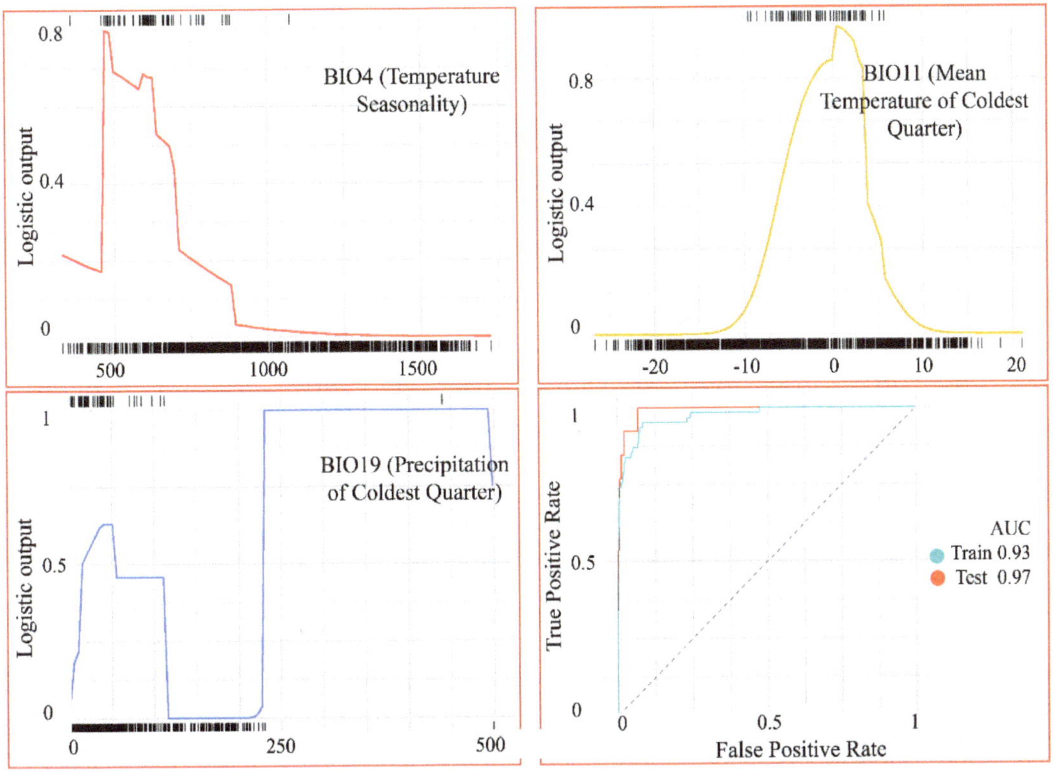

Figure 4. The response of *R. sericea* complex to three climatic variables and ROC curve of the model.

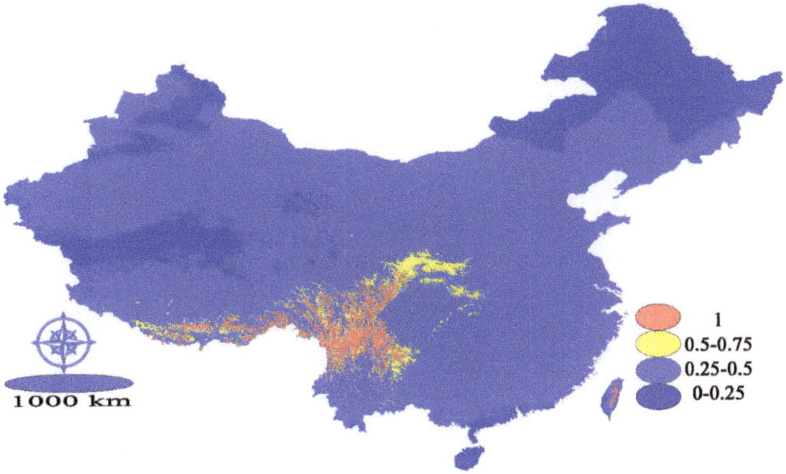

Figure 5. The current potential distribution model of *R. sericea* complex.

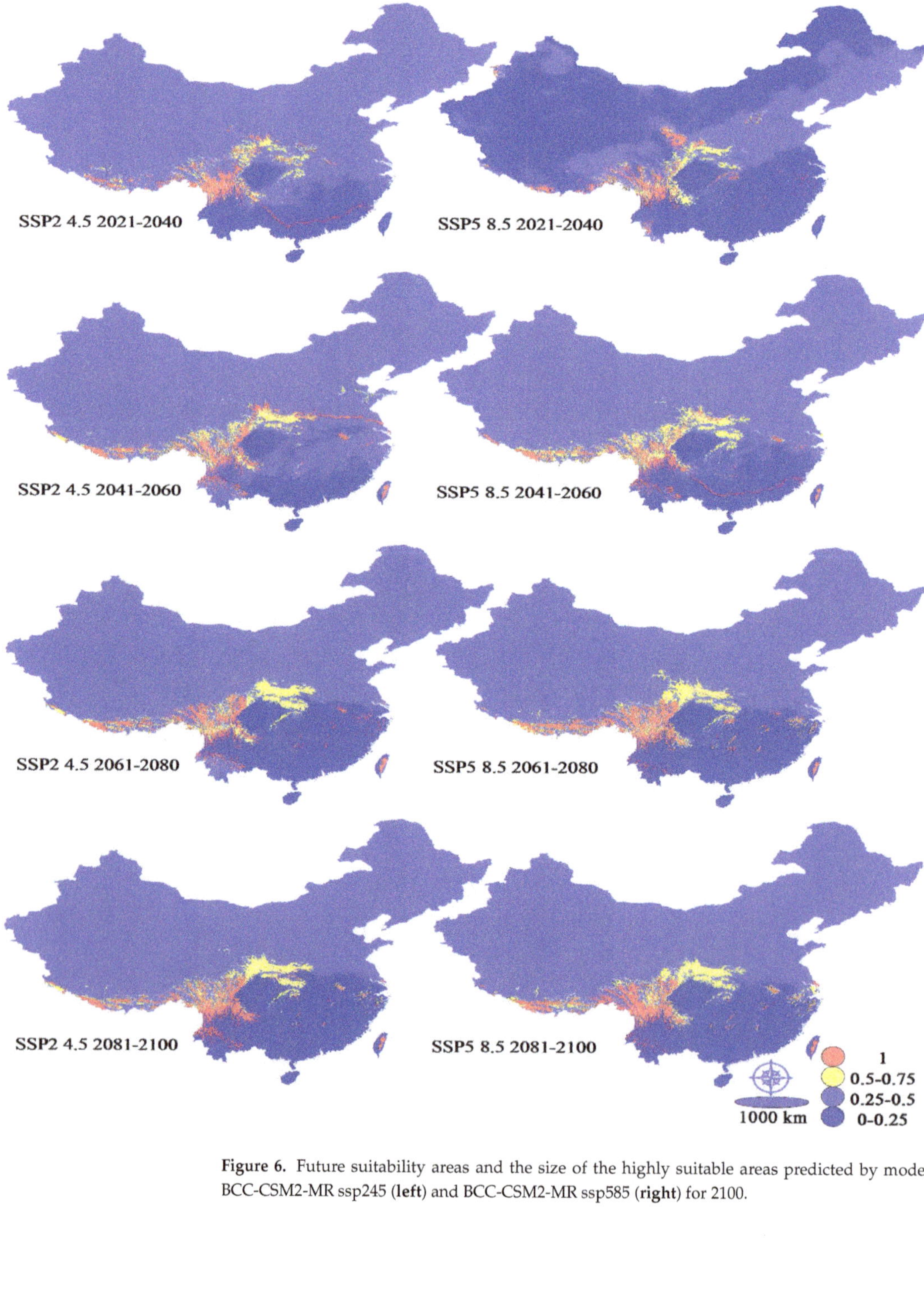

Figure 6. Future suitability areas and the size of the highly suitable areas predicted by models BCC-CSM2-MR ssp245 (**left**) and BCC-CSM2-MR ssp585 (**right**) for 2100.

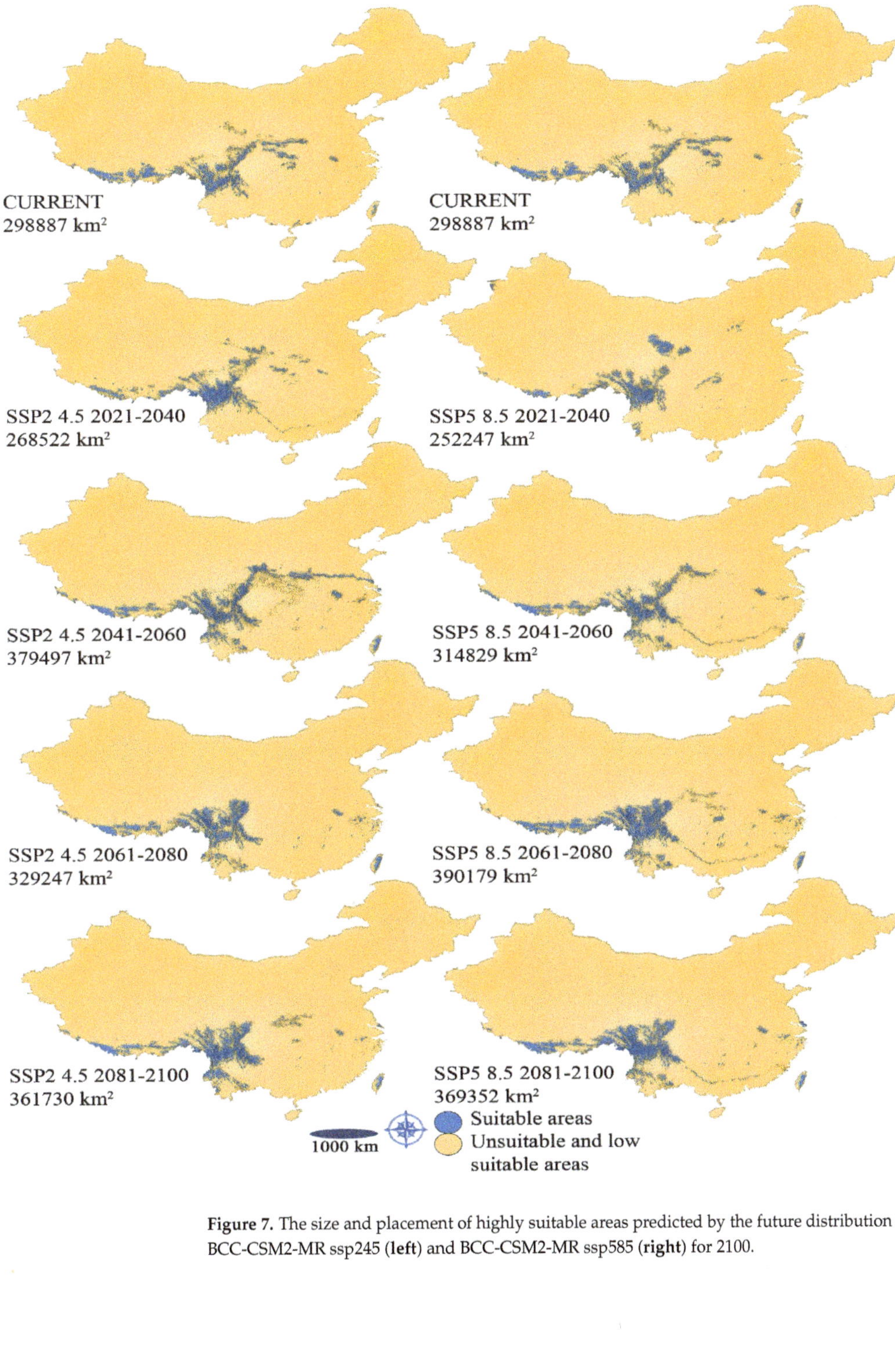

Figure 7. The size and placement of highly suitable areas predicted by the future distribution models BCC-CSM2-MR ssp245 (**left**) and BCC-CSM2-MR ssp585 (**right**) for 2100.

3.3. *Environmental Factors, Morphological Variation, and Correlation*

The ecological species concept of the complex taxa somehow showed variation, and some of the populations occurred parapatrically. Our present work's geographical distribution of RS populations showed a wide distribution range from 1380 to 4168 m in altitude. In comparison, RO has altitudinal variation from 1800 to 4250 m, while RSK has altitude distribution from 2797 to 3851 m. Some RS, RO, and RSK shared the same ecological habitats and grow parapatrically in the study area. We collected each population specimen of RM (2185 m Yunnan, Qujing) and RZ (3007 m Yunnan Diqing), and one population of RMO from an altitude of 3180 m, from Taiwan, Hualian (Table 1, Figure 1). RMO has a unique ecological boundary among the studied taxa and occurs in Taiwan. RZ could also be isolated easily from the rest of the taxa of the complex allopatrically.

The correlation between the quantitative data and 19 bio data is given in (Figure 7). The correlation between the environmental factors and quantitative morphological features shown with single star * correlation is significant at the 0.05 level, while two-star ** correlation is significant at the 0.01 level (Figure 3). The positive and negative correlations between different parameters are shown in (Figure 3). In detail, some of the quantitative characters showed a negative correlation represented with the negative sign (-), while some of the variables have a positive correlation at 0.05 and 0.01 levels. Some variables don't show any significant correlation represented in the Figure without the star * symbol.

4. Discussion

4.1. *Ecotypes Plasticity and Species Boundaries*

Different populations of the *R. sericea* complex were studied here. The individuals' morphological characters showed variation in characters and morphologically distinct taxa. Some populations shared the same environmental and ecological habitat of the complex species. Due to similar characteristics in different species populations and individuals with different characteristics, the complex species had a wide range of character's plasticity (Figure 2A–C). The phenotypic plasticity could activate together through ecotype formation, further increasing suitability. The modest level of phenotypic plasticity may facilitate a population's expansion into novel environments. The trait may place the populations on the slope of an adaptive peak from which natural selection can advance [47]. The RMO adapted to a novel environment and isolated to a new habitat in Taiwan. The limitations to the entirely isolated territory and fundamental asymmetry in migration due to differences in density-dependent viability both redirect selection on a phenotypically plastic trait. Together, these environmental factors might cause the appearance of feature values and genetic constitution that seem specialized to one of the extreme environments of the territory [48]. In the study by Gao et al. [6]), RS and RO were observed parapatrically. However, we observed that the RS, RO, and RSK occur parapatrically, sympatrically, and allopatrically (Figure 1).

Documentation of morphological, environmental, and distribution data provides a basis for the species status of divergent populations of some complex taxa. Conversely, the morphological and ecological characteristics showed that the taxa are distinct taxon based on distinctive features (Table 2). The morphological characters showed that the taxa of the complex are different species. In contrast, predictable morphological features, and environmental and distribution data accept and support different taxon statuses. In our study, the divergent parapatric would stand and be treated as species by default like RMO, RM, and RZ occur parapatrically. Taxon RS has extensive qualitative morphological variation identified in our populations, while the RO has a wide range of characters quantitatively. The PCA analysis showed combinedly that the characters of RO populations are dominated and almost share similar morphological characters with other taxa of the complex (Figure 2A,B,D). Our environmental data showed that the species RMO and RZ have unique environmental characteristics and are isolated from the other taxa of the complex. The factors include low temperature and high precipitation. The morphological characters retain the status of these taxa to distinguish taxon and prove that the taxa in the complex are morphologically different [49–51]. Species of the complex may be a single hyper-diverse taxon or contrasted that the speciation or evolution has happened or is ongoing. If the speciation occurred or is ongoing, we would assume to detect apparent morphological and genetic variations, which may show more robust evidence if speciation is completed, or weaker if the speciation is ongoing or confounded by introgression [52,53]. Overall, however, the morphological characteristics support the status of the species as different taxon.

4.2. PCA Analysis of Environmental and Morphological Characteristics

The PCA analysis of environmental and morphological features demonstrated some significant information about the complex species. We used PCA analysis of ecological features and quantitative and qualitative morphological characters to visualize the relationships and differences among the six species of the complex. For the ecological parameters, we used 2.5 arc minutes resolution ($\sim$5 km^2 at the equator), for which we obtained 19 temperature, perception, and seasonality variables from WorldClim [30,32].

The PCA analysis of only quantitative morphological characteristics showed more variation in the studied specimens of the same species from different populations (Figure 2A and Table 3). The qualitative morphological characters PCA analysis of the studied specimens of diverse populations shows overlap and difficulty in distinguishing the taxa from each other in the studied taxa (Figure 2B and Table 4). We studied the PCA analysis of nineteen bio factors and 12-months temperature and precipitation. Environmental aspects of the studied specimens and populations showed significant importance in differentiating some of the studied taxa (Figure 2C). We also determined the PCA analysis of 19 bio factors, 12-month temperature minimum, maximum, and precipitation combined with quantitative morphological characters (Figure 2D). The PCA analysis of environmental factors and quantitative features of various populations showed the association between the studied taxa (Figure 2D, Table 5).

Table 3. PCA analysis of quantitative morphological characteristics.

PC	Eigenvalue	% Variance
1	475.905	74.668
2	76.9841	12.079
3	36.3351	5.7008
4	13.8951	2.1801
5	10.0186	1.5719
6	7.29266	1.1442
7	5.10194	0.80048
8	2.61096	0.40965
9	1.871	0.29355
10	1.7679	0.27738
11	1.58519	0.24871
12	1.36646	0.21439
13	0.839891	0.13178
14	0.629685	0.098795
15	0.543176	0.085222
16	0.25117	0.039408
17	0.214402	0.033639
18	0.151487	0.023768

Table 4. PCA analysis of qualitative morphological characteristics.

PC	Eigenvalue	% Variance
1	7.45804	31.325
2	3.51155	14.749
3	2.77815	11.669
4	1.5873	6.6669
5	1.55423	6.528
6	0.958856	4.0273
7	0.951377	3.9959
8	0.843049	3.5409
9	0.627619	2.6361
10	0.575717	2.4181
11	0.515355	2.1646
12	0.381654	1.603
13	0.318081	1.336
14	0.253594	1.0651
15	0.238815	1.0031
16	0.197126	0.82796
17	0.182921	0.76829
18	0.169516	0.71199

Table 5. PCA analysis of environmental characteristics of the studied populations.

PC	Eigenvalue	% Variance
1	280,221	93.582
2	13,001.9	4.3421
3	3998.36	1.3353
4	1244.91	0.41574
5	424.43	0.14174
6	254.889	0.085122
7	109.348	0.036517
8	60.992	0.020369
9	46.6265	0.015571
10	28.1597	0.0094041
11	23.9664	0.0080037
12	8.89486	0.0029705
13	6.6421	0.0022182
14	3.56371	0.0011901
15	2.38813	0.00079753
16	1.31154	0.000438
17	0.826218	0.00027592
18	0.770616	0.00025735

4.3. Morphological Characters Evaluation

Morphological characters for species delimitation are usually directly dependent on the field observation and herbarium specimens. These closely interrelated taxa showed dissimilarity in their morphological observation. For the morphological study, we focus on two parts of a plant: (i) the vegetative part and (ii) the reproductive part of the specimens. The vegetative morphology is generally referred to as leaf morphology and is related to different size measures of a species [54]. Morphological features used for taxa delimitation and identification is dependent directly on the studied species specimens.

The present research finding showed that the morphological characters of the *R. sericea* complex sometimes have complicated taxonomic characters, which is challenging to differentiate the complex species. This estimation on the morphological characters and species collection localities, where some taxa co-occur over the geographical regions investigated (Table 1). The study species here are capable of hybridizing with each other. Some of the individuals are commonly possible to assign to species in mixed populations. The species RS and RO in a previous study [6] differentiated through leaflet numbers. The RS showed fewer leaflets, while RO has more leaflets morphologically. Several ongoing or historical evolutionary processes clarify the morphological dissimilarity among entities within the studied taxa. One possibility is that the species of *R. sericea* complex represents a single hyperdiverse species phylogenetically. Due to this, these species display genetic panmixia or high levels of gene flow due to less or more constant contact between the populations [6,55,56]. The hyper-diverse species' morphological characters may be intergrading or demonstrating an approximate relationship with the regional environment [52,53].

4.4. Conservation Strategy

According to Darwin, rarity is a significant precursor of extinction [57,58]. Precise assessments of rare taxon populations and the significant threats are vital to conservation planning and resource allocation for recovery action [59]. As phylogenetically, these species are quite difficult to differentiate, and maybe the species are under ongoing speciation. Our group team has been working on these species for more than a decade [45,46,51,60], and we have information about very few populations, such as the demographic status of RSK, RM, and RZ. The RZ taxon have few numbers of individuals as compared to RSK and RM. As the populations and individual numbers of the one taxon RZ are limited, there is a great chance of drift and inbreeding, altering the genetic structure [61]. Many studies of genetic variation in rare plant taxa have demonstrated that the proportion of polymorphic loci and the number of alleles per locus significantly reduced in small-sized populations [62].

The results of our ENMs indicated that the complex species would have difficulty shortly if climate change drives the taxa into extinction. According to [63], climate change is quantitatively considered in Red list assessments for only a few species. The authors of [64] highlighted that determining the application of SDMs to Red Lists is challenging due to model uncertainties and many biotic and abiotic factors that cannot be studied in such models. They suggested that SDMs and Red List assessments might play a complementary role in conservation actions, such as Red List provides evidence on both current and future risk of extinction for the species, while SDMs warn the magnitude of future extinction risk. Due to the shrinking habitat of the complex in the future climate scenarios, some of the taxa of the complex will lead to extinction due to their rarity and narrow ecological amplitude (Tables 6–8).

Table 6. PCA analysis of environmental and quantitative characteristics.

PC	Eigenvalue	% Variance
1	524,075	95.616
2	17,160.3	3.1308
3	4136.92	0.75477
4	1088.62	0.19861
5	505.171	0.092167
6	373.855	0.068209
7	320.227	0.058424
8	114.599	0.020908
9	72.3684	0.013203
10	61.2497	0.011175
11	44.4784	0.0081149
12	33.4481	0.0061025
13	23.8053	0.0043432
14	22.553	0.0041147
15	12.2862	0.0022416
16	10.5593	0.0019265
17	9.75494	0.0017798
18	7.67523	0.0014003

Table 7. TSS and AUC values obtained from the models implemented in this study.

SSPs and CURRENT SDM	AUC (Train)	TSS (Train)
BCC-CSM2-MR ssp245 2021–2040	0.97	0.90
BCC-CSM2-MR ssp245 2041–2060	0.94	0.83
BCC-CSM2-MR ssp245 2061–2080	0.95	0.82
BCC-CSM2-MR ssp245 2081–2100	0.96	0.83
BCC-CSM2-MR ssp585 2021–2040	0.95	0.81
BCC-CSM2-MR ssp585 2041–2060	0.94	0.82
BCC-CSM2-MR ssp585 2061–2080	0.97	0.86
BCC-CSM2-MR ssp585 2081–2100	0.96	0.83
CURRENT SDM	0.93	0.92

Table 8. Variable importance of the first model carried out with all climatic variables.

Variable	Percent Contribution	Permutation Importance
bio11	28.4441	37.0614
bio10	21.0601	6.4193
bio12	9.9030	2.2631
bio03	8.1910	6.4651
bio02	7.4465	12.7615
bio04	6.7611	0.0000
bio19	3.4775	10.5466
bio01	2.0986	0.0000
bio06	2.0245	0.5266
bio18	1.7853	9.9935
bio15	1.6388	2.0871
bio14	1.6258	4.9419
bio05	1.5583	0.2241
bio09	1.5465	2.8801
bio07	1.3294	0.4519
bio17	1.0934	3.3778
bio16	0.0161	0.0000
bio08	0.0000	0.0000
bio13	0.0000	0.0000

The rare species of this complex should be conserved in natural habitats where the species grow. Direct population management will help conserve these rare species of the *R. sericea* complex in the appropriate habitat. Given current land uses (and other pressures of the Anthropocene), however, human interference may be required to maintain the habitat suitable for conserving these rare taxa. As a result, it is not only species that are conservation reliant, but entire ecosystems and the associated disturbance regimes and ecological succession pathways that define them need to be conserved. However, this requires sufficient data on species' distribution and abundance. The limitation between closely related taxa becomes unclear with hybridization between the same species within the genus and cross genus taxa. Understanding the taxonomy, biodiversity, and conservation biology, the delimitation of species are controversial for biologists, so several species concepts have been evaluated. Reproductive isolation is considered the key standard for delimiting the species [3,5,65].

4.5. Ecology of R. sericea Complex

The *R. sericea* complex showed apparent macro-morphological variation like some other species of different families [66,67]. Our field investigation observed intergrading morphological characteristics; even the populations were geographically intermixed or proximal, except few entities. However, morphological characters only have been inadequately considerable to determine taxonomic problems within the *R. sericea* complex. These characters might characterize acclimatization instead of evolutionary variation. Given limited genetic adaptability among the recognized species and the infrequent individual with intermediate characters, these are particularly factual. So, only through integrating ecology will we imply current or ongoing speciation within the *R. sericea* complex and display assistance for the present taxonomic opinion that the *R. sericea* complex signifies unique, divergent, or diverging entities. The characteristics that differentiate the *R. sericea* complex can have ecological significance as variations alongside the environmental gradient over which these two taxa are diverging.

Morphologically *R. sericea* complex taxa have a smaller number of leaflets or have high leaflet numbers may characterize a variation in temperature, which is higher in the lower elevations of some taxa where the species arises and, therefore, improves the possibility for water loss. Such plants can mitigate water loss in high-temperature environments (niche) with smaller leaf surface areas [68], such as from the smaller number and leaflet size.

5. Conclusions

Through our integrative approach using the field data, ecological factors, and population morphology, we elucidated that qualitative and quantitative characters provide taxonomic descriptions and keys to delimit the taxa. The macromorphological characters showed clear species boundaries, and the taxonomic keys and descriptions provide sufficient evidence as six separate entities. The previous work on the genetics method failed to define species boundaries of the complex. However, species concepts need cohesive evidence such as ecology, morphology, and so on to delineate evolutionarily unique entities since evolution is continuous and leaves different, detectable footprints within different groups of organisms. Additionally, the present knowledge of the conservation status of rare species and the IUCN and its robust network of voluntary experts around the globe is undoubtedly in the best position to guide such work. The present work is based on the collection of samples from more than a decade, suggesting that the species RZ have quite rare populations of individuals and should be conserved. Recognition of unique entities, such as *R. sericea* complex rare taxa, has vital implications for conservation decisions and robust biodiversity estimates. As the one taxon RZ has narrow ecological amplitude, our niche modelling results suggest that the habitat of these taxa is shrinking with the future climate changes. There is a chance of the extinction of the complex's rare taxa, which are morphologically different entities.

Supplementary Materials: The following supporting information can be downloaded at: https://www.mdpi.com/article/10.3390/agronomy12051078/s1. Table S1. The morphological characteristics of different populations of *Rosa sericea* complex from different geographical ranges of Southwest China. Figure S1. The correlation lines among the variables used in the modelling study (Black and red lines indicate a high correlation between variables). Figure S2. The most limiting variables of the species complex distribution in China.

Author Contributions: F.U. and X.G.: Conceptualization, Data curation, Formal analysis, Methodology, Writing—original draft preparation, Writing—review and editing. Y.G. and R.J., Data curation, Formal analysis, Methodology. İ.S. and S.S., Methodology, Writing—review and editing. All authors have read and agreed to the published version of the manuscript.

Funding: This research was supported by the Second Tibetan Plateau Scientific Expedition and Research (STEP) program (2019QZKK0502) and the National Natural Sciences Foundation of China (Project No. 31670192), and National Natural Science Foundation of China 32171605. We are thankful to the Chinese Government Scholarship Program (CSC No. 2018GXZ017484).

Data Availability Statement: All data generated and analyzed during this study are included in this published article.

Acknowledgments: The authors are thankful to the curator and team of CDBI for their support in the examination of samples.

Conflicts of Interest: The authors declare no conflict of interest.

Appendix A

Qualitative and quantitative morphological description of the *Rosa sericea* complex based on Table S1.

Morphological qualitative description of characters

Rosa sericea

Prickles; terete, prickles present, sometimes very broad in pairs below the leaves, dense or scattered, robust to fine, abruptly flaring, and the base is broad, sometimes bristles are very dense. **Phyllotactic arrangement**; not distichous. **Vein on adaxial surface**; mostly sunken and sometimes flat. **Leaf margin**; single serrate. leaf margins 2/3 of apex with serrate, 1/2 with serrate margin were noted. Serrate tooth edge of leaf margin with gland absent. **Thorns**; thorns of petiole and rachis sometimes absent, rare, and sparse were examined. Thorns of midrib sometimes absent, and rare were noted. **Indumentum of leaf adaxially**; pubescent and glabrous. Gland of leaf adaxially absent. **Indumentum of leaf abaxially**; pubescent and glabrous. Gland of the leaf is abaxially mostly absent sometimes rare, sparse, and medium. **Indumentum of stipule abaxially**; sometimes absent, rarely pubescent, pubescent, sericeous, and densely hairy. Gland of stipule, sometimes absent, only on leaf margins, rarely on leaf margin abaxially, sparsely except leaf margin, and densely except leaf margin. **Indumentum of petiole and rachis**; sometimes absent, rarely pubescent, pubescent, sericeous, and densely hairy. Glands of petiole and rachis are sometimes absent, rare, sparse, medium, and dense. Indumentum of the pedicel is sometimes absent, rarely pubescent. **Indumentum of pedicel**; sometimes absent, rarely pubescent, pubescent, sericeous, and densely hairy. The gland of the pedicel is sometimes absent, rare, sparse, medium, and dense. **Indumentum of the receptacle**; sometimes absent, rarely pubescent, pubescent, sericeous, and densely hairy. Gland of receptacle sometimes absent, rare, sparse, medium, and dense. **Sepal morphology**; ovate-lanceolate, abaxially sparsely pubescent or subglabrous, adaxially villous, margin entire, apex acuminate, or acute. **Indumentum of sepal**; sometimes absent, rarely pubescent, pubescent, sericeous, and densely hairy. The gland of the sepal is sometimes absent, rare, sparse, medium, and dense. The number of petals was four mostly few have five. **Stalk**; not inflated. **Hip color**; red or bright to reddish, orange to red or red-purple, immature green or greenish.

Rosa sikangensis

Prickles; terete, prickles in pairs mostly in pairs below leaves, or intermixed sparse, bristles with dense slender. **Phyllotactic arrangement**; not distichous. **The vein on adaxial surface; is** mostly sunken and sometimes flat. **Leaf margin**; double serrate, 1/2 with a serrate margin. Serrate tooth edge of leaf margin with gland present. **Thorns**; thorns of petiole and rachis rare, sparse, medium, and dense were examined. Thorns of midrib were medium and dense were noted. **Indumentum of leaf adaxially**; sericeous. Gland of leaf adaxially absent. **Indumentum of leaf abaxially**; sericeous. Gland of leaf abaxially mostly sparse, medium, and dense. **Indumentum of stipule abaxially**; pubescent, sericeous, and densely hairy. Gland of stipule, sometimes absent, only on leaf margins, rarely on leaf margin abaxially, sparsely except leaf margin, and densely except leaf margin. **Indumentum of petiole and rachis**; sometimes absent, rarely pubescent, pubescent, sericeous, and densely hairy. Glands of petiole sometimes absent, rare, sparse, medium, and dense. **Indumentum of pedicel**; sometimes absent, rarely pubescent, pubescent, sericeous, and densely hairy. Gland of the pedicel is sometimes absent, rare, sparse, medium, and dense. These specimens have more hairs on the leaf abaxial surface as well as on the petiole, and fruits.

Rosa omeiensis

Phyllotactic arrangement; not distichous. **The vein on adaxial surface; is** mostly sunken and sometimes flat. **Leaf margin**; single serrate. leaf margins all serrate, 2/3 of apex with serrate, 1/2 with serrate margin were observed. Serrate tooth edge of leaf margin with gland absent. **Thorns**; thorns of petiole and rachis sometimes absent, rare, and sparse were examined. Thorns of midrib sometimes absent, rare, sparse, medium and dense were noted. **Indumentum of leaf adaxially**; glabrous. Gland of leaf adaxially absent. **Indu-**

mentum of leaf abaxially; glabrous. Gland of leaf abaxially mostly rarely and sometimes absent. **Indumentum of stipule abaxially**; sometimes absent, rarely pubescent, pubescent, sericeous, and densely hairy. Gland of stipule, sometimes absent, only on leaf margins, and rarely on leaf margin abaxially. **Indumentum of petiole and rachis**; sometimes absent, rarely pubescent, pubescent, sericeous, and densely hairy. Glands of the petiole are sometimes absent, rare, sparse, medium, and dense. **Indumentum of pedicel**; mostly absent, rarely pubescent, and pubescent. Gland of the pedicel is sometimes absent, rare, and sparse. **Indumentum of the receptacle**; mostly absent, rarely pubescent, and pubescent. Gland of receptacle sometimes absent, rare, and sparse. **Sepal morphology**; lanceolate, abaxially subglabrous, adaxially sparsely pubescent, margin entire, apex acuminate or long caudate. **Indumentum of sepal**; rarely pubescent, pubescent, sericeous, and densely hairy. Gland of sepal sometimes absent, rare and sparse. **Petals**; number of petals were four mostly few have five. Color of petals mostly white, yellowish and pink. **Stalk**; half inflated. **Hip color**; red or bright to reddish, orange to red or red purple, immature green or greenish. **Prickles**; wing like, prickles present in paired below the leaves with broad base, dense or scattered, abruptly flaring, bristles when present very dense, apex tapering or abruptly.

Rosa morrisonensis

Prickles; stout, prickles in pairs mostly in pairs below leaves, mostly dense sometimes scattered, bristles dense slender. **Phyllotactic arrangement**; not distichous. **Vein on adaxial surface**; mostly flat. **Leaf margin**; double serrate. Serrate tooth edge of leaf margin with gland present. Leaf margin tooth single tooth. **Thorns**; thorns of petiole and rachis not present. Thorns of midrib were absent. **Indumentum of leaf adaxially**; absent. Gland of leaf adaxially rarely. **Indumentum of leaf abaxially**; sericeous. Gland of leaf abaxially sparse. **Indumentum of stipule abaxially**; absent. Gland of stipule mostly absent, sometimes in only on leaf margin. **Indumentum of petiole and rachis**; sometimes absent, and rarely pubescent. Glands of petiole sparse, and medium. **Indumentum of pedicel**; sometimes absent, rarely pubescent, and pubescent. Gland of pedicel sometimes absent, and rare. **Indumentum of receptacle**; absent. Gland of receptacle sometimes absent, and rare. **Sepal morphology**; lanceolate, abaxially glabrous, sometimes sparsely glandular, adaxially densely pubescent, margin entire, apex long acuminate. **Indumentum of sepal**; sometimes absent mostly rarely pubescent. Gland of sepal sometimes absent, sparse and medium. Gland of receptacle sometimes absent, and rare. **Petals**; number of petals were four. Color of petals mostly white. **Stalk**; all inflated. **Hip color**; orange to red or red purple.

Rosa mairei

Prickles; terete, and winglike, prickles present, sometimes very broad in pairs below the leaves, dense or scattered, robust to fine, apex tapering, and the base is broad, sometimes bristles are mostly scattered and sometimes dense. **Phyllotactic arrangement**; not distichous. **Vein on adaxial surface**; mostly sunken and sometimes flat. **Leaf margin**; single serrate. leaf margins 2/3 of apex with serrate, 1/2 with serrate margin were noted. Serrate tooth edge of leaf margin with gland present. **Thorns**; thorns of petiole and rachis absent. Thorns of midrib mostly absent, and sometimes dense present. **Indumentum of leaf adaxially**; pubescent. Gland of leaf adaxially present and dense. **Indumentum of leaf abaxially**; absent. Gland of leaf abaxially dense. **Indumentum of stipule abaxially**; sometimes absent, pubescent, and sometimes sericeous. Gland of stipule, rarely on leaf margin abaxially, and densely except from leaf margin. **Indumentum of petiole and rachis**; absent. Glands of petiole and rachis sometimes medium, and dense. **Indumentum of pedicel**; sometimes absent, rarely pubescent, and pubescent. Gland of pedicel absent. **Indumentum of receptacle**; absent. Gland of receptacle absent. **Sepal morphology**; ovate or lanceolate, abaxially sparsely pubescent, adaxially densely pubescent, margin entire, apex acuminate. **Indumentum of sepal**; pubescent, sericeous, and densely hairy. Gland of sepal sometimes absent, rare, sparse, and medium. Number of petals were four. White color. **Stalk**; not inflated and all inflated. **Hip color**; red or bright to reddish.

Rosa zhongdianensis

Prickles; flat, prickles present, sometimes very broad in pairs below the leaves, dense or scattered, robust to fine, tapering, and the base is broad, sometimes bristles are very dense. **Phyllotactic arrangement**; not distichous. **Vein on adaxial surface**; sunken on adaxial side. **Leaf margin**; double serrate. leaf margins double tooth. Serrate tooth edge of leaf margin with gland absent. **Thorns**; thorns of petiole and rachis sometimes were rare. Thorns of midrib were mostly absent, sometimes rare. **Indumentum of leaf adaxially**; pubescent. Gland of leaf adaxially absent, rare and sometimes sparse. **Indumentum of leaf abaxially**; absent. Gland of leaf abaxially mostly rare, and sparse. **Indumentum of stipule abaxially**; sometimes rarely pubescent, pubescent, sericeous, and densely hairy. Gland of stipule absent. **Indumentum of petiole and rachis**; sometimes absent, rarely pubescent, pubescent, sericeous, and densely hairy. Glands of petiole and rachis sometimes absent, rare, sparse, and medium. Indumentum of pedicel sometimes rarely pubescent, pubescent, sericeous, and densely hairy. **Indumentum of pedicel**; sometimes, pubescent, sericeous, and densely hairy. Gland of pedicel absent. **Indumentum of receptacle**; sometimes absent, and sericeous. Gland of receptacle absent. **Sepal morphology**; lanceolate, 9–12 mm, both surfaces densely pubescent, margin entire, apex long acuminate, caudate. **Indumentum of sepal**; sometimes absent, and rarely pubescent. Gland of sepal sometimes absent, rare, and sparse. The number of petals were four. **Stalk**; not inflated. **Hip color**; dark red, immature green or greenish.

Quantitative discerption of *R. sericea* complex

Rosa sericea

The length ranged from 7–15 mm, but most of the specimens had an average 11 mm of prickles. The width ranges from 7–12 mm. The prickles diameter ranges from 0.9–2.9 cm. **Leaf morphology**; the number of leaflets minimum ranges from 5–11, number of leaflets maximum ranges from 9–15. Length of leaf minimum ranges from 12.5–50 mm, length of leaf maximum ranges from 20–80 mm. Length of first leaflet on apex minimum ranges from 5.12–17.4 mm, length of the first leaflet on apex maximum ranges from 6.71–31.21 mm. Width of the first leaflet on the apex minimum 2.56–6.43 mm, the width of first leaflet on apex maximum ranges from 3.49–9.8 mm. Length of first leaflet on base minimum ranges from 2–7 mm, length of first leaflet on base maximum ranges from 4–18 mm. Width of the leaflet on base minimum 1.5–8.23 mm, width of the first leaflet on base maximum 2–9.78 mm. **Pedicel size**; length of pedicle minimum ranges from 1–27 mm, length of pedicel maximum ranges from 2–41.6 mm. **Sepal size**; length of sepal minimum ranges from 3–19 mm, length of sepal maximum ranges from 6–29 mm. Width of sepal minimum ranges from 1.2–8.7 mm, width of sepal maximum ranges from 1.3–3.5 mm.

Rosa omeiensis

The length of prickles was examined various in different specimens of the same species, variation also have been observed, it was ranging from 3–15 mm, but the average length was between 7–10 mm. The width was ranges from 2–12 mm. Prickle diameter was ranging from 1.5–3 cm. **Leaf morphology**; number of leaflets minimum ranges from 5–11, number of leaflets maximum ranges from 13–17. Length of leaflet minimum ranges from 20–80 mm, length of leaf maximum ranges from 35–115 mm. Length of first leaflet on apex minimum ranges from 5–20 mm, length of the first leaflet on apex maximum ranges from 9–41.1 mm. Width of the first leaflet on apex minimum 2.54–8 mm, width of first leaflet on apex maximum ranges from 3.98–12 mm. Length of first leaflet on base minimum ranges from 3.46–11.12 mm, length of first leaflet on base maximum ranges from 5–17 mm. Width of the leaflet on base minimum 2–6 mm, width of the first leaflet on base maximum 3–8 mm. **Pedicel size**; length of pedicle minimum ranges from 1–21.5 mm, length of pedicel maximum ranges from 2–34.13 mm. **Sepal size**; length of sepal minimum ranges from 3.9–16 mm, length of sepal maximum ranges from 5–23 mm. Width of sepal minimum ranges from 1.89–5 mm, width of sepal maximum ranges from 2–6.4 mm.

Rosa sikangensis

The prickles length was from 7–15 mm, mostly was from 7–10 mm. The width ranges from 2–12 mm width noted. The prickles diameter ranges from 0.8–1.3 cm. **Leaf morphology**; number of leaflets minimum ranges from 7–9, number of leaflets maximum ranges from 9–13. Length of leaflet minimum ranges from 13.41–30.23 mm, length of leaf maximum ranges from 30.12–63.56 mm. The length of first leaflet on apex minimum ranges from 5.12–10.29 mm, length of the first leaflet on apex maximum ranges from 8.11–19.78 mm. Width of the first leaflet on the apex minimum 3.34–5.34 mm, the width of the first leaflet on the apex maximum ranges from 4.51–7.53 mm. The length of the first leaflet on the base minimum ranges from 3.12–6.78 mm, length of the first leaflet on the base maximum ranges from 5.14–11.23 mm. Width of the leaflet on-base minimum 2.34–4.32 mm, the width of the first leaflet on-base maximum of 3.03–6.03 mm. **Pedicel size**; length of pedicle minimum ranges from 3.4–13 mm, length of pedicel maximum ranges from 9.1–32 mm. **Sepal size**; length of sepal minimum ranges from 4.9–14 mm, length of sepal maximum ranges from 7.9–22 mm. The width of sepal minimum ranges from 1.7–4 mm, the width of sepal minimum ranges from 2.9–9 mm, the width of sepal maximum ranges from 2.5–9 mm.

Rosa morrisonensis

The prickles length was from 7–15 mm, mostly was from 7–10 mm. The width ranges from 2–12 mm width noted. The prickle' diameter ranges from 0.8 to 1.3 cm. **Leaf morphology**; the number of leaflets minimum were 7, the number of leaflets maximum ranges from 9–11. The length of leaflet minimum ranges from 15–30 mm, length of leaf maximum ranges from 24–45 mm. Length of first leaflet on apex minimum ranges from 5–8 mm, length of the first leaflet on apex maximum ranges from 7–12 mm. The width of the first leaflet on apex minimum 2.5–5 mm, the width of the first leaflet on apex maximum ranges from 3–7 mm. Length of first leaflet on base minimum ranges from 4–6 mm, length of the first leaflet on base maximum ranges from 5–9 mm. The width of the leaflet on base minimum ranges from 2–4 mm, width of the first leaflet on base maximum 3–5.5 mm. **Pedicel size**; length of pedicle minimum ranges from 4–7 mm, length of pedicel maximum ranges from 5–10 mm. **Sepal size**; length of sepal minimum ranges from 7–14 mm, length of sepal maximum ranges from 8–14 mm. The width of sepal minimum ranges from 2–3 mm, width of sepal maximum ranges from 2.5–3.5 mm.

Rosa mairei

The prickles length was from 7–15 mm, mostly was from 7–10 mm. The width ranges from 2–12 mm width noted. The prickles diameter ranges from 0.8–1.3 cm. **Leaf morphology**; the number of leaflets minimum were 7–9, number of leaflets maximum ranges from 9–11. The length of leaflet minimum ranges from 20–30 mm, length of leaf maximum ranges from 25–40 mm. The length of the first leaflet on the apex minimum ranges from 7–11 mm, length of the first leaflet on the apex maximum ranges from 9–13 mm. The width of the first leaflet on the apex minimum 3–4.5 mm, the width of first leaflet on apex maximum ranges from 3.5–5.5 mm. The length of first leaflet on base minimum ranges from 3–6 mm, length of first leaflet on the base maximum ranges from 5–9 mm. The width of the leaflet on base minimum ranges from 2–3 mm, width of the first leaflet on a base maximum 2.5–4 mm. **Pedicel size**; length of pedicle minimum ranges from 2–7 mm, length of pedicel maximum ranges from 3–8 mm. **Sepal size**; length of sepal minimum ranges from 5–10 mm, length of sepal maximum ranges from 6–12 mm. The width of sepal minimum ranges from 2.5–3 mm, the width of sepal maximum ranges from 3–4 mm.

Rosa zhongdianensis

The prickles length was from 6–8 mm. The width ranges from 4–7 mm width noted. **Leaf morphology**; the number of leaflets minimum were 5–9, number of leaflets maximum ranges from 9–11. The length of leaflet minimum ranges from 19–42 mm, length of leaf maximum ranges from 32–58 mm. The length of the first leaflet on apex minimum ranges from 7–11 mm, length of the first leaflet on apex maximum ranges from 10–14 mm. Width of the first leaflet on the apex minimum 5–7 mm, the width of first leaflet on apex maximum ranges from 7–8.5 mm. Length of first leaflet on base minimum ranges from 5–8 mm, length of first leaflet on base maximum ranges from 7–13 mm. Width of the leaflet on base minimum ranges from 3–4.5 mm, width of the first leaflet on base maximum 4–5.5 mm. **Pedicel size**; length of pedicle minimum ranges from 7–14 mm, length of pedicel maximum ranges from 10–15 mm. **Sepal size**; length of sepal minimum ranges from 9–13 mm, length of sepal maximum ranges from 11–17 mm. Width of sepal minimum ranges from 2.5–3.5 mm, width of sepal maximum ranges from 3–3.5 mm.

References

1. Mayr, E. *The Growth of Biological Thought: Diversity, Evolution, and Inheritance*; Harvard University Press: Cambridge, MA, USA, 1982.
2. Goldstein, P.Z.; Brower, A.V. Molecular systematics and the origin of species: New syntheses or methodological introgressions? In *Molecular Systematics and Evolution: Theory and Practice*; Springer: Berlin/Heidelberg, Germany, 2002; pp. 147–161.
3. Ellstrand, N.C.; Schierenbeck, K.A. Hybridization as a stimulus for the evolution of invasiveness in plants? *Proc. Natl. Acad. Sci. USA* **2000**, *97*, 7043–7050. [CrossRef] [PubMed]
4. Prendergast, J.; Quinn, R.; Lawton, J.; Eversham, B.; Gibbons, D. Rare species, the coincidence of diversity hotspots and conservation strategies. *Nature* **1993**, *365*, 335–337. [CrossRef]
5. Liti, G.; Barton, D.B.; Louis, E.J. Sequence diversity, reproductive isolation and species concepts in Saccharomyces. *Genetics* **2006**, *174*, 839–850. [CrossRef] [PubMed]
6. Gao, Y.; Gao, X.; Harris, A. Species boundaries and parapatric speciation in the complex of alpine shrubs, *Rosa sericea* (Rosaceae), based on population genetics and ecological tolerances. *Front. Plant Sci.* **2019**, *10*, 321. [CrossRef] [PubMed]
7. De Queiroz, K. Species concepts and species delimitation. *Syst. Biol.* **2007**, *56*, 879–886. [CrossRef]
8. Mayr, E. The biological species concept. In *Species Concepts and Phylogenetic Theory: A Debate*; Columbia University Press: New York, NY, USA, 2000; pp. 17–29.
9. Lafuente, A.; Valverde, N. *Linnaean Botany and SPANISH Imperial Biopolitics*; Antonio Lafuente & Nuria Valverde Instituto de Historia; CSIC: Madrid, Spain, 2005.
10. Parkinson, C.L.; Zamudio, K.R.; Greene, H.W. Phylogeography of the pitviper clade *Agkistrodon*: Historical ecology, species status, and conservation of cantils. *Mol. Ecol.* **2000**, *9*, 411–420. [CrossRef]
11. Shaffer, M.L. Minimum population sizes for species conservation. *BioScience* **1981**, *31*, 131–134. [CrossRef]
12. Borchsenius, F. Staggered flowering in four sympatric varieties of geonoma cuneata (Palmae) 1. *Biotropica* **2002**, *34*, 603–606. [CrossRef]
13. Hodel, D.R. *Chamaedorea Palms: The Species and Their Cultivation*; International Palm Society, Universidad Autónoma Chapingo; Mexico Biblioteca Central: Texcoco, Mexico, 1992.
14. Bortolus, A. Error cascades in the biological sciences: The unwanted consequences of using bad taxonomy in ecology. *AMBIO A J. Hum. Environ.* **2008**, *37*, 114–118. [CrossRef]
15. Ely, C.V.; de Loreto Bordignon, S.A.; Trevisan, R.; Boldrini, I.I. Implications of poor taxonomy in conservation. *J. Nat. Conserv.* **2017**, *36*, 10–13.
16. Nixon, K.C.; Wheeler, Q.D. An amplification of the phylogenetic species concept. *Cladistics* **1990**, *6*, 211–223. [CrossRef]
17. Mishler, B.D.; Donoghue, M.J. Species concepts: A case for pluralism. *Syst. Zool.* **1982**, *31*, 491–503. [CrossRef]
18. Weston, M.A.; Clarke, K.; Maguire, G.S.; Sumner, J. Morphological and molecular evidence of population divergence in a widespread shorebird across its southern mainland Australian distribution. *Conserv. Genet.* **2020**, *21*, 757–770. [CrossRef]
19. Yunli, W.; Yangyang, W.; Wenlong, X.; Chaojie, W.; Chongshi, C.; Shuping, Q. Genetic diversity of pumpkin based on morphological and SSR markers. *Pak. J. Bot* **2020**, *52*, 477–487. [CrossRef]
20. Hu, G.-X.; Liu, E.-D.; Wu, Z.-K.; Sytsma, K.J.; Drew, B.T.; Xiang, C.-L. Integrating DNA sequences with morphological analysis clarifies phylogenetic position of *Salvia grandifolia* (Lamiaceae): An enigmatic species endemic to southwestern China. *Int. J. Plant Sci.* **2020**, *181*, 787–799. [CrossRef]
21. McGale, E.; Valim, H.; Mittal, D.; Jimenez, J.M.; Halitschke, R.; Schuman, M.C.; Baldwin, I.T. Determining the scale at which variation in a single gene changes population yields. *Elife* **2020**, *9*, e53517. [CrossRef]
22. Smith, A.L.; Hodkinson, T.R.; Villellas, J.; Catford, J.A.; Csergő, A.M.; Blomberg, S.P.; Crone, E.E.; Ehrlén, J.; Garcia, M.B.; Laine, A.-L. Global gene flow releases invasive plants from environmental constraints on genetic diversity. *Proc. Natl. Acad. Sci. USA* **2020**, *117*, 4218–4227. [CrossRef]
23. Stange, M.; Barrett, R.D.; Hendry, A.P. The importance of genomic variation for biodiversity, ecosystems and people. *Nat. Rev. Genet.* **2020**, *22*, 89–105. [CrossRef]
24. Ivanova, A.A.; Zhelezova, A.D.; Chernov, T.I.; Dedysh, S.N. Linking ecology and systematics of acidobacteria: Distinct habitat preferences of the *Acidobacteriia* and *Blastocatellia* in tundra soils. *PLoS ONE* **2020**, *15*, e0230157. [CrossRef]
25. Hortal, J.; de Bello, F.; Diniz-Filho, J.A.F.; Lewinsohn, T.M.; Lobo, J.M.; Ladle, R.J. Seven shortfalls that beset large-scale knowledge of biodiversity. *Annu. Rev. Ecol. Evol. Syst.* **2015**, *46*, 523–549. [CrossRef]

26. Mace, G.M.; Collar, N.J.; Gaston, K.J.; Hilton-Taylor, C.; Akçakaya, H.R.; Leader-Williams, N.; Milner-Gulland, E.J.; Stuart, S.N. Quantification of extinction risk: IUCN's system for classifying threatened species. *Conserv. Biol.* **2008**, *22*, 1424–1442. [CrossRef] [PubMed]
27. Hochkirch, A.; Samways, M.J.; Gerlach, J.; Böhm, M.; Williams, P.; Cardoso, P.; Cumberlidge, N.; Stephenson, P.; Seddon, M.B.; Clausnitzer, V. A strategy for the next decade to address data deficiency in neglected biodiversity. *Conserv. Biol.* **2021**, *35*, 502–509. [CrossRef] [PubMed]
28. Akçakaya, H.R.; Ferson, S.; Burgman, M.A.; Keith, D.A.; Mace, G.M.; Todd, C.R. Making consistent IUCN classifications under uncertainty. *Conserv. Biol.* **2000**, *14*, 1001–1013. [CrossRef]
29. Gao, Y.-D.; Zhang, Y.; Gao, X.-F.; Zhu, Z.-M. Pleistocene glaciations, demographic expansion and subsequent isolation promoted morphological heterogeneity: A phylogeographic study of the alpine *Rosa sericea* complex (Rosaceae). *Sci. Rep.* **2015**, *5*, 11698. [CrossRef]
30. Hijmans, R.J.; Cameron, S.E.; Parra, J.L.; Jones, P.G.; Jarvis, A. Very high resolution interpolated climate surfaces for global land areas. *Int. J. Climatol. A J. R. Meteorol. Soc.* **2005**, *25*, 1965–1978. [CrossRef]
31. Morelli, T.L.; Smith, A.B.; Mancini, A.N.; Balko, E.A.; Borgerson, C.; Dolch, R.; Farris, Z.; Federman, S.; Golden, C.D.; Holmes, S.M. The fate of Madagascar's rainforest habitat. *Nat. Clim. Chang.* **2020**, *10*, 89–96. [CrossRef]
32. Fick, S.E.; Hijmans, R.J. WorldClim 2: New 1-km spatial resolution climate surfaces for global land areas. *Int. J. Climatol.* **2017**, *37*, 4302–4315. [CrossRef]
33. Phillips, S.J.; Anderson, R.P.; Schapire, R.E. Maximum entropy modeling of species geographic distributions. *Ecol. Model.* **2006**, *190*, 231–259. [CrossRef]
34. Vignali, S.; Barras, A.G.; Arlettaz, R.; Braunisch, V. SDMtune: An R package to tune and evaluate species distribution models. *Ecol. Evol.* **2020**, *10*, 11488–11506. [CrossRef]
35. Zhou, Z.; Bao, W. Levels of physiological dormancy and methods for improving seed germination of four rose species. *Sci. Hortic.* **2011**, *129*, 818–824. [CrossRef]
36. Porfirio, L.L.; Harris, R.M.; Lefroy, E.C.; Hugh, S.; Gould, S.F.; Lee, G.; Bindoff, N.L.; Mackey, B. Improving the use of species distribution models in conservation planning and management under climate change. *PLoS ONE* **2014**, *9*, e113749. [CrossRef] [PubMed]
37. Sari, I.; Kandemir, A. Will *Ferulago glareosa* Kandemir and Hedge (Apiaceae) be extinct in the near future? *Turk. J. Bot.* **2020**, *44*, 427–440. [CrossRef]
38. Guisan, A.; Thuiller, W.; Zimmermann, N.E. *Habitat Suitability and Distribution Models: With Applications in R*; Cambridge University Press: Cambridge, UK, 2017.
39. Allouche, O.; Tsoar, A.; Kadmon, R. Assessing the accuracy of species distribution models: Prevalence, kappa and the true skill statistic (TSS). *J. Appl. Ecol.* **2006**, *43*, 1223–1232. [CrossRef]
40. Warren, D.L.; Matzke, N.J.; Cardillo, M.; Baumgartner, J.B.; Beaumont, L.J.; Turelli, M.; Glor, R.E.; Huron, N.A.; Simões, M.; Iglesias, T.L. ENMTools 1.0: An R package for comparative ecological biogeography. *Ecography* **2021**, *44*, 504–511. [CrossRef]
41. Soltis, P.S.; Gitzendanner, M.A. Molecular systematics and the conservation of rare species. *Conserv. Biol.* **1999**, *13*, 471–483. [CrossRef]
42. Vane-Wright, R.I.; Humphries, C.J.; Williams, P.H. What to protect?—Systematics and the agony of choice. *Biol. Conserv.* **1991**, *55*, 235–254. [CrossRef]
43. Thorne, R.F. An updated phylogenetic classification of the flowering plants. *Aliso A J. Syst. Evol. Bot.* **1992**, *13*, 365–389. [CrossRef]
44. Chase, M.W.; Reveal, J.L. A phylogenetic classification of the land plants to accompany APG III. *Bot. J. Linn. Soc.* **2009**, *161*, 122–127. [CrossRef]
45. Shah, S.N.; Ahmad, M.; Zafar, M.; Ullah, F.; Zaman, W.; Malik, K.; Rashid, N.; Gul, S. Taxonomic importance of spore morphology in Thelypteridaceae from Northern Pakistan. *Microsc. Res. Tech.* **2019**, *82*, 1326–1333. [CrossRef]
46. Fougère-Danezan, M.; Joly, S.; Bruneau, A.; Gao, X.-F.; Zhang, L.-B. Phylogeny and biogeography of wild roses with specific attention to polyploids. *Ann. Bot.* **2015**, *115*, 275–291. [CrossRef]
47. Ghalambor, C.K.; McKay, J.K.; Carroll, S.P.; Reznick, D.N. Adaptive versus non-adaptive phenotypic plasticity and the potential for contemporary adaptation in new environments. *Funct. Ecol.* **2007**, *21*, 394–407. [CrossRef]
48. De Jong, G. Evolution of phenotypic plasticity: Patterns of plasticity and the emergence of ecotypes. *New Phytol.* **2005**, *166*, 101–118. [CrossRef] [PubMed]
49. Cole, S.; Bhardwaj, G.; Gerlach, J.; Mackie, J.; Grant, C.; Almquist, K.; Stewart, A.; Kurz, E.; Duncan, A.; Deeley, R.G. Overexpression of a transporter gene in a multidrug-resistant human lung cancer cell line. *Science* **1992**, *258*, 1650–1654. [CrossRef] [PubMed]
50. Mallet, J. Hybrid speciation. *Nature* **2007**, *446*, 279–283. [CrossRef] [PubMed]
51. Ullah, F.; Ayaz, A.; Saqib, S.; Zaman, W.; Butt, M.A.; Ullah, A. *Silene conoidea* L.: A Review on its Systematic, Ethnobotany and Phytochemical profile. *Plant Sci. Today* **2019**, *6*, 373–382. [CrossRef]
52. Nosil, P. Speciation with gene flow could be common. *Mol. Ecol.* **2008**, *17*, 2103–2106. [CrossRef]
53. Baird, H.P.; Miller, K.J.; Stark, J.S. Evidence of hidden biodiversity, ongoing speciation and diverse patterns of genetic structure in giant Antarctic amphipods. *Mol. Ecol.* **2011**, *20*, 3439–3454. [CrossRef]

54. Hedrén, M. Patterns of allozyme and morphological differentiation in the *Carex flava* complex (Cyperaceae) in Fennoscandiaxs. *Nord. J. Bot.* **2002**, *22*, 257–301. [CrossRef]
55. Hamrick, J.L.; Godt, M.W. Allozyme diversity in plant species. In *Plant Population Genetics, Breeding, and Genetic Resources*; Brown, A.H.D., Clegg, M.T., Kahler, A.L., Weir, B.S., Eds.; Sinauer: Sunderland, MA, USA, 1990; pp. 43–63.
56. Millar, C.I.; Libby, W.J. Strategies for conserving clinal, ecotypic, and disjunct population diversity in widespread species. *Genet. Conserv. Rare Plants* **1991**, *149*, 170.
57. Oostermeijer, J. Threats to rare plant persistence. In *Population Viability in Plants*; Springer: Berlin/Heidelberg, Germany, 2003; pp. 17–58.
58. Darwin, C. *On the Origin of Species: A Facsimile of the First Edition*; Harvard University Press: Cambridge, MA, USA, 1964; Volume 11.
59. Schemske, D.W.; Husband, B.C.; Ruckelshaus, M.H.; Goodwillie, C.; Parker, I.M.; Bishop, J.G. Evaluating approaches to the conservation of rare and endangered plants. *Ecology* **1994**, *75*, 584–606. [CrossRef]
60. Xiao-Mei, W.; Xin-Fen, G.; Zhang, L.-B. A systematic study of *Rosa sericea* (Rosaceae) complex: Are *R. omeiensis* and *R. sericea* conspecific? *J. Syst. Evol.* **2008**, *46*, 919–928.
61. Young, A.; Boyle, T.; Brown, T. The population genetic consequences of habitat fragmentation for plants. *Trends Ecol. Evol.* **1996**, *11*, 413–418. [CrossRef]
62. Luijten, S.H.; Dierick, A.; Gerard, J.; Oostermeijer, B.; Raijmann, L.E.; Den Nijs, H.C. Population size, genetic variation, and reproductive success in a rapidly declining, self-incompatible perennial (*Arnica montana*) in The Netherlands. *Conserv. Biol.* **2000**, *14*, 1776–1787.
63. Akçakaya, H.R.; Butchart, S.H.; Watson, J.E.; Pearson, R.G. Preventing species extinctions resulting from climate change. *Nat. Clim. Chang.* **2014**, *4*, 1048–1049. [CrossRef]
64. Attorre, F.; Abeli, T.; Bacchetta, G.; Farcomeni, A.; Fenu, G.; De Sanctis, M.; Gargano, D.; Peruzzi, L.; Montagnani, C.; Rossi, G. How to include the impact of climate change in the extinction risk assessment of policy plant species? *J. Nat. Conserv.* **2018**, *44*, 43–49. [CrossRef]
65. Lyons, K.G.; Brigham, C.A.; Traut, B.H.; Schwartz, M.W. Rare species and ecosystem functioning. *Conserv. Biol.* **2005**, *19*, 1019–1024. [CrossRef]
66. Ku, T.; Robertson, K. Rosa (Rosaceae). *Flora China* **2003**, *9*, 339–381.
67. Esfandani-Bozchaloyi, S.; Zaman, W. Taxonomic significance of macro and micro-morphology of *Geranium* L. species using scanning electron microscopy. *Microsc. Res. Tech.* **2018**, *81*, 1520–1532. [CrossRef]
68. Givnish, T.J.; Kriebel, R. Causes of ecological gradients in leaf margin entirety: Evaluating the roles of biomechanics, hydraulics, vein geometry, and bud packing. *Am. J. Bot.* **2017**, *104*, 354–366. [CrossRef]

MDPI
St. Alban-Anlage 66
4052 Basel
Switzerland
Tel. +41 61 683 77 34
Fax +41 61 302 89 18
www.mdpi.com

Agronomy Editorial Office
E-mail: agronomy@mdpi.com
www.mdpi.com/journal/agronomy

www.ingramcontent.com/pod-product-compliance
Lightning Source LLC
Chambersburg PA
CBHW040758250926
51850CB00007BA/1027
* 9 7 8 3 0 3 6 5 7 6 7 7 0 *